TAMING RIVERS
Recollections of a Civil Engineer
During the British Raj

Dedicated to
the loving memory of

IQBAL BEGUM, BIBI GUL

our beloved mother and
loving companion of our father

Abdur Rahman Khan, Peshawar, 1979.

TAMING RIVERS
Recollections of a Civil Engineer
During the British Raj

KHAN BAHADUR
Abdur Rahman Khan

OXFORD
UNIVERSITY PRESS

Oxford University Press is a department of the University of Oxford.
It furthers the University's objective of excellence in research, scholarship,
and education by publishing worldwide. Oxford is a registered trade mark of
Oxford University Press in the UK and in certain other countries

Published in Pakistan by
Oxford University Press
No. 38, Sector 15, Korangi Industrial Area,
PO Box 8214, Karachi-74900, Pakistan

© Oxford University Press 2022

The moral rights of the author have been asserted

First Edition published in 2022

All rights reserved. No part of this publication may be reproduced, stored in
a retrieval system, or transmitted, in any form or by any means, without the
prior permission in writing of Oxford University Press, or as expressly permitted
by law, by licence, or under terms agreed with the appropriate reprographics
rights organization. Enquiries concerning reproduction outside the scope of the
above should be sent to the Rights Department, Oxford University Press, at the
address above

You must not circulate this work in any other form
and you must impose this same condition on any acquirer

ISBN 978-969-7342-71-6

Typeset in Minion Pro
Printed on 80gsm Offset Paper

Printed by Delta Dot Technologies (Pvt.) Ltd., Karachi

Acknowledgements
All photographs of dams are courtesy
Pakistan Water and Power Development Authority

Contents

Introduction ix

1. Early Memories — 1
2. Boyhood in Abbottabad — 8
3. Sialkot – Dharamshala – Abbottabad — 15
4. College Days — 21
5. Apprenticeship in Sylhet and Assam, 1914–1915 — 45
6. Railway Project, Bihar, 1916 — 58
7. Homeward Bound — 65
8. Mardan, 1917 — 68
9. Yusafzai Territory — 83
10. Dera Ismail Khan (D.I. Khan) — 95
11. Waziristan, 1927 — 113
12. Across the Frontier into Balochistan — 128
13. Government of India's Inspection — 138
14. Players of the Great Game — 144
15. Political Turbulence — 156
16. Farewell to D.I. Khan — 164
17. Bannu — 175
18. The Second World War, 1939–1945 — 187
19. Peshawar — 194
20. Retirement and Partition of India — 206
21. Bahawalpur — 210
22. Balochistan — 229
23. Irrigation Advisor, NWFP — 240
24. Canal Water Dispute Negotiations — 245
25. Pakistan Delegation's Visit to India — 259
26. Warsak and Mangla Dams — 267
27. The Last Lap — 278

Notes — 283
Index — 291

Introduction

The manuscript of this autobiography of our father, Khan Bahadur Abdur Rahman Khan, was discovered accidentally sometime after his death, while going through his voluminous papers, documents, and personal belongings. Another interesting find in this treasure trove was a complete manuscript of his anthology of Urdu poems dating back to 1909 and right up to the year of his death; it was complete in all aspects and ready for a publisher. There were also several files of his extensive correspondence with newspapers, particularly his contributions in the letters' column of the *Pakistan Times* where, over several decades, he had engaged in lively debates with other learned contributors on a range of subjects that included philosophy, Iqbal's poetry and philosophical thought, cosmology, space, time and relativity, and, of course, many proposals for the development of irrigation system in Pakistan.

In 1946, just before his retirement from the NWFP (now Khyber Pakhtunkhwa) government, as the senior-most non-British Engineer in NWFP, he had sent a comprehensive proposal to Quaid-i-Azam Mohammad Ali Jinnah on the Development of Pakistan, with a focus on NWFP. This proposal, covering 14 pages, was published, in its entirety, in the *Jinnah Papers* (First Series, Volume 3, pp. 547–61). On his death in December 1980, the *Pakistan Times*, in its obituary, eulogised him as the 'Doyen of Pakistani Engineers'.

Our father was a man of great intellect, varied interests, and had a caring and self-effacing nature. He was scrupulously honest in all his dealings, as his story will testify. His life story, which he penned in elegant long hand, covers almost the entire twentieth century, spanning two eras, the first beginning in the reign of Queen Victoria, when Lord Curzon was the viceroy and the new province of North West Frontier was established in 1901, concluding with his retirement from government service in 1946; and the second starting with the Independence of Pakistan and ending in the 1980s.

This is the story of the many challenges and accomplishments of an irrigation engineer, who, through his exceptional abilities, was inducted into the imperial service of Royal Engineers (RE) in 1917 in the NWFP and who courageously served there during the most turbulent period of its history. That was the time when undertaking development work was most hazardous, especially in the border regions of the NWFP. The government and the tribes on the western borders, were in a constant state of armed confrontation, with the government carrying out frequent military campaigns against defiant tribes and mass tribal uprisings, led by the now legendary figures, viz. the Faqir of Ipi (1897–1960) and Haji Sahib of Turangzai (1858–1937). There was the added threat of marauding bands of outlaws and bandits hiding in the lawless tribal territories. Any incursion into the area was thus fraught with high risk. He takes us along on his adventurous assignments in Waziristan and Balochistan where, under the protection of the British-led Indian Army, he trekked in hostile terrain to survey road alignments or dam sites for new irrigation projects. He also gives us a graphic account, during his posting to Dera Ismail Khan as a young assistant engineer, of his epic struggle to save the city from being washed away by the raging flood waters of the mighty Indus River.

He recounts the ups and downs of his career and how it was affected by the deep camaraderie that prevailed amongst the British ruling class when it came to protecting each other's interests, however unjustifiably, especially when these interests were at odds with those of senior Indian officers. At the same time, he acknowledges with admiration the deep sense of commitment and dedication that most of his British colleagues demonstrated in the performance of their official duties. He provides penetrating insights into the styles of British administrative practices and their occasionally unorthodox approaches to dealing with issues of governance. Through interesting anecdotes, he gives us fascinating character sketches of several of his bosses and colleagues and their personalities, idiosyncrasies, as well as their impressive technical knowledge. Many of them eventually became his close friends after having given him a tough time in his working life.

As a young man, during his college years, he got to know many personalities who were destined to achieve fame and prominence in their later lives. He took pride in the fact that Dr Mohammad Iqbal taught him English literature at the Government College, Lahore. Chaudhry Sir Zafarullah Khan, KCSI, was a contemporary at Government College, Lahore. At Aligarh, he came to know Khawaja Nazimuddin and was influenced by the patriotic fervour of the Ali Brothers, Maulana Mohammad Ali and Maulana Shaukat Ali. His friendship with Dr Khan Sahib, Abdul Ghaffar Khan, and Sardar Abdur Rab Nishtar developed during his service. He has given fascinating anecdotes about British officials and other friends and acquaintances, including Sir George Cunningham (ICS), Sir Olaf Caroe (ICS), Sir Ambrose Dundas (ICS), Sir John Wiley (ICS); Col. Noel (IPS), and Maj. Iskander Mirza (IPS). Some of them were of equal rank and seniority to him at that time, with whom he played tennis and weekend cricket. Later, many of these officials rose to positions of eminence, as governors of provinces and as presidents of Pakistan. The hilarious episode of arranging the first visit to Swat state for his friend, Lt. Col. Mohammad Ayub Khan, and his meeting with the venerable Wali of Swat, ironically proved to be a harbinger of the close family relationship that was to come between the ruler of the state and the future president of Pakistan.

The first part of his personal narrative is embedded in the dramatic events that were unfolding in British India during the days of mounting struggle of Mohandas Karamchand Gandhi and the Indian National Congress, in seeking autonomy and independence of India. During those times, the NWFP had a special significance in Indian politics as, being a Muslim majority province, it had become politically aligned to the Hindu-dominated Indian National Congress. He narrates how the Pathans demonstrated their adherence to the Gandhian credo of non-violent resistance when their protesters were fired upon by the British troops and they offered nothing in return but their bared chests and suffered enormous casualties. He recounts how, even after multiple reforms had taken place and limited autonomy had been accorded to the province, the emerging provincial governments in the NWFP, whether they belonged to Sahibzada Abdul Qayyum, Dr Khan Sahib, or Sardar Aurangzeb, were merely a rubber-stamp, without real authority, which continued to be exercised by the British bureaucracy.

Father's retirement from government in 1946, just preceded the Partition of India and the creation of the two dominions of India and Pakistan. For him, it was a period of change and internal turmoil where one had to readjust from the orderly state of governance under the Raj to the chaotic turmoil which followed the division of the subcontinent. His uncompromising integrity and refusal to concede on his principles closed the door to numerous opportunities that arose under the changing norms and values of the new nation.

His post-Independence story provides an interesting backdrop to the events and personalities that shaped the destiny of the new nation of Pakistan. Father was an incorrigible optimist and dedicated to help build and engineer the infrastructure of the new nation. It was, perhaps, with this spirit of optimism that, after retirement from the NWFP government, he accepted a position in the Public Works Department (PWD) of the Bahawalpur state where he arrived in the blazing heat of summer in 1946. He soon got down to work, starting with a formal call on the Amir of Bahawalpur, donned in a red pillbox Turkish cap and carrying out the ritual offering to the Amir of a one Guinea guilt coin, conveniently purchased from the treasury. His tenure in Bahawalpur turned out to be momentous and eventful for, stationed as he was on the border between the emerging states of India and Pakistan, he was witness to cataclysmic events that ensued, and the chaos and bloodshed as a result of Partition. It was in Bahawalpur that he and his colleagues first read the much awaited 'Pakistan Award' and were shocked to learn about the borders of the new state.

His next move was to Quetta as commander and irrigation advisor, Military Engineering Services (MES). As an irrigation expert, he carried out a number of schemes to augment the irrigation resources of Balochistan, including the investigation and design of the Anambar and Bolan dams. A most significant work in this regard, was his epic study on the Quetta water resources. However, the work environment in the MES did not suit him, regarding which, Mr Muirhead, the chief engineer, had lamented: 'Oh, Khan Bahadur, the moral fibre of the country has snapped.' It was, therefore, with a sense of relief that he accepted the offer of Khan Abdul Qayyum Khan, chief minister of NWFP, to return to Peshawar as irrigation advisor for NWFP to work on the Kurram Garhi project. He recounts with justifiable pride that he designed the Kurram Garhi project single-handedly, with only drawing assistance from a draughtsman. This irrigation project, situated in Bannu district of the NWFP, has remained an important landmark in the development of irrigation schemes in the province.

While at Peshawar, the Government of Pakistan nominated him to represent the country as a member of a delegation set up to negotiate the Indus Waters Dispute with India under the aegis of the World Bank. In this connection he first visited the USA for the opening round of the negotiations, which was followed by two visits to India. His interesting travelogue of these visits and the negotiations with the Indian delegation is both entertaining and illuminating. He recounts that he was dismayed at the bad faith demonstrated by the Indians who had gone ahead with the construction of the colossal Bhakra Nangal High Dam on the Sutlej River and the Harike Barrage, downstream of the confluence of the Sutlej and Beas rivers with a gigantic canal taking off to irrigate Rajasthan, while negotiations on the water dispute were still in progress.

In the decade following, father worked as a senior design consultant on the Warsak and the Mangla dams, where he carried out various design assignments related to the construction of the main dams. He notes with regret that Pakistani engineers did not actively participate in the engineering design and construction of the vast Indus Basin Replacement Works, including the Tarbela and Mangla dams and the extensive link canal network, which were implemented in the 1960s. They, therefore, missed the

INTRODUCTION

great opportunity of gaining practical skills and experience in engineering design and construction of large-scale irrigation projects, provided by one of the world's largest engineering projects of the time.

After the passing of our mother in 1976, perhaps to overcome his loneliness, he spent a great amount of time reading and writing and was a frequent contributor to the *Pakistan Times*. Though well in his eighties he still maintained an adventurous spirit and was keen to visit New York to join his son Farid. He wrote to him in December 1980 saying that he was all set to leave, writing again to say that he was postponing his departure until spring when the weather would be warmer. However, this was not to be, as he passed away suddenly one night in Peshawar.

Father's departure was widely condoled by the large fraternity of engineers to whom he had been a mentor, as well as by a wide circle of admirers who were touched by his intellect and his writing. The newspapers eulogised him.

In preparing this autobiography for publication, it was a great pleasure and a labour of love for both of us to compile and assemble hundreds of handwritten pages of his papers. We are grateful to Paull Raza Salim Khan, one of the grandsons and the eldest son of Salim Khan, for his detailed editorial assistance and valuable advice on the structure and format of this work. While working on the draft, Paull remarked, with perhaps some wonder, that he had got to know his grandfather, and so also have many other grandchildren whom he never met. Our younger brother Shahid applied the skills of an architect to prepare the maps for the autobiography, for which we are grateful. We are also grateful to Ayesha and Alia, daughters of Farid Rahman, for their editorial contribution and formatting assistance. For Farid it has indeed been the accomplishment of a major item in his bucket list for which he is most grateful. We would also like thank Farzana for her excellent editing and all the members of the Oxford University Press Pakistan team who have contributed to the publication.

Farid Rahman & Faris Rahman Khan
Islamabad
January 2022

1 Early Memories

I was born on 11 May 1891. In 1893, when I was just over two years old, my mother passed away. I carry a clear recollection of the dull, cloudy day on which she died. I lay in the lap of a woman seated among a silent group of women mourners outside my mother's room. I complained of hunger. A piece of stale bread was placed in my hand. After a bite, finding it distasteful, I let it drop from my hand.

Four men silently entered the courtyard where the women mourners were sitting and went into my mother's room. They came out carrying her bier on their shoulders. The bier was covered with a red cloth. I later learnt that it was the red silk scarf, which she had worn on her wedding day, because she had died young. She was twenty-one years old. 'Where is my mother going?' I cried out in anguish, stretching my arms after the retreating bier. 'She is going to her parents' home,' said the weeping woman who was carrying me. I felt forlorn but vaguely consoled.

Shortly after my mother's death, my grandmother took me to Kohat town to live with her eldest son, my uncle. In those days, railway passengers bound for Kohat were transhipped by boat across the un-bridged Indus river gorge at Khushalgarh. I distinctly remember crossing the river at night. The river water, I noticed, was muddy. For long afterwards, I mistook the square brown patch of land along the railway track, which was lit from the window of my moving compartment, to be still a part of the Indus River channel.

I soon got over the catastrophe of my mother's death and learned to live with her pleasant memories. I came to feel a kind of pride in being called a 'mother-orphan'! Uncle's servant, a tall Pathan from Kohat, often took me to the bustling Kohat bazaar. I toddled alongside him, holding his finger. Heaps of salt about 3 feet high lay for sale in front of the shops. To me, they appeared to be small hillocks. One day, a Pathan soldier with bobbed hair accosted us in the bazaar. He stopped in front of me and gently lifted me up and placed me on his shoulders. I felt as though I was lifted to the roof of a house and cried out in terror on looking down at the pavement whereupon, he gently put me down and went away smiling. Afterwards, I used to relate the incident with great excitement.

Uncle's house had a row of rooms with a veranda facing a large courtyard that opened out into a narrow lane. At the back of the house there was a graveyard. My aunt was a very superstitious lady; she used to say that ghosts prowled about in the graveyard at night. According to her, on many occasions she had heard their screeches as they talked to each other. In my mind, I used to picture their grotesque forms as described by my aunt. In the daytime, I sometimes slowly ascended the flight of steps to the roof of our house and peered down at the graveyard below, but never saw any ghosts. At other times, I would wander out along the lane. Our next-door neighbour, a Hindu gentleman, usually sat on a chair in his veranda facing the lane. As I passed in front of him, he would ask me to stop and would put pieces of candy in the palm of my hand, which I hurriedly grabbed. I remember his acts of kindness to this day.

A few months later, grandmother returned with me to Sialkot. But the return was not to the old two-storeyed mansion in which I was born and where my mother had died. We returned instead to live in a smaller dwelling in the vicinity of the old house. During the time I was at my uncle's place, dramatic changes for the worse had taken place in my grandfather's fortunes. The bulk of his urban and rural landed property in and around Sialkot town and elsewhere, including the 'big house', and also my grandmother's substantial gold jewellery, had been auctioned off to pay his debts to moneylenders. (The big house was purchased by Chaudhry Nasrullah Khan, father of Sir Zafarullah Khan, first foreign minister of Pakistan, who grew up there). Sir Zafarullah was my contemporary at Government College, Lahore, and a good friend. We kept in touch with each other, through correspondence, for some years during the early part of our professional careers.

Hailing from the sturdy Mohmand tribe from Bajaur, my forefather Sultan Mohammad Khan had fought alongside Ahmad Shah Durrani at the battle of Panipat. On Ahmad Shah Durrani's return to Afghanistan, Sultan Mohammad Khan had stayed behind near Islamgarh, in the present-day Gujrat district in Punjab, as a commander of a garrison, where he had also been awarded agricultural property. Grandfather's father, having quarrelled with his brothers, perhaps over inheritance, abandoned his family to seek his fortune elsewhere, and landed in the then bustling town of Sialkot.

My grandfather, Fazal Din Khan, was born in 1845. Even though he was illiterate, he had an impressive career. From a small beginning as a labour contractor, he rose to be one of the biggest masonry contractors of the North Western Railways (NWR), which in the 1860s was being pushed from Karachi seaport to Peshawar on the north western border of British India. The construction contract for the masonry work of the famous Lansdowne Bridge over the Indus River at Sukkur, including its piers, pillars, and other structures. The Sukkur and Rohri railway stations, during the 1870s, marked the culmination of his career. He worked his way along the advancing railway line to construct the Khairpur railway station. His next assignment was to construct the NWR Bridge, known as the Empress Bridge, over the Sutlej River at Adamwahan near Bahawalpur, which was inaugurated in 1878.

Large and bustling labour camps were set up at the construction sites, picturesquely situated on high ground, on the riverbank. As construction progressed slowly, taking many years, these sites took on a semi-permanent character and his regular contingent of supervisors, *mistris* (masons), carpenters, and stone hewers and cutters, settled down to as normal a lifestyle as possible, mingling with hordes of casual labour attracted from the countryside. *Langars* (large community kitchens) provided curried *daal* and *chapatis* for the morning and evening meals. Bazaars of makeshift shops sprang up to cater for odds and ends, and women of easy virtue joined in the loud evening crowds and boisterous melee. Grandfather was a religious man but was fond of poetry and music, and occasionally invited well-known poets and singers from nearby cities to evening soirees, where he entertained the local gentry and public officials.

During that period, shiny silver rupees, packed in sacks, were regularly received at home and stashed away for days in a strong room. Grandfather constructed the two-storeyed mansion in Sialkot town, built many mosques, purchased serais and vast tracts of agricultural land. He also loaded grandmother with jewellery. The eclipse of grandfather's fortunes, which soon followed, was equally dramatic. After building the Sutlej Bridge, he took a large building contract in Bahawalpur state where he made the acquaintance with the Amir of Bahawalpur. His next two big undertakings were the construction of the NWR Bridge on the Tawi River at Jammu, in the Jammu and Kashmir state, and the Alexandra Bridge on the Chenab River near Gujrat, inaugurated in 1876 by Albert Edward, Prince of Wales, later to become King Edward VII. All the three enterprises proved financially disastrous, rendering him heavily in debt to moneylenders. In the end, he declared himself insolvent and the houses and serais and the agricultural land were auctioned off to pay his debts. Worst of all, grandmother's jewellery also went under the auctioneer's hammer!

Grandfather was of medium stature with a ruddy fair complexion. He had a wisp of a beard and thick bobbed hair, dyed in henna. His most prominent feature was a large aquiline nose under a pair of dreamy eyes set deep under somewhat prominent eyebrows. He had a dignified bearing and was often seen talking to himself with hand gestures reproducing the dialogues he had had with prominent personalities, during the heyday of his career. Grandfather would often speak with affection of Mr T.E. Robertson, who was engineer in-charge of the construction of the Lansdowne Bridge at Sukkur. Mr Robertson wore a long beard and enjoyed smoking the *huqqa* (hubble-bubble). Often, he and grandfather would sit on the banks of the Indus River getting into friendly technical arguments such as the quantity of masonry contained in a curved cut-out built on top of the barrage's piers. He would also recall in nostalgic tones the frequent invitations to dinner he would receive from the Amir of Bahawalpur.

Grandfather was fond of stories, particularly from the *Qissa Chahar Dervish* (*Tales of the Four Dervishes*) being read out to him. These stories relate to the adventures of four persons, originally of high rank, but reduced by adversity to destitution, who chanced to meet and fraternise in a serai; each one of them agreeing to narrate the story of his misfortune. One of the dervishes was a prince who, in his narrative, gives a graphic account of the sumptuous dishes that used to be prepared in his royal kitchen. The types of *pulaos* ran into dozens. As soon as the narrative would come to the term *mutanjan pulao*, grandfather would sharply call out, 'Stop! Boys, do you know what *mutanjan pulao* is?' he would ask and then, looking around, would explain that, 'It is not an ordinary *pulao* cooked with rice, meat, and spices. It is much more than that. In addition, various types of nuts and fruits, exotic herbs and other delicacies are added to the *pulao*.' After a brief pause, looking reflectively heavenwards, he would add, 'I partook of *mutanjan pulao* on the *dastarkhwan* (dining cloth) of the Amir of Bahawalpur on many occasions.'

Grandfather usually sat on a charpoy facing the lane and would be found talking to himself. A fakir was his constant companion. Grandmother would loudly scold grandfather for his idleness and would demand that his holy companion leave the house forthwith for he was a nuisance for the household; whereupon, the fakir would start digging up the dirt floor with his stick to unearth some hidden treasure for grandfather while the latter looked on with a tolerant smile. 'Leave him alone,' grandfather would plead, 'He is a man of God.'

Grandmother was a stately woman. She was tall and slender with delicate features, sad-looking eyes, and a fair complexion. Grandfather was scared of her lashing tongue and her taunting allusions to his defeatist attitude towards life and his resolute refusal to resume work to earn a living.

Perforce, she had to fall back on the small monthly sums remitted faithfully by my father and uncle. Their education had been abruptly cut off, before they could complete college, when the crash came in grandfather's fortunes. So, after completing technical training in engineering design, both of them had taken up overseeing positions in the Public Works Department in the frontier districts of Hazara and Kohat, where grandfather had owned property. Grandmother's heart was broken. She acutely felt, in particular, the loss of her gold jewellery. I remember her, in those days, an old lady (she was actually only about fifty) of a somewhat austere temperament, embittered by misfortune. No one dared take liberties with her, except me. She treated my tantrums with indulgence, calling me a poor mother-less waif; she would be perpetually spinning cotton yarn on the *charkha* (a cotton gin or spinning wheel) singing melancholy songs in tune with the hum of the *charkha* and then hack a pull at the handy *huqqa*. The yarn was sent to the local weavers for being converted into coarse long cloth for the boys' clothing because they could no longer afford to purchase fine foreign cloth.

For me, it was hard to come by a pice by way of 'pocket money' in those days. It was a valuable coin. It fetched three mangoes in summer and three juicy lengths of sugarcane in winter. The fruits that were in the bazaar were my means of knowing the change of seasons. Whenever I came in possession of a *paisa*, I fondly scanned the legends embossed on it. On one side was embossed the profile of a beautiful young woman wearing a crown, with a thick braid of her tresses passing below and behind her ear. Grandfather told me that this was our Queen Victoria, who lived in *vilayat* (a country situated thousands of miles across the seas). I felt confused at the thought of 'our Queen' living in a different land, but felt reassured to see her single thick braid of hair passing below the ear like an Indian woman. I felt satisfied that her Indian hairstyle entitled her to rule India.

On other occasions, I came across a relatively new *paisa* showing the crowned and similarly braided profile of a very old woman with withered cheeks and bulging eyes. Grandfather told me that this was the same Queen Victoria who was now quite old, in 1896. Some other coins bore a strange picture of a lion and a single-horned horse, holding a crown between them. Grandfather said those were Company *paisas* issued in the Punjab, after the annexation of the Punjab by the British (in 1849) and during the intervening years. Until the Mutiny (1857), the 'town crier' used to preface his message with a beat of drums and the following preamble: 'The people are God's creation; the country belongs to the King [Mughal emperor of Delhi] and the Authority vests in the "Brave Company" [The East India Company]!'

One day, a Hindu marriage procession passed in the street in front of our house. The procession halted at intervals and the bridegroom's father flung fistful of *paisas* from a bag he was carrying across the bride's palanquin as a gesture of thanksgiving. The swarm of coins was partly grabbed in mid-air and partly on the street by a gang of ragamuffins who followed the procession. After the procession had disappeared around a bend in the street,

I leisurely crossed the roadside and stood in the middle of the deserted street. Lo and behold! A shining *paisa* coin lay on the road in front of me. After carefully looking around to see that no one was watching, I picked up the coin and put it in my pocket

As I grew up, I was left more or less to my own devices. I had forgotten my mother except as a pleasant memory and had little or no contact with my young father. My world thus consisted of my grandmother, aunt, and other female relatives. Although perpetually short of *paisas*, I nevertheless had a sense of dignity, and did not stoop to mean ways for obtaining the coveted coins. In 1897, at age six, grandfather got me enrolled to the first primary class of the Scotch Mission High School at Sialkot. My elder brother and two elder first cousins were already studying in the same school. Shortly after my joining the school, one morning, all four of us were hastily summoned to the house. Grandmother was dying at age fifty-five. She lay on a bed covered with white sheets, in the throes of her last struggle while we stood by, overawed by the grandeur of death. Some women sat near the bed reading passages from the Holy Quran. In the afternoon a large crowd of men assembled in the courtyard of our house. They carried grandmother's bier to the graveyard. Three *maulvis* were hired to read the Holy Quran non-stop day and night for three days by grandmother's grave to prepare her to successfully answer the questions to be put by the angels, Munkar and Nakeer. Tasty meals were sent to the *maulvis* to keep them satiated while engaged in their sacred occupation.

After a few days, grandfather took me to visit grandmother's grave, which lay alongside my own mother's grave. The Shahin-Badshah graveyard was an ancient cemetery sprawling over a vast area. It contained some centuries-old graves, including those of some past celebrities. Whenever I passed by an ornate old grave with an overhead copula, I would ask grandfather whether that was my mother's grave and felt disappointed when he said 'no'. We came upon two earth mounds at the furthermost extremity of the graveyard, one mound fresh and higher, the other older and smaller. Grandfather pointed to the smaller mound, 'This is your mother's grave,' he said, 'and this is your grandmother's,' pointing to the bigger mound. I felt a pang of remorse and grief to see my mother's and grandmother's plain, austere graves, as if they had been of no significance in life. Grandfather asked me to pray for both.

A few months later, my father re-married, for 'there was no one to take care of his two minor sons' (me and my elder brother). One fine morning, father's marriage party consisting of about thirty or forty men assembled in the courtyard of our house. Our family barber prepared a hasty meal of *pulao*, potato curry, and tasty oven-baked bread for the party and they left by train to the bride's (my future stepmother's) house. The next thing I remember is travelling by train with father, stepmother, and brother to Rawalpindi, where father was employed as an overseer on the consolidation of Murree road. In those days (1898), Murree road was almost devoid of roadside buildings. Young saplings, protected by round *pucca*-brick network fences, about 4 feet high, were planted all along the road. I, in my fancy, imagined those fences to be full of clear sparkling water like open wells.

My stepmother was a young girl of somewhat homely appearance. She could not give me a mother's affection, which I craved. I started throwing tantrums and had prolonged fits of crying for no apparent reason. This puzzled my father. I had passed the first primary class from the Scotch Mission High School at Sialkot with some distinction and was a student of the second primary class in Rawalpindi Islamia High School. I also began to lag behind in my studies. Then someone suggested that with the coming of the stepmother I was perhaps missing my own mother. Father took the hint and thereafter started paying special attention to me. He took special interest in my meals and made sure that I finished my daily glass of milk. He also assiduously coached me in my studies, with the result that I passed the second primary class with flying colours and, thereafter, remained at the top of my class right up to matriculation.

2 Boyhood in Abbottabad

Early in the winter of 1899, father was transferred to a non-family station and sent us to live with our uncle Khair-ud-Din Khan at Abbottabad (after whom the present locality of Darul Khair is named in Abbottabad). We travelled from Hasan Abdal railway station to Abbottabad, 44 miles of winding road on a mail tonga—a sturdy, dumpy vehicle pulled by a pair of horses. The coachman sat in the front seat and carried a horn, which he blew to warn off the pedestrians. The tonga had seats for three passengers, one in the front with the coachman and two at the back. There were six stages en route located about 7 miles apart where horses were changed. The passengers' luggage and mailbags were piled on the sturdy roof of the vehicle and the hosteller stood behind, holding on to the back ledge of the roof of the tonga. At the end of each stage, he would take charge of the steaming horses with foaming flanks, which were disengaged from the tonga, and a fresh pair was hitched on, along with a new hosteller. The 'coachman' made the horses continuously run at a smart gallop along the steep winding slopes of the road, which ascended to the high Abbottabad plain.

We arrived at Abbottabad in the mid-winter of 1899–1900. The Abbottabad plain and the surrounding hills were covered with a thick mantle of snow. Thick warm clothes of coarse Kaghan woollen tweed were hastily made available for us and we started playing snowball games with other children. At the end of the winter vacation in early March, I was enrolled in the third primary class at the District Board High School, Abbottabad. Our Hindu teacher, who took my test before admission, was so impressed with my answers to his questions that he asked me to sit at the head of the class. I maintained that position right up to class seven, which I passed in 1904 from that school. In winter, we took our morning meal at 9 a.m. and attended school at 10 a.m. The school closed at 4 p.m. In summer, the school hours were 6 a.m. to 12 noon. In summer, I calculated the school time by noting how far down sunlight travelled along the slope of the Sarban hill. My two cousins were the proud owners of wristwatches, which was not common among youngsters during those days.

My uncle was employed as a district engineer under the direct charge of the deputy commissioner of the district. He designed and constructed schools, dispensaries, village water supply schemes, and flood protection embankments for the district board. He was a man of sturdy build and sported a thick, bushy beard. He was a pious man and regularly said his five daily prayers till the end of his life. He was of a somewhat impressive temperament like his mother (my grandmother) and was extremely fond of good food. Aunt would cook tasty mutton curry and bread fried in clarified butter for him. It was a treat to watch him eat his meal with his beard dripping from his recent bath. For us youngsters, however, there was only ordinary bread and a small portion of the curry on which we fell with voracious appetites. Expensive fruits like mangoes, bananas, and oranges were rare luxuries, but cheap fruits, such as loquats, plums, apricots, pears, and wild fruits including mulberries, ruby-coloured berries, wild figs, and *simbloos*, were always available to us.

There was a large government-owned orchard of apples growing in the Company Garden across the road adjoining our school. One Sunday afternoon, while passing by the orchard, we noticed that no gardeners were in sight. On the spur of the moment, my eldest cousin (aged sixteen) hatched a scheme for stealing apples from the orchard. He planned that he would climb an apple tree, and pass on the plucked fruit to his brother, my younger cousin (aged thirteen), who would hand over the loot to my elder brother (aged twelve) across the low steel wire fence that stood outside the orchard at the roadside. I was told to stand in the middle of the road to report any person approaching from either direction. But the wily gardeners were watching the proceedings from behind some bushes. No sooner had my cousin dropped a couple of apples than two men rushed out of their hideouts and grabbed the younger cousin under the tree with the apples in his hand. They also forced the elder cousin to climb down and got hold of both of them. Then one of the gardeners picked some more apples and, tying them in a larger bundle, placed the load on his head. They proceeded to take my captive cousins to the police station. We two brothers also followed with glum faces.

A neighbour of ours, the Sikh priest of the small Gurdwara situated near our house, spotted us and asked the gardeners to stop and explain where they were taking the boys. The gardeners pointed to the big bundle of apples that we had allegedly stolen and said they were taking us to the police station. Thereupon, we all loudly protested that the gardeners were lying and that we had only taken two apples. This gave our good neighbour an opening to challenge the gardeners. He threatened to accompany us to the police station and state that they were falsely implicating the boys for stealing that big load of apples. This frightened the gardeners and they let us go. But our benefactor gave us a firm lecture on the sin of stealing and made us promise never to steal again.

In the afternoon, after school, we usually played football with a tennis ball against the Gurkha lads of our own age in the beautiful grassy parade grounds of the Gurkha regiment. Our young opponents were tough players and wore sturdy full boots with steel-lined soles supplied by the government while we generally played barefoot and came home with bleeding feet. In the course of time, however, we too became equally tough players. The Gurkha soldiers were noted hill climbers. Annual hill-climbing contests were held among the Gurkha regiments, which attracted huge crowds of spectators from both the cantonment and the city. I remember, on one such occasion, the specified 3-mile route included climbing a steep hill slope to a point 1000 feet high, then a mile-long sprint along the rough craggy top of a hill range, and finally down the slope and back to the starting point. The winner was a young Gurkha soldier by the name of Lachman who climbed the sheer hill slope like a monkey and arrived at the winning post far ahead of the others. Arriving at the winning point he lay down exhausted. Thereupon, his British Company commander started briskly massaging his thighs and calves. This gesture made a great impression on the public.

On a dull cloudy afternoon after a tiring full day's school session, I was the last to leave the classroom for home along with two classmates who had also lingered behind. Walking in a leisurely and erratic fashion, my companions started playfully flinging small pieces of gravel, which lay on the roadside, at each other. A small piece accidentally struck an English gentleman's back, as he was walking by. He quickly turned around as the two culprits ran and hid behind the tall hedge of the adjoining Company Garden. The gentleman looked irresolutely at the escaping figures and then his glance fell on me. I was standing at the edge of road and had taken no part in the gravel flinging game. The English gentleman beckoned me with his stick to come to him. As I had nothing to fear, I readily went to him. He noted down my name and class and walked away.

This incident took place on a Saturday afternoon. On the following Monday, the whole school was gathered in the school compound. The headmaster and the entire teaching staff stood in front of the rows of students. The headmaster called my name and ordered me to come forward. I was publicly caned by him for throwing a stone at the deputy commissioner. I went home weeping and told the tale to my uncle. White with rage, my uncle grabbed a sturdy staff and went with me to the headmaster and demanded an explanation for the outrage. The headmaster was all apologies but said that he was helpless in the matter. He had to carry out the deputy commissioner's orders! Uncle was dumbfounded and could do nothing but silently return home. It was unthinkable to protest against this act of a British deputy commissioner. The deputy commissioner was evidently out to assert his authority. The ethics of the matter were apparently of no concern to him.

Years later, in 1937, as a member of the Indian Service of Engineers, I was working as executive engineer in the Bannu PWD. One day, while poring over an old voluminous file relating to proposals by various authorities for development of irrigation and agriculture on the tract commanded by the indigenous Kachkot and Landidak canals, I came upon a note on the subject by a revenue commissioner of the NWFP. The views expressed in the note were of a somewhat mediocre and impracticable nature. I was not astonished to see the signature at the bottom of the note to be that of the same deputy commissioner who had made an undeserved example of me.

One fine morning in late January 1901, we were playing marbles on the road near our house. A news hawker passed by, shouting: 'Queen Victoria dead, her elegy, price two paisa.' We were already familiar with eulogising poems, some of which we had learnt by heart, concerning the great Queen. These poems were composed by a poet of somewhat mediocre talent whose chief qualification lay in his claim to be a scion of the late Mughal emperors of India. So, his poems were included in our Urdu courses of the third and fourth primary classes. One of the poems, I remember, was in celebration of Queen Victoria's Diamond Jubilee in 1887. The other one was an elegy on the death of Prince Victor, the grandson of the Queen, during the 1890s. A few lines from this elegy are reproduced below from memory:

> 'O Fort William! Why is thy flag this day at half-mast?
> 'O India! Why art thou plunged in despondency?

In 1902, a contingent of Boer prisoners of war arrived in Abbottabad. They were lodged in tents, in a prison camp surrounded by a barbed wire fence, which was located at the foot of the hills near Kakul village. We often visited the Boer camp on Sundays and noticed that the Boer prisoners were treated leniently. They were not as fair as the British, having fairish tanned complexions, but they were said to be 'white men' and I was puzzled that even white men could be made prisoners. The Boer prisoners were adept at fashioning inkpots, paperweights, etc. from the nearly soft limestone rocks using only penknives as working implements. They offered their handicrafts for sale to the public through the barbed wire fence, while the guards looked on without interfering. As prisoners, the Boers were polite to the point of servility. They were eventually released and they left for their homes.

One afternoon in 1903, at the close of school, sweets were distributed among all the students. The occasion was the coronation of King Edward

VII in England. The next day was declared a school holiday. School sports and races were held in the parade grounds. One of the races was a 100-yard handicap race, in which older boys were placed in back rows, and the younger ones in front. I also took part in the race and was put in the front row. I was a fast runner, and was leading the race, when I was tripped from behind by a Hindu boy and fell prone only a few feet short of the winning rope. And so, the Hindu boy won the race. There were cries of 'shame' from the spectators.

After a short while, the boy's handicap race was repeated. I ran to take part in it. The teacher of a rival school, who was in charge of arranging the boys in rows according to their sizes, recognised me as the boy who had almost won the previous race, and placed me in the last row. When the second race started, boiling with rage, I ran as never before, and left all the competitors far behind and easily won the race. There was a shout of applause from the spectators. The prize was three brand new silver one-rupee coins bearing the bald effigy of King Edward VII. Rs 3 was a great deal of money for a schoolboy in those days.

Years later, I learnt that a parallel durbar had been held by Lord Curzon, viceroy of India, at Delhi, on the occasion of the coronation of King Edward VII, which, in its splendour, eclipsed the coronation ceremony in London. Lord Curzon, clad in imperial regalia and followed by British Indian troops in splendid uniforms, had ridden at the head of an elephant procession, followed by an equally imposing retinue of heads of Indian states. The durbar ceremonies included a ball in the splendid audience hall of the Mughal emperors, in Delhi's Red Fort, in which Lady Curzon also took part. Urdu poet, Akbar Allahabadi, immortalised the splendours of the ball in a poem; extracts from his verses are quoted below (from memory):

> *A ball has been held in the Imperial Fort.*
> *Every Imperial Hall has been illuminated,*
> *The splendours of the Imperial past have been revived.*
> *It is the talk of the town;*
> *That the Lady Curzon danced in the hall,*
> *Dazzling all onlookers.*
> *That every other woman looked on her with envy;*
> *She suddenly burst into the Hall, a shining vision*
> *Resplendent in her golden dress;*
> *The sinecure of all eyes, right up to the Heavens,*
> *Although a celestial dancer herself,*
> *Venus lacked her grace.*
> *The ball was a model of heavenly 'Indra's' pleasure fest,*
> *And pursuit of pleasure continued to the early morn.*

You may believe it or not,
We have only heard these tales.
Those who were there would know better
This picture is the creation of a futile imagination.

The poet was, of course, taken to task by the government. It is said that one of the objectives of King George V's splendid coronation durbar that was held later in Delhi in 1911 was to outshine Lord Curzon's durbar of 1903.

Uncle had designed and constructed a village irrigation watercourse taking off from a perennial stream. The work involved some half a mile length of rock-cutting in the head reach of the watercourse. The watercourse head, however, got heavily silted during the summer flood causing serious interruption in irrigation of the fields. The landowners concerned complained to the deputy commissioner and made allegations against uncle's integrity as having fraudulently permitted the contractor to execute less than the designed depth of excavation in the rocky reach of the watercourse. Uncle's boss, the British deputy commissioner, being a layman, readily believed the landowners' allegations. At first, he threatened to dismiss uncle, but on second thoughts reduced uncle's salary by half.

And thus, an era of austerity suddenly descended upon the family. Barley was substituted for wheat for our daily bread. Beef and cheap goat trotters were purchased instead of the expensive mutton. The trotters, moreover, did not require the expensive clarified butter for cooking. Austerity became the rule for all other daily necessities of life. This state of affairs lasted for about six months. One day the deputy commissioner mentioned the matter to a garrison engineer (a fellow Britisher) of the Military Engineering Services who volunteered to check uncle's work and suggest measures for improving the working of the watercourse. The garrison engineer's inspection completely exonerated uncle. The deputy commissioner, to his credit, forthwith restored uncle's pay with retrospective effect. Our days of adversity were over.

In 1904, I was a student of class seven. One day, while we were taking our English lesson, an old European gentleman walked into our classroom. As our class stood up as a token of respect to him, the visitor asked our teacher as to what lesson was being taught and was informed that it was an English lesson and the subject was the fall of Constantinople by Edward Gibbon.[1] The visitor exhibited visible interest in the subject and asked the teacher to ask someone to read out the opening paragraph. Our teacher beckoned me. I stood up and read out the first two paragraphs. 'Eenuff, Eenuf,' exclaimed the gentleman. He then asked me whether I could draw on the blackboard the sketch of the city of Constantinople and the environment as shown in our textbook. I fancied myself rather good at drawing and readily drew the sketch of the city and the Bosporus, the Golden Horn, the Sea of Marmara, and the coast of Anatolia. He smilingly corrected the Anatolian part of my sketch, remarking that the Anatolian coastline was much longer than I had drawn.

After enquiring my name from the teacher and noting it down in his pocketbook, our visitor said goodbye and left as he had come. Shortly after, the headmaster burst into our classroom and asked whether the visitor had disclosed his identity. Our teacher replied in the negative. Further enquiries revealed, that he was the distinguished archaeologist Sir Aurel Stein, who had recently returned from the Sinkiang province of Central Asia (China) after excavations of the sand-buried ruins of Khotan and had since been appointed as education officer by the government of the NWFP of India. He had paid a surprise visit to our school in that capacity and by chance had entered our class!

I lived for five years in my uncle's house at Abbottabad. During this period, my education was laid on firm foundations. Abbottabad was a beautiful place to live in. Viewed from the top of the Brigade Circular hill, which we frequently climbed, Abbottabad town and its cantonment presented a breath-taking view of red gabled houses amidst clusters of dark green pines sprawling over large verdant areas of the 'Rush' plain. In spring, the hedges were decked with dense clusters of wild roses, and the air carried the scent of white and yellow jasmine, honey suckle and *attar* roses. Plum, apricot, loquat, apple, and pear trees were ablaze with white and pink blossoms, and fresh buds peeped out of the bare branches of maples, poplar, and oak trees. The forest-clad heights of the Thandiani range, in the east, were covered with a mantle of snow and the majestic Kaghan snows stood in the far north; pink in the morning sun, they appeared celestial.

The summer monsoon rains of July and August turned the hill slopes a vivid green hue with grass that was harvested in autumn and stored as winter fodder for the cattle. Autumn was perhaps the most beautiful season of Abbottabad, with the leaves and the grasses on the hill slopes turning purplish crimson, and clear azure skies, and transparent air bringing the forests on the hill slopes and the distant snows in sharp focus. Winter in Abbottabad had its own charm since the surrounding hills received heavy snowfall. The Abbottabad plain itself received moderate amounts of snow two or three times in the season, usually followed by days of balmy sunshine when we romped around in the snow and played with snowballs. There were crusts of ice on the surface of springs and brooks. The air was crisp, cold, and exhilarating. I had climbed all the accessible peaks around Abbottabad and tasted the wild fruits of the hill slopes, the *simbloos*, the *singlees* (local berries), wild figs, and prickly pears.

3 Sialkot – Dharamshala – Abbottabad

In 1904, my father returned to Abbottabad after a lengthy period of service in distant places. His last assignment, as supervisor on the construction of the Quetta–Nushki Railway in Balochistan, carried a high salary, enabling him to save a sizeable amount of money. He had resigned from that post after a long bout of typhoid fever at Nushki, which had left him weak, dispirited, and discouraged. He resolved to turn a new leaf and start a steel trunk manufacturing business in Sialkot. And so, the family returned to Sialkot, and we settled down in grandfather's small house. Father's company made a promising start. The quality of our product was superior to that of our numerous competitors and purchase orders started to pour in. But he was an honest man, and a novice in the devious ways in which business was done. Local postmen were bribed to deliver the postal orders for the supply of steel trunks addressed to father's firm to his rival companies, mainly owned by well-established Hindu manufacturers. They would respond to these orders saying father's company had gone out of business and dispatch a steel trunk of their own on approval basis. Their tactics, not known to father, led to the ruin of his enterprise within a few months and a total loss of his investment.

The next year, in 1905, a notable event took place in Sialkot. Mirza Ghulam Ahmad, the founder of the Ahmadiyya movement proclaimed his much-publicised 'message of peace' to non-Muslims (namely Christians and Hindus) in the Raja Mandi Serai of the city. Long before he delivered his message, there was great excitement in the streets, both among Muslims and non-Muslims. Mirza Ghulam Ahmad claimed, on the authority of the Quran, that Jesus had not been crucified and had died a natural death. But since a messiah was expected by almost all religions, he, Mirza Ghulam Ahmad, was the messiah as prophesised. His claim to prophethood antagonised most of the Muslims, who previously had lauded his services to Islam in his religious debates and encounters with Christian missionaries and Hindus.

Some years before his claim to prophethood, Mirza Ghulam Ahmad had served as a *munshi* with my grandfather, while the latter was the construction contractor of the NWR Bridge over the Chenab River between Wazirabad and Gujrat stations. Both my father and grandfather were present in Sialkot on the occasion of Mirza Ghulam Ahmad's delivery of his 'message of peace'. At that time, I was fourteen years old and an interested listener; I remember father asked grandfather, as to what he thought of the 'character' of Mirza Ghulam Ahmad, who had served with him for many months. Grandfather pondered for some time and recalled an incident which he described as follows: 'On one occasion, Mirza Ghulam Ahmad remarked: "Khan Sahib, God Almighty has been kind to you" [meaning he was well off in a worldly sense]. However, I am always worrying about my future and have to do something about it.'

My father was a scholar of Persian and Arabic, and had received sound religious education before he joined an English medium school. He turned to me and said: 'I am convinced that Mirza Ghulam Ahmad is not a prophet as he claims, but he rendered valuable services to Islam in successfully countering the Christians' and Hindus' attacks on our religion. So, we shall go to Raja Mandi, and hear what he has to say. And you will also have an opportunity of seeing this remarkable man.'

I accompanied my father to Raja Mandi to hear Mirza Ghulam Ahmad's message. It was a hot sultry afternoon. The dirt courtyard of the serai was chock full of people of all religions. The air was thick with dust. We could only secure standing room at the edge of the crowd. Presently, Mirza Ghulam Ahmad arrived in a 'tum-tum' (a one-horse carriage of primitive construction) accompanied by three or four companions. He was clad in a long white robe that fell well below the knees over a wide white pyjama, and wore a voluminous white muslin turban. Mirza Ghulam Ahmad was of medium height and build with a remarkably fair complexion; he had bobbed hair and a thick beard, both dyed red. His companions were clad in similar attire. Owing to the deafening noise of the crowd, we could hear nothing of Mirza Ghulam Ahmad's 'message'. The next day it was out in the newspapers. The message primarily was a repetition of the invitation of peace extended to the Jews and Christians, in the Holy Quran, with the added offer to the Hindus that he would acknowledge Sri Krishna as a prophet of God at par with other prophets, provided the Hindus acknowledged him as the 'promised messiah'. The Hindus, of course, rejected the offer as evidenced by hostile posters, which appeared on the walls of the town the very next day, under the caption 'Challenge! Challenge!'

In 1906, father got a job as an overseer in the PWD at Dharamshala[1] hill station for work on the reconstruction of the Dharamshala cantonment which had been devastated by a severe earthquake a year earlier. I was left at Sialkot as a boarder in the Government High School, where I was a matriculation student. That year, the British Crown Prince George and Princess Mary paid a visit to India and, in due course, visited Jammu, the winter capital of the Kashmir state, as guests of the maharaja of Kashmir. A magnificent camp was set up in the Satwari Gardens on the outskirts of Jammu for the imperial guests.

One Sunday afternoon, a fellow boarder and I went to Satwari, by train to see the splendour of the much talked about Satwari camp. While we were standing by the side of an ornate, flower-decked road in the Satwari Gardens, a phantom drawn by a couple of magnificent black horses approached in our direction. The road was deserted. Prince George and Princess Mary were enjoying a leisurely drive in the gardens, with the heir apparent of the maharaja of Kashmir, a lad six or seven years old, was seated between them. Just as the carriage was passing in front of us, Princess Mary kissed the little lad. They passed by, sublimely unconscious of the presence of the two boys standing by the roadside gaping at them.

One day, during our English lesson with the headmaster, a distinguished looking Englishman, entered the classroom unannounced. The headmaster and the whole class stood up. The topic under study was the famous oration on 'the quality of mercy' by Portia, in Shakespeare's play *The Merchant of Venice*. The visitor asked if anyone knew this passage by heart. From a sign from the headmaster, I stood up and fluently recited the whole passage from memory. The visitor appeared pleased; I later learnt that he was Sir Francis Younghusband, the then resident in Kashmir, and I felt proud of the fact that I had earned approval of a distinguished person.

I passed the summer vacation of 1906 with the family at Dharamshala cantonment. It was a beautiful hill station situated on the grassy lower slopes of the magnificent snow-clad Dhauladhar range of the outer Himalayas, which commanded a grand view of the plain of Kangra for hundreds of miles. The pony track from Shahpur in the plains to Dharamshala was lined with a dense forest of magnificent trees and undergrowth, thick with medicinal herbs. Some Gurkha regiments were stationed there. I was on friendly terms with the Gurkha lads of my own age. We played marbles with buttons as stakes. I, being fairly adept at the game, had my pockets full of buttons.

After the failure of his industrial enterprise, father considered his PWD overseer's job at Dharamshala as a lucky break and went about his duties with great enthusiasm. He was required to lay down the alignment of a cart road for wheeled traffic along the hill slopes, at a uniform gradient from Dharamshala to Matour in the plain, with a vertical drop of 3000 feet over a distance of 12 miles. His boss, the sub-divisional officer (SDO), despite father's opposition insisted on modifying a certain length of the alignment, which had been proposed by father. W.S. Dorman (executive engineer) inspected the alignment for final approval. When he reached the spot where the alignment had been altered under the sub-divisional officer's orders, he stopped short, looked at father and said, 'Which idiot of an overseer has bungled this part of the alignment? Why shouldn't the overseer be dismissed?' Father looked at the SDO who remained mute. All of a sudden father lost his temper and threw down the drawings he was carrying. Mr Dorman abruptly turned around and started walking down the alignment as if nothing had happened. Realising the gravity of his impulsive act, father picked up the drawings and started walking behind the officers. The rest of the inspection passed off without incident. A few days later, an official letter was received from the executive engineer confirming father in the post of senior overseer in the PWD, which meant a great deal to him. He had, henceforth, security of employment with prospects of further promotion and a pension. It transpired that at the end of his inspection, Mr Dorman had taken the SDO aside and questioned him about the overseer's strange conduct. And the SDO had probably admitted his responsibility in the matter.

At the end of the summer vacation I returned to Sialkot and back to a boarder's life. The messing charges in our boarding house were only Rs 2 per month. The curry supplied by the cook consisted of *daal* only, without any frying medium such as *ghee* or oil. The majority of the boarders came from rural areas, and they brought with them supplies of *ghee* to add to their curries. I had no such resources and so passed my year in the boarding house in relative austerity. In March 1907, I passed my matriculation examination from the Government High School, Sialkot, with distinction, standing first in Sialkot district and securing a merit scholarship. With a light heart, I went to join my family at Kangra, where father had since been transferred. There was great jubilation in the family at my success, and father took me and my elder brother (a student of mechanical engineering at the Bombay Technical Institute) on a pleasure trip to Dera Gopipur on the Beas River, a famous fishing station situated on the boundary of Kangra and Hoshiarpur districts of the Punjab province.

On the way, we visited the famous Hindu temple of Jwalamukhi, situated on a small rock outcrop. The 'idol' of the temple was a natural sulphur flame over which the small temple edifice had been erected. It was claimed that the flame remained lit and was never extinguished. We also noticed natural deposits of sulphur lying among the dark rocks of the hillock. It was said that Emperor Akbar had paid a visit to the temple and had presented an offering of a golden umbrella and a charter granting the revenues of a number of surrounding villages to the temple priests. The British government had also honoured Akbar's charter and the priests were still enjoying the revenues. Dera Gopipur was a pleasant little town situated on the right bank of the Beas River. At that point, the river flows with a rapid current, in a steep gravely plain. It has a width of about 150 yards between high vertical banks. Dera Gopipur is famous as a fishing resort. We purchased a thirty-pounder Mahseer fish.

The town of Kangra was a pleasant place, situated in a fertile plain. Its rice crops were watered by small perennial streams that were fed by Himalayan snows. The Buriganga hill torrent flowed nearby in a deep gorge in a bed of large boulders and, in those days, was teeming with Mahseer and other game fish. The town of Kangra was built around its famous temple. Nearby were the ruins of the once imposing Kangra Fort with its stone ramparts extending over a wide area. A copious spring exists within the fort area. Sultan Mahmud of Ghazni plundered both the fort and the temple during the course of his campaigns in about AD 1010. I spent a pleasant time in Kangra taking long solitary walks around the town and countryside, enjoying the magnificent views of the majestic Dhauladhar range stretching for hundreds of miles in an east-west direction a few miles away to the north. I longed to join a college, but father had other plans for me. He planned a business career for me and so I was sent to Abbottabad to serve a term of apprenticeship in the general store run by my two first cousins. The intention was that I would then open my own business in that town and would be joined by father on his retirement

My cousins ran a large general store in the Abbottabad bazaar. Along with two shop assistants, I opened the shop every morning at 6 a.m. sharp and remained on duty until 10 p.m. Our noon meals came from home. The servants swept the shop's floor while I dusted the goods and stacked them in the shelves and was also in charge of the till. Our military customers were chiefly Gurkha women who purchased glass bangles. The Indian soldiers (both Hindu and Muslims), with their meagre pay, could only afford to purchase such trifles as thread, needles, and buttons. Goods were sold in our shop at reasonable and fixed prices, which attracted regular customers. My cousins arranged more capital and decided to add books and shoes to the store provisions. The elder cousin took me along on 'wholesale' purchasing missions to the towns of Karnal, Ambala, Delhi, Agra, and Kanpur. At Karnal and Ambala, we purchased cheap books and shoes of local indigenous manufacture. At Kanpur we purchased superior quality stuff from two British firms of boots and shoe manufacturers. The bigger of the two firms, Messrs Cooper Allen & Co., sold their slightly defective manufactured goods at half price. My cousin purchased a lot of the defective stuff and sold it at full price at his shop.

I spent a year at my cousins' shop, but a business career did not appeal to me. When there were no customers around, I would read books. John S.C. Abbott's *The History of Napoleon Bonaparte* was, in those days, my favourite reading. I would sometimes even neglect customers while poring over some book. On one or two occasions, I sold some articles below their approved sales price on the customer's plea of poverty. On one such occasion, my magnanimity was detected by my elder cousin who got furious and wrote to father asking him to withdraw me from the shop as I was unfit for doing business. On my part, I beseeched father to allow me to join a college. Father acceded to my request. I had passed my matriculation examination with distinction the previous year and was admitted, in June 1908, to the intermediate class of the prestigious Government College, Lahore.

4 College Days

Government College, Lahore (1908–1910)

Captain E.W.C. Sandes, RE, our professor of civil engineering at the Thomason Civil Engineering College at Roorkee, some years later remarked, in the course of a lecture on buildings, that the building of the Government College, Lahore, was a classic example of a faulty design; it was more like an elegant cathedral than a college. Yet it was a noble Gothic structure, exquisite in design, as well as construction, with an imposing tower capped by a lofty spire, which housed an astronomical telescope. The college grounds had a set of tennis courts and a cricket pitch for net practice. The front lawn was the famous 'Oval', accommodating a full-scale hockey or football ground where we played league matches. The college's regular cricket ground lay in the extensive Chauburji ground situated a mile away.

I was able to gate-crash into the Government College's cricket First XI. I started attending net practice uninterrupted, doing only fielding for a few days. One day towards the close of the evening practice the captain invited me for some batting. I promptly started punishing the college's top bowlers, hitting them all over the ground. Impressed, the captain invited me to come to practice the next day. On the following day, I was told to put on the pads first and the captain, himself a fast bowler, along with five spin bowlers started bowling to me in right earnest. I made a polished batting display all around the wicket by playing both on the back and front foot, executing a variety of strokes. On the third day, Mr Jones, professor of chemistry and the president of the cricket club, watched my play and nodded his approval. I was taken into the First XI and allotted one-down position in the batting order. Incidentally, I had been a member of the School Cricket XI from the fifth primary class onwards.

At Government College the principal and all the professors were British, and the assistant professors and lecturers were Indians; of the latter only one was a Muslim. The Muslims formed only a quarter of the student body, which was predominantly composed of Hindus and Sikhs. The latter considered the prestigious college their own special preserve. A veiled sense of rivalry prevailed between the two communities. Keen competition existed between the Hindus and Muslims in the academic and sports spheres. Personally,

I felt impatient to match my mettle against all competitors, the majority of whom had also passed their matriculation in the first division. To me it was a pleasant dream to have returned to academic life after the dreary year spent as a shop assistant.

The college had a single hostel for the boarders which was called the Quadrangle. This was a single-storey square structure with two wings consisting of dormitories for junior students, and the other two wings had small cubicles for seniors. The Muslim students, about forty in number, comprised about one quarter of the hostel population. There were separate dining halls and kitchens and communal baths for the two communities. Messing was under the private management of students. Our usual meals consisted of mutton and vegetable curry and *chapatis*, with the addition of 'fashion' once a week of *pulao* and rice pudding. Our messing costs amounted to about Rs 9 a month, which was reportedly the costliest amongst local colleges. A Hindu sweetmeat shop and a Muslim fruit seller's shop were located within the hostel compound. On summer mornings, while on our way to the college at 6 o'clock, we would have a glass of cool diluted milk at the Hindu sweetmeat shop; the owner kept these glasses in rows on separate tables for Hindus and Muslims. A glass of cool diluted milk cost one *anna* (one-sixteenth of a rupee). That was our sole breakfast in summer. During winter we had our brunch before attending college at 9 a.m.

A few weeks after I joined the college, a simmering hostility between the Hindu and Muslim students flared up in the Quadrangle hostel where the two communities lived in close proximity to each other. Some Hindu students of the hostel reported to the college authorities that two Muslim boarders had committed an unnatural offence. The principal, Mr Robson, rusticated the Muslim boys in question on the verbal evidence provided by the Hindu students. Since the charges were believed to be false, as the Muslim students accused of the offence bore good character, most of the Muslim boarders as well as day scholars went on strike to protest the unjust decision. We held our daily meetings in the office of a local Urdu daily newspaper presided over by the editor of the paper. Eventually a delegation of influential Muslim citizens of Lahore intervened with the principal, and the two rusticated students were reinstated.

A Muslim boarder took his revenge on the Hindu boarders who had reported against the Muslim boys, in a bizarre fashion. There were rows of common corrugated iron latrines erected at a short distance away from the Quadrangle for the use of the boarders. At rush hours, rows of sweepers were in attendance behind the latrines who removed the used iron pans from behind and quickly cleaned and replaced them for the next visitor.

The Muslim boarder in question, bribing the sweepers, disguised himself with a dirty turban, and holding a brush and a can full of coal tar, took up his position among the row of sweepers. As soon as one of the marked Hindu boarders entered a latrine, the compartment was signalled to the Muslim avenger who silently replaced the sweeper of that compartment and liberally smeared the victim's bottom with a thick coating of coal tar through the aperture below the latrine. This process was repeated in the case of all the accusers.

There were also bitter sectarian squabbles between the Muslim student boarders themselves. Some Muslim boarders belonged to the Ahmadiyya sect, which believed Mirza Ghulam Ahmad to be the promised messiah. Whereas, the Muslim boarders of the Sunni sect, who were in the majority, believed that prophethood had concluded with Prophet Muhammad (PBUH) and that Mirza Ghulam Ahmad's claim was false. The brain of the Ahmadi group was Chaudhry Zafarullah Khan,[1] the son of a Sialkot lawyer who had purchased my grandfather's big house in auction. So Zafarullah Khan was brought up in the same house in which I was born. He had matriculated the same year (1907) as I from a different school but was one class ahead of me in college as I had spent one year in my cousin's shop. We remained on friendly terms over the years.

The brawn of the Ahmadi group was Chaudhry Mohammad Sayal, a BA student, hailing from the rural areas of Jhang district. The Sunni champion was Malik Barkat Ali, a first-year student, from the rural areas of Gujranwala district who was a match for Sayal in physique and strength. One summer morning, these two stalwarts were seen in a grim, silent fistfight in the bathroom, landing blows on each other, fit to fell an ox. This was evidently their way of settling their religious differences. After both had inflicted enough punishment on each other, they silently left the bathroom.

The Ahmadi–Sunni differences came to a head on the arrival, in Lahore, of the old and ailing Mirza Ghulam Ahmad in the summer of 1908. He was put up in a room on the first floor of a follower's house situated on Railway road. Close by, on a vacant lot, some Sunni *maulvis* set up their headquarters in a bid to draw the Mirza into a public debate. The most prominent among them was Pir Jamaat Ali Shah of Sialkot, a divine, noted more for his piety than learning. Mirza Ghulam Ahmad, however, was past having any dialogues, being in the throes of his last illness. His spiritual successor Maulana Nur-ud-Din, a hakim, held talks on the ground floor of the same building. We sometimes attended these meetings. On one occasion a student asked the Maulana whether he had ever had communion with God. 'Yes, I talk with Allah every day,' replied the Maulana. 'But have you seen

God,' persisted the student. In reply Maulana Nur-ud-Din enigmatically responded, 'Can you see Time?' A few days later Pir Jamaat Ali Shah in the course of his usual evening sermon loudly challenged Mirza Ghulam Ahmad to a religious debate within the next twenty-four hours. 'Otherwise, Allah Himself would decide the issue.' The next morning Mirza Ghulam Ahmad died. The incident gave a tremendous boost to Pir Jamaat Ali Shah's reputation as a 'saint'.

The Ahmadiyya community was stunned by the death of their 'prophet'. A directive was hastily circulated amongst them not to enter into religious controversy with non-Ahmadis even under the gravest provocation. A host of satirical elegies were published and had brisk sales. The mortal remains of Mirza Ghulam Ahmad were conveyed by train for burial to his native town of Qadian in Gurdaspur district. I went to the Lahore railway station to see the scene of the train's departure. The Pathankot Express was standing on platform number one. Mirza's bier had been placed in a goods wagon, which was attached to the Express. A first-cum-second class bogie and also a number of third class bogies were reserved on the train for the Ahmadis who were accompanying the bier to take part in the funeral rites at Qadian. Maulana Nur-ud-Din was sitting in a third class carriage with downcast bloodshot eyes heavy with grief. Mirza Bashiruddin Mahmud Ahmad, Mirza's eldest son, a fair complexioned and well-dressed young man in his early twenties, sat in a first-class compartment, busily thrusting out his hand for being kissed by a long queue of Ahmadis on the platform.

Government College, Lahore, was not only the most elite institution in the province, it had a well-deserved reputation for the high standard of education imparted there. Principal Robson always took English lessons for the incoming first year students himself during the first few weeks. He advised the students to come prepared for the next day's lesson after consulting dictionaries for different word meanings. Our regular professors of English language also followed this practice. The mathematics professor always gave us homework in continuation of the day's lesson and insisted on its regular compliance. There were well-equipped chemistry and physics laboratories for practical work, along with lectures on theory delivered by highly qualified British professors. On Saturdays, there were group meetings in which the students read out essays, recited poems, and so forth. I belonged to the group of Mr A.S. Hemmy, professor of physics.

In 1909, Dr Mohammad Iqbal (aka Allama Iqbal) was appointed professor at the Government College, to teach philosophy to the senior classes and English poetry to intermediate classes. He came to the classroom immaculately dressed in a dark English suit and wore a Turkish Fez. During lectures, he kept a serious mien and seldom smiled. I remember only one occasion on which he exhibited signs of emotion. While teaching Alfred Tennyson's beautiful little poem titled, *Break, Break, Break*, Dr Iqbal remarked with some emotion that only a poet could appreciate the feelings of another poet! He was in his early thirties but, owing to his habitually serious temperament, looked older than his age.

An Urdu poetry competition was held annually in the Government College, Lahore, carrying Lala Jiya Ram Trust's first prize of Rs 50, willed by the late Professor Jiya Ram (1888–1907) of the college, and a noted scholar of Urdu literature. The competition was held in the college hall, with the principal in the chair. The college staff and the students were given leave to attend the competition. Interested students from other colleges also came to attend the event. Each competitor recited his poem before the audience. In the end a panel of judges from amongst the college staff declared the prize-winning poem.

In 1910, I took part in this competition with an Urdu poem titled *A Morning in Kangra*. The panel of judges consisted of Dr Iqbal and assistant professors Lala Ruchi Ram Sahni and Mr Ram Chandra. Dr Iqbal declared my poem as the winner, but Lala Ruchi Ram Sahni disagreed and declared the poem titled *Naya Kaaba* (*New Kaaba*) by my friend and classmate Abdul Qadir to be the best poem. In bantering verse, Abdul Qadir had rebuked Iqbal for censuring the Hindu Brahmin for his worship of 'stone idols' instead of 'mother India' in one of his recently published Urdu poems titled *Naya Shiwala* (*New Altar*). In my poem, I had recalled in nostalgic verse Sultan Mahmud's invasion of Kangra early in the eleventh century. The controversy over the winning poem caused a minor sensation. In the midst of the subdued hubbub, Principal Robson rose and asked Dr Iqbal as the principal judge to declare the winner. Dr Iqbal got up and recited the first four lines of my poem before the audience, with their English translation, and declared my poem to be the winner.

One Sunday morning, I was chatting with two classmates in my cubicle in the Quadrangle when, to my amazement, my father walked in unexpectedly. My companions, guessing that the visitor was my father, rose up and respectfully greeted him. When they left, father asked many questions about them. I informed him that one was the son of an inspector of schools, and the other that of a deputy commissioner. Furthermore, both were good sportsmen, and held high positions in the class. Father then asked some more searching questions about my other friends and about our college life in general. We had lunch together which I procured from our kitchen. After lunch, father took me to the Anarkali Bazaar and to my further amazement, ordered a new suit of clothes for me, and also bought me some new shirts and a pair of new shoes. 'Since you are in good company here, you need some decent clothes,' he casually remarked. He left in the evening. It transpired that my elder brother who had paid me a visit earlier had reported to our father, that I was keeping bad company of rich boys and was neglecting my studies. And so father had come to make personal enquiries!

Our university's intermediate examinations were held in March 1910. One episode during the examination stands out in my memory. The physics practical examination of the candidates from all local colleges was held in the spacious zoology hall of the Government College. It was a two-hour long process involving the performance of three practical experiments and writing out the results. The invigilator was a Hindu gentleman from some other college. I completed my work in 45 minutes and, after going over the results of the experiments for another 15 minutes, rose to hand in my paper. The invigilator hurriedly came to me and enquired whether I wanted paper, pen, or ink. I said I wanted to hand over the paper. Seeing that I was a Muslim, he smiled and said, 'There is no hurry; there is still one full hour left, sit down and do your best.' I sat down with a calm composure, more so, as some other students had begun to cast amused glances in my direction. Going over my answers for another 15 minutes I finally rose up. The invigilator came to collect my paper with the remark: 'All right, if you must hand over.' After giving him the paper, I collected my belongings and walked towards the exit door, while the invigilator was idly going through my paper. Before I left the room he called me back, 'Which college do you belong to?' he enquired. 'I belong to this very college,' I replied. 'That explains it,' he said. 'You can very well go; your paper is one hundred per cent correct.' I passed the intermediate examination, with English, Science, and Mathematics as compulsory subjects, attaining eighth position in the University of the Punjab, and was awarded a government scholarship. In the optional subject of Persian language, I attained the first position in the university.

My father, Mohammad Din Khan, had started his engineering career as an overseer in the 'subordinate' ranks of the Public Works Department. By sheer dint of hard work, honesty and perseverance, he rose in official position and personal prestige. He would undoubtedly have risen further to an officer's position had it not been for his untimely death at the young age of forty-three. He had great ambitions for me to enter the Superior Engineering Services by competing for selection in the highly competitive Thomason Civil Engineering College at Roorkee in the United Provinces. There was a substantial government scholarship available for Muslim students of the Muhammadan Anglo-Oriental (MAO) College, at Aligarh (later Aligarh Muslim University) for entering the Roorkee Engineering College. Entry into the engineering college was through a tough all-India competitive examination. I, however, resolved to try my luck and decided to join the Aligarh college for my BA degree, as a preliminary step towards the ultimate goal of an engineering career in the prestigious Indian Service of Engineers (ISE).

The Muhammadan Anglo-Oriental (MAO) College, Aligarh (1910–1911)

The newly constructed Minto Circle hostel, in which I was housed in a dormitory along with students from outside the province, was situated in a wilderness in a barren plain, a mile away from the college buildings. Only one quadrant of the 'Circle' had been constructed in which we were housed. On moonless nights, the hostel would plunge in darkness due to lack of outdoor lighting. On those nights, the howling of jackals and the distant ghostly boom of cannons, allegedly emanating from the direction of the ruined Aligarh Fort, situated a couple of miles further away from the hostel, rendered the surrounding wilderness gloomy and depressing. A bloody indecisive battle had been fought between the British and the Maratha defenders of the fort a century earlier in which the British were said to have fared the worst. Ghostly scenes of that battle were said to be seen at the fort every evening ever since. A number of expeditions of students led by professors had visited the fort to pinpoint the source of the sounds of cannons, but when they arrived at the fort the sounds appeared to come from somewhere else. These ghostly sounds were also heard regularly, in the evening all over the college campus. Their source, however, remained a mystery.

Unlike the noble structure of the Government College, Lahore, the Aligarh college building was a great disappointment. It was a long featureless barrack-like structure with a gabled-roofed central hall, flanked on both sides by rows of twin lecture rooms with verandas on both sides. On one side of the building was the imposing college mosque. Prayers were compulsory for students, but defaulters were treated leniently. The college building was enclosed by two students' hostels, the Sir Syed Court and the Syed Mahmud Court. Both were one-storey quadrangles, the college building forming their fourth common side.

The Syed Mahmud Court was the most coveted hostel and housed studious scholars. The Sir Syed Court housed students mostly from affluent families and lacked the academic atmosphere of the Syed Mahmud Court. After about a fortnight, I got thoroughly sick of the Minto Circle hostel, and asked Dr Ziauddin, professor of mathematics, who allotted rooms to students in hostels, to shift me to the Syed Mahmud Court. I hinted that I longed to return to the Government College, Lahore, where I had been promised the secretaryship and later captaincy of the cricket club in my BA final year. Since I had passed the intermediate examination with distinction, Aligarh college obviously could not afford to lose me. So, I was promptly allotted accommodation in the Syed Mahmud Court. My roommate was M.M. Sharif who later took his MA degree in philosophy from University of Oxford, and was, for many years, professor of philosophy at the University of the Punjab, Lahore.

Meals consisting of mutton curry, *chapatis* and a small dish of *daal* were provided by the college authorities through a contractor. The quality of food was much inferior to that at the Government College, Lahore. The mutton curry dish prepared at Aligarh, chiefly for the students from United Provinces (UP), was full of stinging red pepper and was hardly edible. A substitute dish termed the 'patients' dish' was prepared for weaker palates. It was a pale and unpalatable concoction, served with equally unappetising oven baked bread which tasted of washing soap. So, I and other fellow students from the north were perpetually famished. There were, in addition, also 'fashion' dishes of mutton *pulao*, and a delicious bread pudding once a week. Our breakfast consisted of a cup of milk with a couple of biscuits. MAO college at Aligarh was the premier Muslim educational institution in India, founded by Syed (later Sir Syed) Ahmad Khan in the 1860s, shortly after the Indian Mutiny of 1857 (aka Sepoy Mutiny or First War of Independence). Its mission was to introduce the depressed Muslim community of India to modern English education, which they had hitherto shunned as an insidious tool of the infidel usurpers of the Mughal throne. It was also to help them compete with the Hindu community which had readily adopted the language of the rulers and were rapidly leaving the Muslims behind in securing government jobs and patronage. The college was run on the lines of the English public schools, by securing, in the initial stages, the services of some eminent English educationists. A measure of freedom was allowed to the students of the college which was unknown in the government-run institutions.

One programme that became a custom was the 'mud riot' which took place during the first heavy shower of the summer monsoon that usually occurred each year in June. That day was declared a college holiday and the students smeared each other with mud; new students being the main victims. In June 1910, the year of my joining the college, the students proved so rowdy that the honorary secretary of the college, Nawab Viqar-ul-Mulk, forbade the holding of mud riots in future. Another distinctive feature was the college uniform. The students were expected to wear dark *sherwanis* (tight fitting long coats buttoned up to the neck), red fez caps, and white *pyjamas* or *shalwars*. This distinctive uniform of Aligarh students was considered a hallmark of good education and good manners.

I was cordially invited to join the College Cricket XI but decided to join the riding class instead, which I figured would prove more useful in my prospective engineering career. The college maintained a stud of thirteen horses in the riding club. The captain of the riding class was a senior student, and our coach was a retired non-commissioned officer of a cavalry regiment. The training started with the 'master mare', a gentle creature, which seemed to sense the novice mounts on her back. While riding the master mare, the students learned the correct method of mounting a horse and the correct way to sit. As they became adept in riding, the students progressively rode more spirited ponies, ending with a rough pony which had the propensity to grab the bit in his teeth and run away with the rider back to the stables. The riding course covered seven months and included learning to trot, canter, gallop, and charge in formation. It also included show jumping and tent pegging.

A lad from the 'English House' hostel of the Aligarh Collegiate School joined the riding class. The English House was reserved for sons or relatives of Muslim rulers of Indian states, and from other important families. The new entrant hailed from East Bengal. On one occasion, after he had acquired fair proficiency in riding, he was allotted the docile 'master mare' to ride that day. In protest he burst into tears. I have a clear recollection of the lad due to that trivial incident. That lad was Khawaja Nazimuddin, later to become the governor general of Pakistan, during 1948–51.

I had arrived at the college rather early, and for the first few days was the sole chemistry student in the BSc class. One day, while conducting practical experiments in inorganic chemistry, I proceeded to elaborately wash the pestle and mortar before starting the experiment. Professor Allah Bakhsh, under whose personal supervision I was conducting the experiment, grew impatient at my extra precautionary measures, grabbed a pestle and mortar and without washing it dropped the dry compounds in it and then poured some hydrochloric acid over them and started briskly pounding the mixture. He was suddenly overpowered by fumes and fainted, dropping down in a swoon. Alarmed, I rushed for help to the professor of physics in the next room, who hastily administered some ammonia and other restoratives to the unconscious Professor Allah Bakhsh and brought him around after half an hour's struggle.

The mathematics periods under Dr Ziauddin, a noted mathematics scholar, proved discouraging. Firstly, our mathematics class was overcrowded, and besides, I found it difficult to understand the doctor's pronunciation in the noisy classroom. With similar conditions in some other classes and the heavy workload of five tough science subjects, I got thoroughly discouraged.

My roommate M.M. Sharif, an arts student himself, also kept constantly needling me that, as I had stood first in the Persian language in the University of the Punjab, an arts career as a teacher was best for me. So, with considerable qualms of conscience I took up the arts subjects of History and Persian language in place of those for a BSc degree. I wrote to my father about my change of subjects, which must have pained and shocked him, for it was his cherished ambition that I should pursue a career in engineering, but in his reply, he supported my decision.

During the summer holidays of 1910, I went home to Abbottabad. One day at noon, I was walking down the cantonment road carrying an umbrella, when a British military officer and his wife approached from the opposite direction; my cousin, who was with me, hastily told me to close my umbrella. With innate pride and being conscious of the fact that I was a BA student, I refused to yield to the servile custom of closing one's umbrella in the presence of a Britisher. When the couple came abreast of me, the military officer, livid with rage, snatched the umbrella from my hand and threw it in the gutter and after swearing at me went away. His wife kept looking on without a word of protest against this act of wanton aggression.

A similar incident occurred at Abbottabad a few days later with a couple of distinguished Indians—a Muslim judge of the Lahore High Court, and a leading Muslim lawyer. Both were later knighted by the British Indian government. The pair was walking along the same cantonment road when they were accosted by a British military officer. The latter expected them as usual to greet him but they took no notice. 'What is your name?' demanded the military officer, arrogantly. 'My name is Shah Din,' he said. 'But you should also have said, Mr Justice of the Lahore High Court,' added the lawyer companion of Justice Shah Din. Taken aback, the British officer said, 'Good morning,' and walked away.

In winter, distinguished Indian personages were invited to visit the Aligarh college. In the winter of 1910–11, Dr Iqbal came and recited an Urdu *ghazal* in the college hall to a crowded audience of students. The UP students, whose mother tongue was Urdu, were critical of Iqbal's alleged misuse of Urdu idiom.

Another distinguished visitor from the Punjab was Maulana Zafar Ali Khan, who delivered a religious lecture in the college hall. It was customary for college students to be permitted to assemble in the college hall to listen to these lectures by eminent personalities. The personality of Maulana Zafar Ali Khan did not strike me as very impressive. Occasionally, our visitors were eminent non-Muslims as well. During the same winter, we listened to a lecture from the distinguished Indian nationalist leader, Sarojini Naidu. She was a youngish, good-looking lady, elaborately dressed in a silk sari and wore expensive jewellery. She delivered her message of Indian nationalism in impassioned tones, in faultless English.

Our constant visitors, however, were the Ali Brothers from Aligarh. Maulana Mohammad Ali had just returned from the UK after taking his MA degree from Oxford, and had just started his, later famous, weekly journal *Comrade* from Aligarh. A joke about him was going around in the college. He had received a 'telegram' with the abbreviated address: 'Mohammad Ali Oxon'. The local telegraph clerk translating it in Urdu for the benefit of the Urdu-knowing postman, who had to deliver the telegram, had written on the envelope 'Mohammad Ali, Bullock Owner'. Both the brothers had a lively sense of humour. The elder brother, Maulana Shaukat Ali, in particular, often regaled the college students with lectures laced with hilarious jokes.

Maulana Mohammad Ali, more than anyone else in India, during the course of the long struggle against British Imperialism in India, was responsible for relieving the fear of the British rulers from the minds of the Indians. Brought in chains before a British judge for publishing his epic edition, 'The Choice of the Turk' in the *Comrade*, on the entry of Turkey on the side of Germany in the First World War (1914–18), he sat down on the floor of the court-room disregarding the judge's order to stand up. When asked by the judge whether he pleaded guilty or non-guilty, he said he did not recognise the authority of the court to judge him. The judge acidly remarked, 'You perhaps do not have the courage to acknowledge your guilt,' to which Maulana Mohammad Ali retorted, 'My courage is not in question, my courage is not on trial!'

All India Competitive Examination

The students of the UP had an overwhelming majority in Aligarh college; they tended to look askance at students from the Punjab and the NWFP for their 'lack of sophistication'. There was no love lost between the two groups. There was keen rivalry in the academic and sports fields as much as in all other activities of college life. In the final examination of the third year (BA), which was conducted by the college authorities, I topped the list in my class of 180 students and was declared the 'senior scholar', a position of considerable prestige and some privileges. I, thereby, incurred the resentment of some UP students and professors.

In obedience to my father's wish, I consented to try my luck, just once, for the All India Competitive Examination for entry into the civil engineers' class of the Thomason Civil Engineering College at Roorkee, due to be held one month later. The selection was highly competitive and, in that year, only twenty candidates were to be selected from all over India. I obtained leave from college and went to Roorkee for intensive preparation and for taking the examination there. Twelve other students from our college, mostly graduates, had already gone to Roorkee ahead of me for the same purpose. They had rented a bungalow in the cantonment and engaged a cook for preparing meals. I joined them and was allotted the tiniest room, which I shared with a Punjabi candidate. All the other candidates hailed from UP. Sumptuous meals were prepared for us. Our mess manager was a burly student from UP with an overbearing temper who fancied himself as a pugilist and a 'muscleman'. He regarded me and my Punjabi companion with open disdain and covertly instructed the cook to serve us last and with the dregs of the curries and the daily concoction of milk, almond, cardamoms, and poppy seed, to keep our brains fresh. I protested against the discriminatory treatment being meted out to us, but the mess manager remained rude and uncompromising.

The next day I heard the bully in the next room hurling obscene abuses at someone in a loud voice. Out of innocent curiosity, I went in and asked him politely who had displeased him. 'It is you, who else,' he shouted. 'You had the cheek last night to object to my messing,' he added, using expletive language. White with rage, I struck him a blow on his right temple. He swayed drunkenly and dropped down in a dead faint. I felt scared lest the fellow might have died, but after a time he recovered consciousness and was led tottering by his friends to the adjoining room. I and my Punjabi companion were later advised to shift elsewhere. Luckily, the tenant of an adjoining bungalow, a district judge on hearing what had happened, readily offered to take us both in. We enjoyed his hospitality for the next fortnight till the end of the examination.

The 'muscleman' went through the examination with a black eye. As he was the college's star candidate, I feared that if he failed the examination it would be attributed to our fight. When I returned to the college after the examination, many students and professors shunned me. After a few weeks, unofficial information regarding the result of the competitive examination was received. Only one successful Muslim, from all over India figured in the list of the first twenty candidates selected that year for admission to the civil engineers' class, and that happened to be myself. All the other twelve candidates from Aligarh college, including the 'muscle-man', had failed to qualify. With the receipt of this news, the atmosphere in the college dramatically changed. I was congratulated by everyone whom I met. The vice principal, Dr Ziauddin (the principal was Mr J.H. Towle, an Englishman) stopped on way to the lecture room to congratulate me. 'You have upheld the honour of the college,' he gravely remarked, adding, however, in regretful tone, that the 'muscleman' had failed. He also failed the following year, and adopted another profession.

The long summer vacations commencing in mid-July were drawing near when I had to say goodbye to the Aligarh college and join the Thomason Engineering College at Roorkee. The annual college week was celebrated at Aligarh before the commencement of the summer vacation. There was a grand college dinner and the round of sports and other festivities throughout the week. I wrote to my father and elder brother, inviting them to visit Aligarh during the college week. They came, and I took them to meet Nawab Viqar-ul-Mulk, the revered honorary secretary of the college. Nawab Sahib treated my father and brother with marked courtesy. He said, 'We are proud to have your son as our pupil. Your son is our son. I assure you we shall arrange a government scholarship for him for the duration of his studies at the Roorkee Engineering College.' He called his personal assistant and had invitations for the annual dinner and other functions of the college week issued to my father and brother. Furthermore, they were to be the guests of the college during the celebrations.

Thomason College of Civil Engineering, Roorkee

'Gentlemen, if you talk politics in this institution, out you go!' This was the gist of the principal's preliminary address to the new first year civil engineers' class. This speech was delivered in the upper subordinate hall, the most spacious and the best equipped hall in the college, reserved for its upper subordinate students, who were mainly European sergeants from the army, and who after qualifying filled the subordinate ranks of the Military Engineering Services. In comparison, the lecture rooms of the civil engineers' class were unimpressive as only a handful of Europeans and Anglo-Indians managed to get into the civil engineers' class through the tough competition entrance examination.

The Thomason College of Civil Engineering, Roorkee, that claimed to be one of the best institutions of its kind in the world at that time, was a great institution sprawling over a vast area, to the extent of many square miles, with its noble college building, workshops, laboratories, the professor's bungalows, cricket, football and hockey fields, and an eighteen-hole golf course. The college offered courses in civil, electrical, and mechanical engineering to engineer, upper subordinate, and lower subordinate classes.

The civil engineers' class had a three-year academic programme. The first year covered most of the theoretical mathematics, including calculus and differential equations, mechanical and laboratory work, drawing, practical classes in the blacksmith's, carpenter's, and moulding shops, and the college's engine room; as well as practical use of dynamite and demolition course and theory of surveying. The second year course included applied mechanics, several civil engineering subjects and a five-week survey camp for tachometric plane table and contour survey in a highly cut up area full of deep nullahs near Roorkee. The third year course included the entire remaining course in civil, electrical and mechanical engineering and, during its last three months, preparation of a complete project, including survey design and cost estimates. The third year project was considered to be the unique feature of the college. It was claimed that the projects prepared by the students were complete and of such high standard that they could be handed over to construction agencies. Different projects were assigned to the third year class each year such as a large building structure, a hydroelectric project, or a railway project to be prepared at an actual site.

The principal, Lt. Col. E.H. Atkinson, CIE, RE, was a true aristocrat. He did not take classes and was a strict administrator and disciplinarian. At the outbreak of the First World War he was called back to England, promoted to the rank of a major general and was appointed chief engineer of Britain's first army corps which initially landed in France to fight against Germany. All the professors and assistant professors of all the three classes of the college were Europeans; the Indians only held the posts of demonstrators. The engineers' class students were allotted bungalows, on the average of four students per bungalow. We, the six Muslim students in all the three civil engineers' classes, lived in two bungalows. We made our own messing arrangements. The Indian civil engineer students were expected to dress in European clothes and behave like European officers and were forbidden to mix with students of the subordinate classes who would ultimately have to serve with them in a subordinate capacity in the PWD. This was in accordance with the British Indian government's general policy of keeping the 'officers' and 'subordinate' ranks apart in all the services; the former class was meant for Europeans, and the latter created for the Indians.

Marks allotted for work were cumulative; final results were declared on the aggregate marks obtained during the three years. These included marks for the three final examinations and for daily class work, and also marks for sports and general deportment, including dress and general conduct. It was compulsory to pass the examinations for each year. Four out of the five Europeans who had entered the civil engineers' class with me in 1911, failed at the end of the first year. They had to leave the college. Out of the twenty students admitted each year in the civil engineers' class, the first seven on the list of the finally qualified candidates were recruited to the PWD on a permanent basis in the Indian Imperial Service (converted in 1920 to the Indian Service of Engineers). The rest also usually got jobs in the PWD as temporary engineers on equivalent status and were also permanently absorbed in the PWD on the opening of new development projects.

The Roorkee town was called by European residents as 'the heavens in India'. Situated only 20 miles from the foothills of the Himalayas, it enjoyed a pleasant climate. The summer monsoons set in in June following a brief spell of summer; the winters were mild and bracing. The countryside was fertile and produced all kinds of cereal crops and fruits, particularly an excellent variety of mangoes and lychees. The magnificent Upper Ganges Canal ran along one side of the city. Designed and constructed by Sir Proby Cautley, KCB, in 1843–5, it was one of the earliest and most spectacular irrigation works built by the British in India. It irrigated a large tract of the Ganges–Jumna Doab. The canal took off from the Ganges River at the point where the crystal-clear azure current of the river emerged from the thickly wooded Himalayan foothills near the town of Haridwar, an important Hindu pilgrimage site. It had many spectacular cross drainage works built in the 20-mile reach between Haridwar and Roorkee. At Roorkee town, the canal had a surface width of 200 feet with a depth of 13 feet, and a discharge of 8000 cubic feet per second.

Our college had a boating club on the canal bank for the civil engineers' class students. Admission was only after a rigorous swimming test. The boats in use were the 'doubles', 'singles', and the 'mess tours', the latter for holding the annual regattas. Coolies were provided for steering the 'doubles' for the convenience of fresh students. The small singles boats were piloted by the students themselves. We generally rowed about a mile downstream of the canal in midstream, and then rowed slowly upstream close to the canal bank. This afforded us enough exercise. Promotion from the double to the singles was made after a rowing test. Admission to the mess tours was open only to students after a very stringent test of rowing proficiency. One evening, I boarded a singles boat alone and had rowed a few yards downstream when an English teenage girl appeared on the canal bank and hailed me to come back and take her in. Knowing the strict rules against taking in non-swimmers, especially in the flimsy 'singles' boat with a few inches of overboard—and a British teenager, at that—I refused and kept rowing downstream. She, however, ran along the bank after me and shouted, 'Oh please, please, do take me in for a ride in your beautiful boat.' And so, I took her in and started rowing downstream.

It was, apparently, a new experience for her and she was thrilled at the novel adventure and shivered to see the racing current a few inches from the rim of the boat. 'Oh my!' she exclaimed at intervals. She had heard of the disaster of the Titanic, a star British liner on her maiden voyage to the United States of America which had only recently sunk (March 1912) with thousands of passengers on board after colliding with an iceberg at night. Recalling the ghastly details of the tragedy, she said, 'What would happen if this boat overturns?' 'Don't worry,' I replied. 'I could swim to the bank to safety, along with you.' She felt reassured.

One late evening, I was returning from the canal to the college after my usual rowing exercise in the singles boat. It grew almost dark under the shade of the mango trees planted along the canal bank. Suddenly, a tall British soldier of the Roorkee cantonment emerged from the shadows and blocked my way. He was big and tall and towered over me. 'Cough up,' he said, pointing to my pockets. Knowing perfectly well what he meant I said innocently, 'I do not understand you, what do you want?' Hearing me speak English he left me alone in disgust and, swearing at me, walked away.

The Delhi Durbar

In the Christmas of 1911, the memorable Delhi Durbar was held for the coronation of King George V, as the Emperor of India. That gorgeous pageant was designed to eclipse Lord Curzon's Delhi Durbar of 1903 as well as to reproduce the fabled splendour of the great Mughals. We were given leave from college to see the historic event at Delhi. A beautiful city of tents had sprouted up on the banks of the Jumna River with a carefully laid network of roads covered with *surkhi* (red brick dust) and bordered by beautiful strips of delicate green grass and flower beds in full bloom; rows of full-grown ornamental trees had been replanted there with consummate skill. Geometrically aligned blocks enclosed self-contained tent-towns to accommodate the 600 odd independent rulers of Indian princely states and their retinue, who had arrived to pay homage to their suzerain.

A magnificent royal tent was erected as the temporary residence of Emperor George V and Empress Mary. Close by, a royal canopy was erected and a gilded throne placed on a raised dais under the canopy for the coronation ceremony. The pedestrians were permitted to roam about on the roads of the royal tent city and view its wonders. The coronation of Emperor George V took place amidst scenes of brilliant splendour with the feudatory chiefs and rulers of the Indian princely states resplendent in their bejewelled dresses, in attendance. The star attraction of the durbar for the general public, however, was the grand procession in which the king emperor, accompanied by his subject chiefs and rulers and their select army units, would pass along a 3-mile route from the Durbar Hall to the Red Fort of the great Mughals. Tier upon tier of wooden benches were erected all along the route to seat the public, through tickets. We purchased five-rupee tickets for back seats at a point of vantage along the route, behind the Badshahi Masjid, and occupied our seats in freezing cold, long before dawn. In the faint light of early morning, noticing that many expensive front seats were vacant, we gradually shifted forward until we occupied the most expensive ones, nearest the road.

It was a fine day. Shortly after sunrise, the processions started arriving, with the minor chiefs and their picturesque retinues forming the vanguard. They were followed in strict order of precedence by bigger rulers seated in phaetons and followed by their mounted and foot army units in picturesque national uniform with pendants flying. Whenever a Hindu ruler passed in front of the crowd the Hindus clapped their hands; the Muslims did the same for the Muslim rulers. The Muslims lustily cheered when the carriage of the nawab of Bahawalpur, then a lad of seven or eight years passed in front of us. Among the subject rulers, the last to arrive in the procession were the maharajas of Baroda, Mysore, and Kashmir, and the very last being the Muslim nizam of Hyderabad, ruler of a territory greater than France or Germany.

About midday the four-hour long procession of the feudatory chiefs and rulers came to an end. It was followed by the British Indian troops and by selected contingents from British regiments stationed in India. In their wake followed a couple of horsemen of superb physique, clad in smart dark uniforms with shining breast plates and shining spiked steel helmets, and mounted on magnificent black chargers. Behind them rode an inconspicuous figure clad in an identical dark grey-black uniform, but without the breast plate and steel helmet. He rode a small black Arab pony wearing an ordinary military hat. All eyes were riveted on the pair of magnificent horsemen and their ordinary looking follower passed unnoticed. However, since I was sitting in the front row no more than 20 feet away from the road, when he passed in front of me I at once recognised King George V, and clapped my gloved hands emitting only muffled sounds and shouted, '*Badshah*!

Badshah!' (The King! The King!). I was, however, promptly silenced by the row of security men in plain clothes who were streaming through the tiered rows of spectators, keeping abreast of the King. Thus, a perfect camouflage had been devised for the King's safety in the protected procession.

Very few spectators had, however, noticed the king in the procession. Rumours, therefore, went around in the town that the king emperor had not been in the procession. The gossip was promptly contradicted, and it was announced that on the following day the king emperor would appear in the *jharoka* on top of the wall of the Red Fort where the Mughal emperors used to give *darshan* to their subjects. The public was invited and big crowds saw the king emperor in the *jharoka* and paid homage to him. It was announced that the king emperor had commanded the shifting of the capital of British India from Calcutta (now Kolkata) to Delhi, as a coronation gift to the people of northern India. It was originally planned that the new capital would be located in the large open space between the present old Delhi of the Mughal Emperor Shah Jahan and the Jumna River, in which the temporary Delhi Durbar town had been located. But the plans were hastily changed, and the proposed new capital was to be located among the old monuments and ruins of the abandoned older Delhi of the Mughal Emperor Humayun, on higher ground about 4 miles away from the present old Delhi and further away from the Jumna River.

The oldest town of Delhi, the 2000-year-old Indraprastha, was built 25 miles away from the banks of the River Jumna and remained the capital of various Hindu dynasties. In the twelfth century AD, the Muslim king Qutubuddin Aibak founded the first Muslim town of Delhi at the site of Indraprastha and made it the capital of India. He erected the famous Qutub Minar which stands to this day. Since then, various Muslim dynasties built their new capitals in Delhi shifting it nearer to the River Jumna. The previous old Delhi of the Muslim era is the present old Delhi built by the Emperor Shah Jahan in the seventeenth century. There, thus, remained only a small space between old Delhi, and the River Jumna, for the British to build their capital.

But there was an old legend to the effect that when the town of Delhi reached the banks of the River Jumna the then ruling dynasty would be overthrown. That legend was revived and was the talk of the town when the British Indian government's plans to build the New Delhi near the banks of the River Jumna became known. In view of the currency of the legend, the proposed New Delhi was planned well away from the river among the ruins of Emperor Humayun's long abandoned capital. It was said that the step had been taken to forestall any possible psychological effect of the legend on the people of India. But sceptical Indians attributed the change of site for the proposed capital as a sign of superstition.

That was my first visit to Delhi as a young man of twenty. I had a powerful emotional reaction on seeing the splendid Muslim monuments of Delhi—the Red Fort with its matchless marble palaces, their walls inlaid with semi-precious stones; the gems of Moti Masjid; the Badshahi Masjid of Delhi; Emperor Humayun's tomb; the Qutub Minar; and the beautiful little tomb of Princess Jahanara Begum—the Emperor Shah Jahan's favourite daughter—bearing the inscription:

> *Baghair subza na poshad kase mazar mara*
> *Ki qabr posh ghariban hamin gayah bas-ast*
>
> *Let nothing but green grass cover my tomb,*
> *for this is the best grave cloth for the poor in spirit.*

It was hard to believe that the creators of those masterpieces, which, after the lapse of many centuries, were still good as new, were no more! I left Delhi with moist eyes, absorbed in the above reflections. Strangely enough, the Delhi Durbar, in comparison, had made a much lesser impression on me.

One bright Sunday morning, I was taking a walk in the spacious playgrounds of the college when I stopped to see a cricket match in progress between the second and third year engineering classes. On spying me, the captain of our third year team came running to me and asked me to hurry to the hostel and come back in cricket gear as they were short of a player and wanted me to play. Our side was batting and making a poor show of it. I went in eight down and punished the bowlers, scoring fifty-five runs in half an hour that sealed a win for our class. Luckily, for me, Captain E.W.C. Sandes, RE, professor of civil engineering and president of the college sports, was watching the game. He called me over and enquired about my previous cricket experience and then asked the team's captain why I had not been included in the First XI. The captain, who was a Hindu, stammered a reply, 'Sir, we know he is a good player, but we want bowlers, we already have good batsmen in our side.' Captain Sandes walked away with a grim expression on his face.

I was the only Muslim student at Roorkee in a class of twenty, in which the majority were Hindus. Since the coveted 'guaranteed' appointments in the Indian PWD, open to the top seven in the class, depended on the combined marks obtained for daily class work, as well as in the annual final examinations, there was fierce day-to-day competition even among the

Hindu students. In the first year, the entire field of pure mathematics was quickly covered in the daily lessons. Unfortunately, I had given up BSc and had taken arts courses in my BA at Aligarh which proved a terrible handicap at Roorkee. I could not master the subject from the cursory Roorkee lessons and fared very badly obtaining only 33 per cent marks in the subject, which were just pass marks. Moreover, our mathematics professor was said to be in the habit of repeating his daily lessons in mathematics verbatim from year to year, and also prescribing identical questions for daily class work. It was also believed that in his question paper for the annual final examination he very often repeated his questions.

This fact was known only to some astute Hindu students, so they had secured exercise books on class work for the first year, from the outgoing students, and thus they practically knew beforehand the items the professor would be taking up on a particular day. My competitors thus secured 60 to 90 per cent marks in pure mathematics, leaving me hopelessly behind from the outset in the race for a guaranteed appointment in the PWD.

Five British colleagues were also in the same boat. In the final count, at the end of the first year, four out of five failed to qualify. In his annual report the principal remarked that in the final examination of the first year class, Abdur Rahman and Brown had just scraped through. One of the four failed Anglo-Indians (pure blooded Englishmen born in India) was taken into the electrical engineering class in the second year at the Roorkee college. I did much better in the subjects of applied mechanics, theory of structures, and other civil engineering subjects. I had returned from the brink and had gained confidence in my academic ability. The five-week survey camp in the second year comprised tachometric plane table contour survey of the cut up and broken tract of about a square mile area in the vicinity of Roorkee town. The professor of the surveying staff toured the area from time to time and evaluated the field work done by each student at the site. On checking my work, he was surprised to see me surveying and plotting on the plane table not only the deep drainage lines in the middle of the gorges, but also their irregular top edges on the ground surface. I stood second in the class in that survey job.

Reaching home for the summer vacation in July 1913, at the close of the second year in college, I found my father down with typhoid fever. I and my elder brother lovingly nursed him and sought help of the best available doctors. However, despite our stringent precautions he had a couple of relapses of three weeks each, which prolonged his illness to over two months. He became extremely weak. In the meantime, my vacations came to an end and I suggested that I would take leave from college for a year for nursing him. But my father refused to consider the suggestion. He said, 'You have your own life to live, and have nothing to do with my illness.' Saying this, he blessed me and, extending his hand, bade me goodbye! With tears in my eyes I parted from him. A fortnight later, I received the fateful telegram announcing his death. My father's death was a terrible shock to me. I hurried back home to Haripur near Abbottabad, where he was buried. I, along with my elder brother and grandfather, visited his grave. We prayed for him. Grandfather said his back was broken on his favourite son's death, at the early age of forty-three. How I loved my father's grave and blessed all the inmates of the graveyard in which he was buried.

He was the best of fathers: always considerate to his wife, our stepmother and to us, his children, and to our step-brothers and step-sisters. He never uttered a harsh word on our misdeeds. His silent disapproval was punishment enough. In spite of his relatively junior position in the PWD, he was respected by his British superiors, and was also held in respect by fellow-citizens, irrespective of their status, wherever we had lived. He was of a saintly disposition, never hurt anyone's feelings, and never quarrelled with anyone. Before joining an English missionary school, he had received a sound religious education and was a Persian and Arabic scholar. He had a small library of books on Islamic theology and history in Urdu, Persian, and Arabic languages. I benefited from the Urdu and Persian collection and became well versed in Islamic history during my teen years.

Luckily, I had enough income from my government scholarship and other arranged resources, at the Roorkee college, without having to rely on my father's finances. So, I did not feel any financial embarrassment on his death. The college courses during the third and final year were mainly on civil engineering subjects in which, to the amazement of my competitors, I started to forge ahead. A fellow top-ranking student remarked within my hearing one day that it was good that I had not studied higher pure mathematics, otherwise I could have topped the class.

The Annual College Boating Regatta

The annual college regatta was a notable event in Roorkee. Large city crowds as well as the British gentry from the Roorkee cantonment assembled on the canal bank to witness the mile race downstream between the third and second year civil engineering students in the college mess tours. The boats had four rowers each and a cox. All were expert rowers selected after vigorous tests. The boats were of light construction and, when not in use, were hung from the low sloping roof of the college boating shed on the canal bank.

They were carefully disengaged from the roof of the shed and carried by the rowing crew and gently placed in the water. Each boat was 63 feet in length and only 3 feet wide, with about a 3-inch free board.

The leader of our third-year class team in the 1913 annual race was Wood of the third year electrical engineering class, our former colleague of the civil engineers' class, who had failed in the first year. He was our 'stroke'. I occupied the bow position. The race started with a pistol shot. Captain Sandes rode on a fast motor bike along the canal bank to keep abreast of the fast racing boats, to see that there was no foul play en route. Ours was a strong team and we were confident of easy victory. When only halfway down the race tract, we had left our opponents two or three boat lengths behind. At that point our leader, Wood, suddenly started 'digging' his oar which had the effect of sharply retarding our speed. We shouted 'Wood-steady!' but he completely ignored us and kept on digging. Meanwhile, our opponents' boat started forging ahead, and soon overtook us. Realising that Wood's behaviour was deliberate we stopped rowing and shouted: 'You traitor!' We rowed the boat to the canal bank; Captain Sandes had stopped and seen it all. Our second year opponents in the meantime comfortably completed the course and won the uncontested race.

The matter was reported to the principal, and was considered in a meeting of British professors, especially called by the principal. No action, however, was taken against Wood, and the second year class was declared the winner. It turned out that Wood nursed a grudge against our class for having failed in the first year, and he also wanted the second year class team to win which had two Anglo-Indians in its crew. This incident brought to the surface the covert tension that existed between the small number of Anglo-Indians and the overwhelming majority of the Hindus in the engineers' class. The Anglo-Indians could not compete with the graduates and MAs, who were the pick of the dozen or so Indian universities, who entered the civil engineers' class at Roorkee college.

The maximum age prescribed for admission in that class was twenty years and an affidavit about the date of birth was required to be furnished by each candidate along with his application, for admission into the competitive examination. It was surmised, however, that most of the Hindu students of our third year class were actually over age and had furnished false affidavits regarding their ages. Confidential enquiries were carried out by the government about the actual ages of all the students of our third year civil engineer's class from their school and college records. It was found that only myself and Brown, the lone surviving Anglo-Indian in our class, had filed true affidavits. All the other students, on the other hand, had furnished false affidavits and were above the prescribed maximum age on entering the college. Show-cause notices were issued by the college authorities to all the Hindu students of our class to explain why they should not be removed from the college. The Hindu students collectively hired the services of an eminent Hindu lawyer, Sir Tej Bahadur Sapru, who asserted in his plea on behalf of the students that the ages given in their affidavits were correct, and that the entries in the school and college records were unreliable. He put the onus of proving that the latter were correct, on the government. The matter also took on a political complexion since the Hindu students came from different provinces of India. So, the matter was dropped, but for the future it became the rule that for purposes of entry into government service, the age declared by the students at the time of their matriculation examination would only be accepted as correct.

One day, our class came out of the mechanical workshop and one of the Hindu students, Gauri Dutt Pandey, who had topped the list of successful candidates in the competitive examination for entry into the civil engineers' class, and was expecting to secure a permanent appointment after qualifying, was walking alongside me with a slight limp. The civil surgeon of Roorkee, a British doctor who also looked after the health of the college students, happened to be passing by and noticed Pandey limping. He stopped him and enquired what the trouble was. Pandey could not give any definite cause for his problem. The civil surgeon, thereupon, told Pandey to attend his office the next day. Pandey's examination revealed that he was suffering from tuberculosis of the bones in an advanced stage. He was ordered by the college authorities to leave the college within twenty-four hours. He pleaded in vain that the final qualifying examination was only three months away and that he was expecting to secure a permanent appointment. However, he had to leave immediately. Within three months he was reported dead.

For the final project, we were allotted a railway project, which was for the design of the proposed (5–6 feet) broad gauge branch line from Deoband railway station on the main line of the North Western Railway, about 30 miles away from Roorkee to the town of Purkazi 12 miles away. The main features of the project were the junction and terminus stations at Deoband and Purkazi; and bridges on two major drainage crossings of the Kali Nadi and the Sila Nadi, two large tributaries of the Ganges River. The work involved surveys, design, calculations, preparation of drawings, estimation, and specifications. The project was to be submitted in a complete form, such that it could be handed over to a contractor for construction. Lots were drawn for pairs of distinctive colours for survey flags for each student.

My colours were green and black, over which I superimposed four white crescents. Distinctive flags were necessary to enable our professor of surveying to identify the different students working in the field while checking their work at site. A survey camp was established at Deoband. Surveying instruments, field books, survey poles, flags, and other materials were supplied. Each student was strictly required to work independently without consulting his colleagues. For messing arrangements, I was paired with Banerjee, the sole Bengali Hindu student in the class. The field work was expected to be completed in five weeks. While camping at Deoband, I took the opportunity to pay a visit to the old renowned Muslim school of theology at Deoband on a Sunday and paid my respects to the saintly head of the institution. On his suggestion, I also visited another saintly teacher hailing from Kashmir, who was generally engaged in devotions in his small bachelor's cell in his spare time, after teaching classes. Muslim students of all ages came to Deoband from all over India to study at that famous seat of religious learning. The curriculum of studies at Deoband, I was informed, was essentially the same as at the famed Nizamiyya University of Baghdad in the eighth century AD during the time of Caliph Haroon-al Rashid. With youthful exuberance I suggested that the English language and science be added to the theological curriculum, little realising that the two do not mix in the curricula of traditional religious *madrassa* education.

I had just completed surveying the alignment down to Kali Nadi, the first major drainage crossing, where the land from both sides sloped sharply down towards the torrent, across a shallow valley cut up by transverse drainage lines. Both the Kali Nadi and Sila Nadi torrents had their source in the lower Himalayas and had small perennial flows but passed large floods during the summer monsoons. Viewed from the banks of the Kali Nadi, the country between the flanking ridges of high ground presented a nightmare of ravines and gullies. Suddenly, a figure appeared on the ridge behind me on the Deoband side. Recognising my flags, he hurried down and waited. He was utterly bewildered to see the desolate country around Kali Nadi and did not know what to do. He was a Hindu student from Sialkot district and so taking pity on him I gave him instructions for selecting a site for the torrent crossing. He was among those who got a permanent appointment.

According to the final results I had secured seventh position in the railway project, and eleventh position as a whole in the class. I thus missed a permanent appointment but became eligible for a year's apprenticeship in a government department. I went home to Abbottabad. A few weeks later, I received an official letter from the Government of India asking me to report to Sylhet, in Assam province, for a year's practical training in the PWD, Buildings & Road branch of that province.

5 Apprenticeship in Sylhet and Assam, 1914–1915

Tracing a straight west-east route to Sylhet across the railway map of India, parallel and adjacent to the foot-hills of the Himalayas in the north and after checking train connections of the different railway systems along the selected route, I started my week-long railway journey from Havelian railway station, near Abbottabad to Sylhet, across the entire breadth of India. I broke journey en route for twenty-four hours, at the beautiful city of Lucknow. A policeman on duty outside the Lucknow railway station suggested that I could stay at the Darwaza Band Serai while in the city. The serai was a spacious quadrangle of small single storey rooms with verandas, with access from the main road through a huge gate. A number of 'Bhatiari' women of all ages were in charge of sets of rooms, who let out the rooms to visitors for a small charge. As soon as we entered the serai my baggage was seized and pulled in different directions by half a dozen Bhatiaris each of who shouted: '*Mian*, come to my room.' I selected the room of the oldest among them, to the chagrin of the young ones. Vendors of sweets, fruit, and *sanday ka tail* (lizard oil) offered their wares in alluring manner to the visitors (local application of lizard oil was supposed to bolster masculine virility).

Lucknow was a beautiful town with the Gomti River daintily flowing through it. It was the last refuge of the Muslim civilisation in India after the decline of Delhi. The most impressive Muslim monument in Lucknow was the audience hall of the Oudh kings with a flat arched roof, in brick, of 84-foot clear span, with an ornate ceiling; it was indeed a marvellous piece of engineering. There were numerous other monuments of the Muslim era, more ornate, but lacking the dignity of the Mughal architecture of Delhi, Agra, and Lahore. The next day, I boarded the slow-moving train of Bengal North Western Railway, for a tiring thirty-six-hour journey to the railway's eastern terminus station of Kaichar. I had to say goodbye to the familiar food of meat curries, *chapatis*, and *pulao* at Lucknow. From then onwards, *puris* fried in *ghee* and pickles and vegetables were the only fare purveyed on the train. The plain country along the northern border of India running parallel to the 'Terai' of Nepal was intensely green. We crossed numerous tributaries of the Ganges swollen with monsoon floods flowing southwards. The largest of them was the Ghaggar River, which appeared to have greater discharge than the Ganges itself.

From Katihar to Parbatipur was a further twelve-hour exhausting train journey. We (I and my servant whom I had brought along from home) reached Parbatipur railway station the next evening dead beat, and I promptly fell fast asleep on the benches of the sheltered platform. The next train to Lalmonirhat was due at 2 a.m. A sudden deafening noise—that of the rain falling on the roof of the railway shed—woke us up. It was exactly 2 a.m. by my watch. Our train was standing on a platform to be reached by an over-bridge. We grabbed our luggage, crossed the over-bridge and scrambled on to the first carriage we came across, and the train started moving! We reached Lalmonirhat in the morning at 6 a.m. The train stopped in front of a small single room with a low front veranda, which served as the platform. This was the Lalmonirhat railway station. There was standing water all around the train and the station's veranda was submerged. We had to wade through ankle-deep water to reach the room. It was still raining.

Wet, hungry, and miserable, we waited for the next train which was to leave at 9 a.m. It was an express train that started from Calcutta and was reported to be full of pilgrims bound for the famous temple of Kamrup for the annual pilgrimage. A train arrived and to our astonishment it had plenty of room. So, we found comfortable seats on it. The train started. A Bengali gentleman sitting in the same compartment asked me in English where I was going. I said, 'Sylhet.' 'But this train is going to Cooch Behar, due north,' he said. 'Your train, going east, has not yet arrived.' We were in a panic. Luckily, after moving only a furlong or so the train stopped. We threw down our luggage, got off the train and, grabbing the luggage, hurried back to the station to find the long pilgrim train standing on the platform. The pilgrim train was overcrowded. There was only one inter-class compartment on the train for which I held a ticket. It was a large compartment with seats all around and an open space in the middle. The seats were all occupied by Bengalis from Calcutta, and the centre space was covered by mounds of luggage. A hefty looking Bengali was standing at the only entrance to the carriage, barring the way of passengers from entering. I surveyed the situation from the running board. The sentinel at the closed and barred carriage door scowled at me. I dumped my luggage inside the compartment on the piles of the pilgrims' luggage and climbed in through the window. The burly gentlemen guarding the window cried out in pain and quickly stepped aside as I inadvertently stepped on his bare feet with my heavily booted foot.

I told my servant to find any room that he could in a third class compartment. Looking around I was met by hostile glances from all directions. A teenage Bengali girl was the sole female passenger in the compartment. Two sturdy looking Bengalis, holding large staffs in their hands, quietly took their seats on each side of the girl, and looked belligerently at me as if ready to repel any assault I might carry out on the girl. I later realised that my all-too-obvious Pathan appearance had triggered the above precautions. An hour later, everyone purchased *puris*, pickles, and vegetable curry for their breakfasts from the vendors on a large station. Being famished, I also bought food and, finding a temporarily vacated seat, sat down, and had my breakfast.

At 2 p.m., our train arrived at Amin Gaon station on the right bank of the Brahmaputra River, and unloaded all the passengers and luggage under a long shed. The passengers had to cross the river in a steamer and board the train waiting on the far bank, bound for the Kamrup station and the town of Gauhati further upstream. Swarms of coolies, all Muslims, wearing *dhotis* and red caps, handled the passengers' luggage. Piling my luggage on the platform, I walked the entire length of the train and back, calling out my servant's name in every compartment. He was nowhere to be found. Realising that he had been left behind on the Lalmonirhat station, my immediate worry was to have my luggage conveyed to the steamer moored on the riverbank. The coolies, not understanding my language, shied away from me. Exasperated, I grabbed a coolie by the neck and compelled him to carry my luggage to the steamer. Luckily, the same coolies accompanied the passengers on the steamer across the river and loaded their luggage on the train on the far bank before being finally paid. The Brahmaputra River, which at that point was still in the upper reach in Assam, had a mighty current. It was 4 miles wide and of great depth, swollen with the monsoon floods. The swift flowing current was confined within high banks. The opposite bank was not visible from our side. A number of Hindu male and female pilgrims took a dip in the holy river (it is considered the son of Brahma, the creator who is one of the three gods of the Hindu trinity) before boarding the steamer. The young female bathers stood practically nude but unconcerned in their muslin *dhotis* rendered transparent by their dip in the water.

An hour later, the train reached the Kamrup station on the Brahmaputra River, where all the pilgrims disembarked. An almost empty train reached the town of Gauhati, a few miles further up the river, where I got down. A midnight train from Gauhati was to take me through the last leg of my railway journey. Gauhati, the headquarters of a district in Assam, was a pleasant station. The roads were surfaced red with powdered burnt brick, and detached cottages, with white-washed walls and red gabled roofs, were bordered by trim grassy lawns. Pretty Assamese girls clad in fine muslin *dhotis* and close-fitting short blouses, with elaborate hairdos festooned with strings of flowers, strolled about on the roads. The pretty scene was in striking contrast to the drab atmosphere of our NWFP towns where girls of well-to-do families usually stayed indoors and only came out heavily veiled. Surprisingly, most of the big stores in Gauhati belonged to Muslims, descendants of the ancient Arab seafarers, who had spread over most of the Far Eastern countries in the Middle Ages. Tasty Muslim dishes of chicken curry and *pulao* were available in the Muslim restaurant, which I fell upon with relish, being famished for want of decent food since many days.

The train for Karimganj station arrived punctually at midnight. My missing servant alighted from the train, escorted by two policemen. I claimed him, showing them his ticket which was with me, to their satisfaction. Early the next morning, I noticed the train passing through thickly forested, hilly country of upper Assam, the home of the wild elephant. The pungent smell of dense green vegetation made me giddy. We were passing through an alternating chain of tunnels across spurs and lofty viaducts across the side valleys. This was the famous tunnel section of the Assam Bengal Railway. On a small wayside station, Assamese women, with distinctly Mongolian features and wearing sarongs, offered bananas and guavas for sale. Being hungry, as I didn't have breakfast, I purchased fruit worth two *annas* and in exchange received a very large bunch of bananas and about five kilos of guavas. After having my fill, I reluctantly threw the rest out of the train. At noon we reached the railway station of Karimganj, on the Assam Bengal Railway, my final destination, where I alighted. The hilly section was left far behind and we were now in a pleasant plain of green grass and rice fields.

The town of Karimganj was about 3 miles away from the railway station and was reached by a narrow gauge that the railway constructed for the convenience of the passengers. I left the servant with the luggage at the station and boarded the narrow-gauge train for town in quest of my friend, Ali Ahmad, an assistant engineer in the PWD and a former senior colleague at Roorkee, who had likewise been drafted to the Assam PWD by the Government of India two years earlier and was posted at Karimganj.

Karimganj, a tehsil headquarters of the Sylhet district, was a bustling town situated on both the banks of the Surma River at the junction of the Noti Khal stream with it, the latter splitting the town in half. Double decker steamers from Calcutta cruised up along the Surma River to Karimganj and beyond, carrying passengers and goods both ways. One had to ferry across the Surma and Noti Khal rivers in order to reach different parts of Karimganj town.

The population was Bengali speaking and was predominately Muslim. The principal shops in the town were also owned by Muslims. Since no one understood Urdu, or my English pronunciation, I had considerable difficulty in locating my friend's house. Enquiries from an English-speaking Hindu clerk at the local post office only elicited the reply: 'PWD? No letters for PWD.' Overhearing our conversation, a young Muslim man, who coincidentally was a clerk in the PWD, offered to escort me to Ali Ahmad's house. Ali Ahmad, a devout Muslim who sported a carefully trimmed beard, was having a session with his barber while seated in the pleasant sunny lawn of his cottage. His young wife observed strict *purdah*. He fed me lavishly and since I had to report to the executive engineer, PWD, Sylhet division the very next day, he advised me to start by boat down the Surma River for Sylhet, the very same evening. He also wired Maulvi Qudratullah, a prominent businessman and landlord of Sylhet town, to meet me at the Surma River landing at Sylhet the next morning. Sylhet was 50 miles downstream from Karimganj. The boat hired for me was a tiny dinghy manned by a single boatman, with a small round reed shelter in the middle for sleeping. The boatman smoothly paddled the boat down the river all night while we lay in blissful slumber. In the morning, at sunrise, we reached the Sylhet landing on the Surma River. On the high riverbank, stood my host Maulvi Qudratullah with a one-horse phaeton carriage to receive me.

Sylhet

Before twelve noon that day, I reported for duty to Mr A.P. Mullick, executive engineer, PWD (B&R) branch, Sylhet division, and was relieved to learn that my pay had started from that date. Mr Mullick was a Bengali Hindu gentleman of liberal views, without a trace of communal bias, as it turned out later. I received instructions from him to take notes on the Police Chummery building which was under construction. It was an earthquake-proof single-storey structure with 4.5-inch thick burnt brick walls built in a framework of vertical and horizontal steel joists and a gabled roof. Earthquake-proof construction for buildings was adopted in Sylhet town after the devastating earthquake of 1895.

Sylhet was a beautiful town of red-brick-powder paved roads, detached cottages and bungalows situated on high ground on the banks of the Surma River. There were no patches of bare ground anywhere. Besides the red-brick surfaces, the ground was either grassy or almost wet with rain in summer or, with dew in winter, teeming with stinging insects and pools of water. Ground water was within a foot or two of the surface. Small pools of clear water, 3 or

4 feet deep, served as 'baths' and as drinking and household water supply. The Sylhet district, politically a part of the Assam province, was ethnically a part of north Bengal. Bengali was the local language of the mixed Hindu and Muslim population of the city. Muslim upper classes also spoke Urdu.

The citizens, both Muslims and Hindus, had fairer complexions than those of the people of southern Bengal. Besides, there were fair-looking Manipuris of Mongolian origin and Khasis of a mixed Mongolian and Caucasian race. The countryside raised two or three rice crops a year, usually flooded during summer monsoons by heavy floods coming down from the neighbouring Khasi and Jaintia hills, which also contained the world's wettest spot, the Cherrapunji station. Villages were located among thick groves of bamboo and betel nut palms on natural laterite outcrops which nature appeared to have thoughtfully scattered about on the plain. My landlord, Maulvi Qudratullah was a middle-aged man of about fifty-five, with a nut-brown complexion and a pockmarked face, with a grey head and beard. He owned three cottages on a main road. I occupied the middle one; on each side were Hindu tenants, a professor and a civil servant. In my front lawn, there were four small graves of Maulvi Qudratullah's four sons who were killed under the debris of his collapsed house in the 1895 earthquake. That event had overnight turned Qudratullah's hair grey. Since his first wife was then overage, he had since married two younger women, both of whom, however, couldn't give him any children. So, he was abnormally attached to his sons' graves.

He lived with his three wives in a house, heavily guarded by thick hedges, adjoining our cottages. It was customary for upper class Muslims, even in normal circumstances, to have more than one wife, one of whom would be the head wife, and the other younger and favourite ones, were usually taken from poorer families. On the other hand, the Hindu families usually had one or more young widows on their hands who never remarried. In Bengal, in those days, young girls reaching puberty were usually married off. Thirty-year-old women were considered to be old. The usual diet of both the Muslims and Hindus was coarse rice, fish, pulses, and mustard oil. Well-to-do Muslim families sometimes added chicken curry. The sweet dish was almost invariably slices of bananas with milk. Milk cattle were, however, small and of poor yielding variety. Dutch canned milk, both skimmed and creamy, was readily available in shops. Large quantities of fresh fish of many varieties from the Surma River were sold in the busy fish market every morning. The delicious *hilsa* fish was also plentiful and cheap. Fish was usually cooked with gravy and was eaten with boiled rice. Fried fish, which consumed a lot of mustard oil, was considered a luxury. Banana of various varieties, large and small, and also pineapples were cheap and abundant. I usually purchased a large bunch of bananas and hung it from a beam in the veranda, eating the ones that ripened daily. Wheat flour and mutton were also available but at a cost.

A few days after settling down, a Hindu gentleman who introduced himself as a retired sub-inspector of police paid me a visit. He offered to supply Manipuri, Khasi, or if preferred, local Hindu and Muslim girls as mistresses, at reasonable rates. He casually remarked that it was customary for single 'officers' posted in Sylhet to hire mistresses, quoting a recent instance of a young British engineer recently posted in the PWD, who had hired a Manipuri girl through him at Rs 15 per month, maintaining her in the out-houses of his bungalow. To his incredulous stare and evident disappointment, I told him I was not interested.

I made acquaintance of my two Hindu neighbours—the professor and the civil servant. The professor was an enlightened Hindu gentleman of liberal views. I also developed friendly terms with a number of officials and members of some old aristocratic families of Sylhet. One of the latter had an ancestral library of rare manuscripts in Persian and Arabic, the value of which the young aristocrat did not appear to appreciate. I borrowed two rare manuscripts from his library, the *dewan* of the Persian poet, Hafiz, and the *masnavi* of Jalaluddin Rumi. I discovered that Bengali was a soft and gentle language with no swear word. I also liked Bengali music, with its haunting refrains and softly undulating metres which appeared to be in harmony with the soft and beautiful environment of Bengal.

By and by, I received instructions to move out in the district on inspection tours of roads and bridges under construction, and thus had many opportunities to visit the rural areas and see the village life of Sylhet district. The rural population was predominately Muslim. The villagers lived in reed huts plastered with mud on the inside and huddled together on laterite mounds amidst fiercely competing vegetation of bamboo canes, and betel nut palms, and thick elephant grass with a labyrinth of narrow spongy footpaths. This was also the domain of wild animals, including elephants, wild boar, sloth bear, different varieties of deer and leopards. The Royal Bengal Tiger also abounded and roamed freely and frequently terrorised the scattered village population. Mr Mullick, in his early briefings, had cautioned me to be careful in my tours, in the countryside specially, advising me against travelling after dark and always touring in the company of some local staff members. I took this advice somewhat in the spirit of a careless young man, for which I had to pay a price. My stay in Assam was not without its adventures with wildlife; three incidents are worth recounting here.

It was the winter of 1914–15 and I was assigned the job of constructing a solid steel pile bridge across a large unbridged drainage gap on Mile 14 of the Sylhet-Cherrapunji road. Cherrapunji station was situated about 20 miles further on the road among the Khasi hills. The surrounding plain, a part of the Sylhet district, was devoid of habitation for miles around. It interspersed with occasional vivid green patches of dry ground dotted with stately hollong trees, and also with tall, slender betel nut palms. The low, densely wooded Khasi hills ran parallel with the road a few miles away. It was a deceptively beautiful country, infested by snakes, and was home to the wild buffalo, wild boar, and the tiger.

Encounter with a Man-Eating Tiger

I had set up temporary quarters in the Companiganj rest house, which was built on a low, natural laterite mound on the far bank of a small, navigable river that was crossed by ferryboat, 3 miles further, on the same road. I did the 6-mile round trip to the site of the construction work and back, daily on a push bike. There was always a regular trickle of pedestrian and elephant-borne traffic on the road; therefore, in the daytime, there was usually not much danger from wild animals, except that there had been reports about the presence of a man-eating tiger in the vicinity, that had killed a woman who was out collecting firewood and attacked a party of sadhus on their way to a nearby shrine, severely mauling two persons before being shooed off by the clamour created by the rest of the party.

News about the man-eater had created terror among the neighbouring villages and my servant was particularly concerned about my safety. One afternoon, while busy disbursing wages to the labourers, at the site of the work, I noticed with a start that the sun was about to sink below the horizon. My assistants expressed concern at the prospect of my having to return to the rest house in the evening, since the road was considered unsafe after dark. I was urged to spend the night at the labour camp.

Unfortunately, at that time there were five cases of suspected cholera in the camp, probably due to drinking unclean water from the large scour-hole in the drainage channel which teemed with live, as well as putrid fish, which had been trapped in it since the recession of floods of the previous summer monsoons. So, weighing the comparative risks of catching cholera, or a possible encounter with the man-eating tiger, I decided to take the latter risk, and to return to the rest house despite strong protests from my staff. As I left, I was cautioned to keep the bicycle bell ringing all the time. Viewing the rapidly fading evening sky with growing concern, I quickly started on my return journey, bumping along on the rough road embankment as fast as I could, while continuously ringing my bicycle bell, and peering right and left through the flanking jungle at the marshy terrain below, my nerves taut, and my senses alert to spot danger from any quarter. The road ahead was completely deserted.

In the fading evening twilight, the road appeared to be a long, narrow tunnel hemmed in by tall reeds and jungle, gently curving away to the left in the form of a gigantic arc, about 2.5 miles in length. Arriving safely at the far end of the arc without incident, I felt a profound sense of relief. A sharp turn to the right around the nearby bend ahead led to the remaining short, straight stretch of the road to the riverbank, and safety. No sooner had I turned the bend that I saw, a huge tawny beast, a tiger, sprawled right across the roadway, 150 yards or so ahead. It was of enormous size, completely blocking the narrow road on top of the high embankment. The sound of my bicycle bell must have alerted the tiger from far for it was looking in my direction.

When I saw the tiger obstructing my way, I had a sinking feeling in the pit of my stomach but immediately started looking around for means of escape. The obvious course seemed to be to quickly turn around while still riding the bicycle and run and get back behind the bend of the road which I had passed moments ago. Common sense, however, warned me that the width of the roadway was too narrow, and its dirt surface too rough to allow a successful turning manoeuvre. The attempt would surely have resulted in my sliding down the slope of the 10-foot high embankment into the marsh below, which was of uncertain depth—a frightening prospect! There was too, the further grim possibility of the tiger following me into the marsh and grabbing me there, a helpless victim. The only possible alternative was to get down from the bicycle, turn around and remount and flee backwards. Instinct warned against that step as well. The tiger was steadily watching me while I was inexorably drawing nearer to it. The moment I would have set foot on the ground, the beast would have reached and grabbed me in a few quick bounds.

All these options and their dim chances of success had crossed my mind with lightning speed. Then the hopelessness of my situation suddenly dawned on me and engulfed me in despair. Curiously, I had no feeling of fear or terror, except deep regret and despair at my approaching and untimely end! Luckily, these feelings instantly gave way to a frenzy of resentment and rebellion against my impending fate. 'Oh, no, no. Providence will surely come to my rescue!' Buoyed up with that feeling, and heedless of the consequences, I bent low over the bicycle handles, putting in my last ounce of effort in increasing its speed! Violently ringing the bicycle bell with nervous energy,

I made a reckless charge at the crouching beast! Involuntarily, I raised my eyes. In a mighty spring, the tiger executed a wide arc high up in the air across the roadway and dropped down with a loud splash in the marsh below beyond the toe of the embankment. A split second later, I crossed the spot where it had been lying.

I was suddenly drenched in perspiration. My legs were leaden. My arms and hands were limp. With stupendous effort I made the run, arriving at the riverbank 200 yards ahead. It had grown almost dark. The riverbanks were deserted. Only the ferryman was standing on the far bank awaiting my return. After repeated shouting, he reluctantly rowed the ferry boat across, but only after making sure of my identity: He greeted me with the query: 'There is a huge tiger lying on the road a short distance away holding up traffic since noon. It is the dreaded man-eater. Did you see it?'

Snakes Galore

During the summer following my adventure with the man-eating tiger, exceptionally heavy summer monsoon rains in the Khasi hills during early July brought floods of an unprecedented magnitude, which spread across the wide plain of Sylhet in a continuous sheet of water several feet deep. It completely submerged the streams, rivers, and the Sylhet–Shillong road, although the latter was on an embankment several feet high. I was assigned the tasks of preserving flood marks on tops of the trees and telegraph posts along the road during the peak stages of the flood. I left Sylhet in the morning accompanied by a personal servant in a small boat, manned by a single boatman. Skirting the high ground around the Sylhet town, via the Surma River, and then entering a small tributary, we at last reached the open flooded country into which the tributary suddenly disappeared. We took a course parallel to the Sylhet–Shillong road, keeping on the upstream side of the road while daubing paint at flood level on tops of the telegraph posts and on trees, as we rowed along.

The road was completely submerged under 4 feet deep water and, being on an embankment, acted as a weir, with a continuous waterfall along its lower edge, which emitted a dull roar. The site of falling water was visible for miles along the straight length of the road. Now and then, a violent disturbance in the surface of the placid flood waters disclosed the presence of a river underneath. On noticing it, we would make a laborious detour upstream, to avoid being sucked down through the unbridged gaps across the road. It was an eerie feeling to realise that a navigable river lay concealed under the agitated flood waters beneath the boat!

We had not rowed far when I noticed a zigzag disturbance on the surface of the water rapidly approaching our boat. Presently, a huge cobra reared its head and tried to swarm up into the boat. Frantically, I grabbed a bamboo pole and lashed at it, until it lay limp on the surface of the water. In the meantime, my servant was similarly dealing with another intruder on the other side of the boat. I was contemplating the strange scene, lost in reverie, when a shout from the servant made me aware that another king cobra had eluded our vigilance and had clambered up inside the boat, and was trying to hide among the heap of bamboos lying within in the boat. Luckily, the snake was too scared to put up a fight. We killed it and threw it in the water. Until 4 p.m., when we anchored our boat against the veranda post of the Companiganj rest house, it had been a nightmare journey spent in lashing out at the intruding snakes, daubing paint for flood marks, and negotiating tumultuous submerged streams. In all, we had accounted for thirty-nine snakes, of which three had succeeded in getting into the boat. The snakes, as we discovered, were highly venomous cobras and kraits along with some rat snakes and some common vine snakes.

The Companiganj rest house was situated on the far bank of a stream, on top of a natural outcrop of laterite, about 10 feet higher than ground level. It had a further 4 feet high wooden plinth above the top of the laterite mound. The winter water levels of the adjoining stream channel were usually about 20 feet lower than the ground surface. It was nevertheless a navigable stream. River steamers from Calcutta passed up and down the stream during the winter season carrying loads of the famous Sylhet oranges and lime, from the lime kilns in the Khasi hills to the Calcutta market. During the flood, water levels had risen to within 6 inches of the plinth level of the rest house, thus rising about 34 feet above the winter water levels in the stream! That flood was probably of the magnitude of once in a century. To the best of my recollection, it accrued on 13 July 1915.

A Cruel Custom

'Eli, eli, eli,' shouted the crowd of people around as the leopard, goaded by the stab of the lance which had drawn blood, lashed back at its assailant but fell back helplessly after its vain assault against the stockade which was encircling it on all sides. Its assailant had thrust his spear in the leopard's body from outside the stockade, through its intestines. The bamboo and cane stockade in which the leopard was confined, but left free to roam about, was about 20 feet in diameter and 10 feet in height. It was elastic and yielding, but strong enough to withstand the savage rushes of the infuriated beast.

Half a dozen men carrying long spears encircled the stockade and stabbed the leopard from all directions with shouts of 'Eli, eli'. The delighted crowd greeted each rush of the leopard at its cruel assailants outside the stockade. At last, exhausted by its numerous impotent rushes against the stockade, and profusely bleeding from many parts, the tortured creature lay down on its back in the middle of the stockade, and looking heavenwards uttered piteous moans. I could stand the scene no longer and, feeling nauseated, left in disgust. The creature was later tortured to death. The leopard in question had carried off some goats and young cattle, and perhaps also some human babies from the outskirts of Sylhet town. It had somehow been trapped and captured and was tortured to death in the customary fashion described above. Executions of captive leopards within a stockade by spear stabs was a gala affair and was witnessed by crowds of people of all walks of life in Sylhet in a holiday mood. Amongst the crowd of people round the stockade I was shocked to recognise the venerable head of the aristocratic family, from whose library I had borrowed two rare manuscripts.

* * * *

Towards the end of my apprentice year at Sylhet, I was permitted by the Railway Department to take notes on the construction work of the Railway Bridge on the Kushiyara River at Fenchuganj. The river was navigable with a large centred span to admit passage of double-deck steamers. The brick piers of the central span rested on well foundations and were laid in deep running water. Muck and shingle were thrown in the river at the spot of the proposed pier until the mound was well above the water surface. The well curb was then placed on the dry surface of the mound and well sinking started, using steam driven centrifugal pumps and crab dredgers. There were smaller spans on brick piers likewise founded on well foundations, on either side of the main span which were built on the riverbed in winter. Work was started from the west bank of the river. The west abutment was also constructed in the dry river against 20 feet high earth embankments. Work had been completed on the west abutment and also on the two adjoining piers in the dry riverbed and plate girders were placed on the two completed spans.

A few days later, however, the west abutment suddenly sagged, imparting a horizontal thrust to the plate girders resting in it. The thrust was transmitted to the next pier, which in its turn also tilted and passed on the thrust to the second pier, like a falling house of cards. As a result, the west abutment and the two adjoining piers collapsed and the plate girders resting on them also fell down. Expert investigations as to the cause of the disaster disclosed that the soft peaty riverbed had settled under the heavy load of the earth embankment. The latter also had settled with the sinking riverbed, and in the process the embankment and other structures had overturned. A few more bridge spans were consequently added to the structure. Mr Boyajian (executive engineer) a naturalised Englishman of German origin, in charge of the work, was replaced by an experienced Muslim engineer, who completed the job without further mishap.

Completing my apprentice year in the Sylhet PWD, in late 1915, I was granted a practical training certificate by the PWD of Assam government, along with a personal testimonial by my boss, A.P. Mullick (executive engineer). I returned home via Calcutta which was the shortest route, by the fast Calcutta Mail of the East India Railway, running between Calcutta and Peshawar. At Calcutta, I stayed at a small but clean hotel on the Chitpur road. The daily charges for a large furnished room with attached bathroom and four decent meals were three-and-a-half rupees. I stayed for a week in Calcutta, enjoying its sights, principally the Chowringhee, the 'finest street in Asia', flanked on one side by the Eden Gardens; Sir Stuart Hogg Market; the Zoo; and the colossal Indian Museum. I was surprised to see a large scale geological cross section of the Salhad valley (near my home at Abbottabad), hung on a wall of the geological hall of the museum. I saw my first cinema show in Calcutta, *The Neptune's Daughter* (1949), showing a lady in bathing costume, diving in an elaborate swimming pool.

I had collected a lot of cane furniture at Sylhet, which I booked for home at the huge Howrah railway station at Calcutta. The process of booking the luggage and having it loaded speedily and safely in a goods train bound for the NWFP was facilitated under the guidance of an experienced coolie (porter) by greasing the palms of half a dozen clerks for the amount of four *annas* each. I left for home by the Calcutta Mail. My compartment was crowded with Marwari merchants—men and women bound for their homes in Marwar in north western India. The women wore massive gold bangles which covered their forearms and ankles, in addition to rings, bracelets, nose rings, etc. Each woman thus carried a weight of many pounds in gold. Both men and women wore dirty clothes and were unprepossessing in appearance. At first, they were alarmed at my entry into the compartment. Pathans were evidently looked upon with suspicion and as a potential threat. Eventually, the Marwaris reconciled to my presence, even as a possible asset in case of an assault on their gold-laden women-folk from any quarter en route. Thanks to the skill and efficiency of the railway porters at the Howrah station, aided by a generous tip of eight *annas*, my cane furniture reached in perfect condition at my home.

6 Railway Project, Bihar, 1916

After returning home from Sylhet in the autumn of 1915, and spending a few months with my family, I became restless with inaction and started looking for a job. In response to an advertisement in the *Pioneer* of Allahabad, for an assistant engineer required for the survey of a branch line of the Dehri–Rohtas Light Railway in Bihar, I applied for the job and was promptly offered it.

Dehri-on-Sone, the headquarters of the Dehri–Rohtas Light Railway was a pretty little town situated on the right bank of the Sone River, on the Grand Chord Line of the broad gauge East Indian Railway. The Dehri–Rohtas Light Railway brought limestone from the Rohtas hills, 30 miles away, to the lime kilns in Dehri-on-Sone. My assignment was to carry out surveys for a new 15-mile long branch line from Tilouthu station on the Dehri–Rohtas Railway, to the town of Sasaram, a place of considerable commercial importance. Rohtas was also famous for the extensive ruins of the Rohtas Fort (or Rohtasgarh, situated in the Sone valley, Bihar) built by ancient Hindu rajas, which was entered through a stratagem and then ransacked and overrun by a local Pathan soldier of fortune, named Farid Khan, in sixteenth century AD. Farid Khan extended his domain from his Rohtas base in Bihar and later, by defeating the Mughal Emperor Humayun, became the Emperor of India, taking the title of Sher Shah Suri. He was a great administrator and reformer. He also introduced in India the system of measurement of cultivated land. The credit for the system of collection of land revenue in India based on accurate measurement of land and seasonal crops, that has since been followed by his successors to date, goes to his genius. He also constructed the 1500-mile long caravan road from Peshawar to Calcutta, whose alignment is essentially followed by the modern Grand Trunk Road.

My first survey camp outside the town of Dehri-on-Sone was in a village called Hurka, 5 miles away from Dehri. The summer was intolerably hot. Hot air seared the body like red hot coals even in the early hours of the morning and even within the shelter of my new Swiss Cottage tent. I, therefore, hastily moved into an incomplete mud hut built outside the village, which I and my survey *khalassis* shared with the local *patwari* of the village. The *patwari* was a pious Hindu who regularly performed his early morning devotions after a bath and a drink of hashish prepared by pounding dry hashish leaves with water. I informed him that in my district of Hazara, the hashish plant was in abundance and grew wild. His mouth watered. He could hardly believe that such a heavenly spot existed on earth. The nights being hot, we slept in the open space outside the hut which was enclosed by a low mud wall without a gate. Some cattle also walked in at night and lay down sharing the courtyard with us. One night we were aroused by loud cries of the cattle and by their wild stampede. A leopard had carried off a calf from inside the compound where we were sleeping. However, life in Dehri-on-Sone also had its pleasant interludes. In the balmy winter of 1915–16, the 16-mile long artificial lake, created by the anieut (weir) built across the 2-mile width of the Sone River, from bank to bank, was a haven to swarms of wild fowl, after their long and exhausting autumn trek from the polar seas. On Sundays, I and my surveyor used to go out for day-long duck shooting in the lake, hiring a small dinghy paddled by a single boatman, and taking our lunch along. We would lie behind a camouflage of thorns and bushes placed above the rim of the boat, enabling us to approach within gunshot of the birds, which did not seem to mind the boatman.

One evening, returning after a particularly successful day with a bag containing a ten-pounder wild goose, two fat Brahminy ducks, some pintail and mallard, and a dozen river snipe, we were accosted by my boss, Mr S.G. Reilly (executive engineer), who was returning empty-handed, cycling along the river bank. He was amazed to see our bag. Despite his mild protest, I pressed our wild goose on him, to share which he invited his British friends to a feast the next day. My assistant surveyor's wife, in strict *purdah*, was expert in roasting ducks. Another subordinate of mine was a young Hindu Brahmin, permanent way inspector. He used to invite me to lunch now and then. His young wife, also in *purdah*, turned out delicious meals of *puris* and vegetable dishes. But she would not touch the dishes containing the remnants of my meals for washing until a dog had lapped them clean!

My next camp lasting six months was at the large village of Dewadand situated at the foot of a low forested spur of the Rohtas hills. The village had about a thousand houses. The inhabitants were Hindus, the majority being of the low Koeri caste. The landowners were high-caste Rajputs. There were a few families of humble Muslim weavers, as well. The land revenue of Dewadand and some other neighbouring villages was assigned, since the time of the Mughal Emperor Farrukh Siyar (r. 1713–19), to a Muslim *jagirdar*, who occasionally visited the village to collect his dues. The village had good quality land on which mainly wheat and fodder crops were grown in winter, and rice, linseed, and sugar cane in summer. The area was not irrigated by canals. The winter crops were raised on the winter rainfall.

The summer monsoons brought copious rainfall which was collected in an artificial tank and carefully distributed amongst the cultivators for watering the summer crops. Crushing and sale of linseed oil was done by the villagers as a cottage industry.

On Sundays, I would sometimes climb up the top of the neighbouring hill to enjoy the fine view of the surrounding country. The majestic Sone River with the broad silvery ribbon of its 2-mile wide channel completely confined between two parallel banks, flowed parallel with the direction of the hill range a few miles away, offering an awe-inspiring view. The hillside consisted of large pieces of rock, cut up into numerous fissures and caves partly hidden under bushes and tall grass and reeking with the smell of wild beasts. For that reason, as a precautionary measure, I invariably took along with me one or more of my survey *khalassis*, carrying long spiked bamboo poles. In fact, the whole countryside around was inhabited by both tigers and leopards. One Monday morning, my survey *khalassis*, who belonged to the neighbouring village, arrived late for work for the reason that a tiger had been lying, in broad daylight, on the footpath leading to our village.

I set up my quarters in the village 'guesthouse' situated just outside the village dwellings. It was a mud hut with two small rooms on the ground floor, one of which served as the kitchen. The other housed my survey staff. An apology of a room on the second floor served as my bedroom. Living was cheap in Dewadand village. Mutton was non-existent, while the Hindu population did not raise chickens and eggs. Only pulses, potatoes, sweet potatoes, and a few vegetables were available in addition to cheap wheat flour, milk, and clarified butter. Mangoes, of the sucking variety, were cheap and plentiful. I would purchase a basketful of mangoes for the paltry sum of eight *annas*, and bury them in wheat chaff to ripen, and ate a dozen or so that ripened daily.

An Amateur Physician

At first, my Hindu neighbours were highly suspicious of me, and my presence in the village. Gradually, however, their suspicions were allayed, partly due to propaganda in my favour by my survey staff who were all Hindus from the neighbouring villages, and partly by the extraordinary circumstance of my turning into an amateur doctor. The railway authorities had supplied me with some common medicines, such as quinine and aspirin tablets, etc. for my survey staff. The news about my store of medicines spread amongst my neighbours and requests for medicines started pouring in. Gradually, the rumour spread in the whole village that I was a doctor and, in particular, a specialist in children's ailments, as I had a special weakness for small children. Each morning, before starting for survey work, I had to attend to a small crowd of patients, men and women, some of whom arrived with their babies and young children, waiting outside my quarters for treatment. I was at my wits' end about how to deal with complicated ailments. The villagers would not take my protests seriously that I was not a doctor. During peak periods, the daily attendance of my patients was about fifty.

At that time, I got hold of a medical book titled *Secret Remedies: What They Cost and What They Contain* (1910), issued by the British Medical Association, in which treatment of a large number of common ailments was described for the benefit of the general public. I selected a few well-known medicines and sent their prescriptions to a firm of chemists in Calcutta and placed an order for preparation of large amounts. I also obtained antiseptics and medicines and dressings for treating wounds, boils, eye and ear infections, and medicines for relieving toothache, etc. I appointed one of my intelligent survey *khalassis*, named Tilak, to attend to women and tackled men and babies myself. The supplies from Calcutta were purchased from my personal resources.

As a layman doctor, I recollect two spectacular cures. One midnight, the son of a neighbouring Hindu villager, a sweetmeat seller, woke me up and said his father was in great pain. 'What is his problem?' I asked. 'His windpipe is choking due to acute pain,' he said. Guessing that the pain might be of a rheumatic nature, I gave the boy a generous portion of Zam-Buk* which I had procured, from Calcutta, and instructed him to gently rub it on the affected part followed by hot fomentation. The next morning a large tray of freshly prepared sweetmeat was sent by the grateful sweetmeat seller whose life I had been instrumental in saving.

Another morning, on a palanquin carried on the shoulders of four sturdy men, a female patient in purdah was brought in by her husband, a Rajput landowner, from a village 7 miles away. The young husband gave me, in strict confidence, a long list of his wife's ailments, but being a proud Rajput, he declined to allow me to even have a look at her. Under the circumstances, I expressed my inability to prescribe any medicine. After a long discussion, he permitted his wife to expose just one of her hands outside the palanquin. I at once noticed the pallor of her hands and fingernails and concluded that she was suffering from anaemia. Her husband had also told me that she suffered from an irregular monthly cycle as well. I prescribed the then well-known iron and quinine pills for her alleged anaemia and the better-known drug Aletris Cordial for the menses troubles. A few weeks later, the grateful husband came all the way from his village to say that my prescription had done his wife a lot of good!

* * * *

There were four or five houses of Muslim weavers in the village. One Sunday morning, I was invited by a weaver to his house to meet a *maulvi* from Sasaram town who had come to Dewadand village on his annual visit to perform some religious ceremonies for the Muslim weaver community. Mystified as to the nature of the religious ceremonies, I willingly accompanied the weaver to meet the religious man. One of the religious services rendered by the *maulvi* to the Muslim weaver community was to loudly utter the words, 'In the name of God, God is great', three times and to blow three times on the knife used by the weavers for slaughtering animals for their food. This served them for a year. He also said prayers on their behalf, which also sufficed for them for a year. The weavers gave the *maulvi* gifts of grain and cash. The *maulvi* told me frankly that Islam had only a precarious hold on the ignorant Muslim minorities living among the Hindu majority in those areas. He became interested in me on learning that I had a good salary. He said 'A respectable young man of means like you can afford to have three wives. You are leading a lonely bachelor life. I can arrange your marriage with a virgin or widow, according to your choice from amongst the poor, middle class or aristocratic Muslim families of Sasaram as desired.' Moreover, he added, "Marriage with a widow would be an act of charity.'

Bands of Hindu minstrels regularly toured the rural areas of Beharat at the time of gathering the harvest. They performed dances and sang lilting religious songs and received gifts of grain in return. A party of minstrels also arrived at the important village of Dewadand at the end of the autumn harvest. They gave some excellent performances in the open ground in front of my camp and also pitched their tents nearby. On the evening of the day the party of minstrels was to arrive, I was visited by a village elder. He remarked that, owing to my exemplary character and conduct during my stay at Dewadand village, I had earned the respect of the village elders. However, being a young man away from home for many months, I would naturally be feeling sex-starved. So, they had decided to arrange a mistress for me from amongst the young girls of the minstrels' party for the duration of their stay in the village. He further remarked that the charges would be moderate; however, I declined the offer with thanks.

This rather amazing incident, apparently, showed that I had gained the goodwill of the proud Hindu village elders. One afternoon, I was intrigued by the inviting smell of fried mutton coming from my kitchen. The cook informed me that the welcome gift of mutton was from a Hindu neighbour. Further enquiries disclosed that the meat was part of a goat that had been ritually sacrificed by the villagers for the local deity, an idol of the goddess Kali at the village temple. Now eating meat sacrificed before idols is strictly forbidden in Islam. So, I asked my servant to throw away the meat in question. The news of my act leaked through my Hindu survey *khalassis*. A few weeks later, on the occasion of the Muslim festival of Eid-ul-Azha, I sacrificed a fat sheep as a customary sacrifice in commemoration of Prophet Ibrahim's (AS) offer of sacrificing his son, Ismail, before Allah. I sent some meat to my Hindu neighbours, and village elders. It was unanimously returned with the message that since I had returned their meat owing to religious scruples, they were returning my meat on the same grounds! And so, I left for my next camp at Tara Chandi with mixed feelings.

Tara Chandi village was situated 5 miles farther along the proposed railway line. It was also situated at the foot of a small rocky spur of the Rohtas hills on the far side. The weather having somewhat cooled down, I pitched my tent under the shade of a mango grove outside the village. The place was considered unsafe because of close proximity of the wooded hillside which was known to harbour wild animals. I kept my lantern burning brightly throughout the night within my carefully closed Swiss Cottage tent as a precautionary measure. In the mornings, on more than one occasion, I was assailed by the fresh pungent smell of wild animals, possibly tigers or leopards, which had been prowling round my tent during the night. On a sunny Sunday afternoon, after a leisurely lunch, I was relaxing in my camp chair under the mango trees, which at that late season were devoid of fruit, when a marriage party of Hindu villagers, about twenty in number, carrying a bride's litter (carrier) enveloped in a red cloth, halted under the mango trees a few paces from me. Two or three members of the party promptly left for the nearby hillside to collect dry cow dung and sticks for lighting a fire.

A shallow pit about 3 square feet was dug in the ground, in which the cow dung and sticks were piled up and a fire was started. A sack of wheat flour was unslung and kneaded into dough in a large iron basin. Salt and small slices of onions were mixed with the dough which was formed into round balls of the size of a large fist, one for each member of the party. The dough balls were thrown into the glowing cow dung fire along with an equal number of large round eggplants. When the dough balls and the egg plants were judged as properly cooked a ball of cooked dough and an eggplant was extracted by each member of the party. Each person ate his meal separately in his own brass vessel, which all travellers used to carry during journeys.

That was a marriage feast. The usual morning and evening meal of the villagers was boiled barley ground into flour and kneaded into salted dough. Mouthfuls of the dough were pushed inside the mouth deftly with the thumb of the right hand, to be swallowed.

My surveys soon reached the outskirts of the proposed terminal station of Sasaram, with its groves of Toddy palms and mango gardens. The Toddy palm tree yielded large quantities of a sweet and pleasant-tasting milky fluid every day, which spontaneously fermented with the heat of the day, and was sold as cheap liquor, usually at the site by the owners of the Toddy gardens. Boiled grams fried in mustard oil and highly seasoned with salt and red chillies were the usual accompaniment of the Toddy drink. Customers purchased liberal supplies of both these delicacies for a paltry sum of money and lay down, for hours, under the Toddy palms in blissful stupor. I rented a house in Sasaram town for myself and my survey staff. Being the resting place of Sher Shah Suri, Sasaram had a large Muslim population, including some well-to-do middle-class and old aristocratic families of landowners. I called on a number of Muslim notables of the town and was hospitably received everywhere. The mausoleum of Sher Shah Suri was an impressive plain stone structure in Pathan style of architecture, a forerunner of the later, matchless Mughal architecture in India.

7 Homeward Bound

My survey completed, I returned to my headquarters at Dehri-on-Sone to complete the plotting work. I had been away from home for about a year and was very homesick. Since I had done the assigned task and the survey had been completed, I sent in my resignation. Mr Reilly was well satisfied with my work, and he urged me to stay on by refusing to accept my resignation, but I was adamant, and so I left for home. However, on my way back, I resolved to try for a new job before meeting my wife, Khanum Begum, and family.[1] Hence, I broke my journey at Lahore, where I learnt that the job of a lecturer in civil engineering had fallen vacant at the Osmania University, in the Muslim state of Hyderabad Deccan, a place even farther away from home than Bihar. Yet, I set about finding ways and means of securing the lecturer's job. Stopping on the way at Lahore, I visited Dr Mohammad Iqbal, who expressed great pleasure on learning that I had graduated in engineering from Roorkee. I secured a letter of introduction from him, for the finance minister of the state, and started for Hyderabad Deccan.

The Muslim state of Bhopal in central India fell in the way, where an old Roorkee colleague was working as forest engineer to the state. I broke journey at Bhopal and paid a surprise call on my friend. He was residing in a suite of a vacant palace of the ruler of Bhopal. On his pressing invitation, I stayed with him for a week, which happened to be the annual week of festivities in connection with the birthday celebrations of the Begum of Bhopal, the lady Muslim ruler of the state. My friend secured invitations for me to all the week's functions, including the formal durbar of the Begum of Bhopal Sultan Jahan, who was an elderly widow. In the absence of a 'formal dress suit' required for the occasion, I attended wearing a dark English suit. The garden party was held in the beautiful grounds of the Taj Mahal palace. Four-year-old Princess Abida Sultan, the heir apparent of the Bhopal throne, was brought by her nurse to the party.

Bhopal was a beautiful town spread along the shores of an artificial lake on low green hills. It was an island of Muslim culture in Central India. The majority of the state's subjects were Hindus. Bidding goodbye to my friend, I resumed my journey for Hyderabad Deccan. At Manmad Junction of the Great Indian Peninsula Railway, a short distance shy of Bombay (now Mumbai), I boarded the Nizam's Guaranteed State Railway which would take me to the state's capital, my destination. My sole fellow passenger in the second class railway compartment was a *taluqdar* (commissioner) of the Hyderabad state. On learning that I was going to Hyderabad, he informed me that the town was in the grip of a plague and everybody was living out in tents. He advised me to postpone my visit. Thoroughly alarmed and repenting my foolhardy adventure in quest of a job so far from home, I quickly decided to get down at the next station of Aurangabad and return home.

As the return train to Manmad was not due until the evening I had time to visit the mausoleum of the Mughal Emperor Aurangzeb and his favourite wife [Dilras Banu Begum] at Aurangabad, as well as the ruins of the town of Daulatabad. Emperor Aurangzeb's mausoleum was an imitation, on a small scale, of the famous Taj Mahal at Agra, but it inspired feelings of awe and reverence for the great Mughal emperor who ruled for fifty years over an empire greater than British India—from Afghanistan to Cape Comorin and from Bombay to Assam. Daulatabad, also known as Deogiri, of medieval India, was invaded by Sultan Alauddin Khilji of Delhi in the thirteenth century, from where the Sultan had carried away elephant loads of gold. A few centuries later, Sultan Mohammad Tughlaq of Delhi transferred his capital from Delhi to Deogiri, which had a more central location for governing India, and renamed it Daulatabad. The project failed because the sultan also insisted on transporting the bulk of the population of Delhi to the new capital. Daulatabad town was built around the slopes of a natural outcrop of rock rising sheer to a height of about 200 or 300 feet from the surrounding flat plain. The top of the hill was surmounted by a strong fort, providing a magnificent view of the surrounding country for miles around. The ruins of the fort and the town were in a tolerable state of preservation. I spent the night at Manmad station in a waiting room, which I shared alone with a young Maratha girl. She was traveling with a servant, who was not entitled according to the railway regulations, to stay in that waiting room. The night was bitterly cold and while I slept soundly under a warm blanket, the poor girl having no adequate covering shivered with cold throughout the night. I felt sorry for her, but under the prevailing social circumstances could not offer her any assistance.

Early the next morning, I arrived on the platform with my luggage, awaiting the arrival of the third class express from Bombay. When the train arrived, it had only one inter class compartment which was occupied by a number of British Tommies. There were two or three vacant seats, for which a Hindu gentleman and I were the only candidates. The Hindu gentleman was immaculately dressed in European clothes. The coolies first deposited the gentleman's stylish luggage in the compartment while the Tommies looked on. Then the gentleman himself entered. As soon as he got in the train he was seized by the Tommies and thrown out of the compartment on to the platform. His luggage was thrown out after him. Angry, bewildered, and rubbing his bruised parts, the gentleman walked away and soon returned with the station master, an elderly Anglo-Indian (Eurasian). The Tommies greeted the station master's protests with obscene swear words. He walked away frustrated advising the Hindu gentleman to go to some third class compartment.

I had been a silent spectator to this drama. The Tommies next turned their attention to me. I asked the coolie to deposit my luggage in the compartment, the Tommies looked on as before. Then I entered the compartment. Nothing happened. I took a seat. No one took notice of me. The Tommies grew busy with their breakfast of huge loaves of bread, and large mugs of hot tea which was brought for them in a huge kettle. I again broke journey at Bhopal to pay a farewell visit to my friend. He again detained me for a week, and unsuccessfully tried to find a job for me at Bhopal. One day, at Bhopal, a palmist called at my friend's house and offered to give a reading of our fortunes. He was a young, uncouth lad belonging to the wandering Raul tribe of the Punjab, who are professional fortune tellers. I offered him my hand to read. He predicted, among other things, that I would get a good job within a month. I left Bhopal and broke journey at Roorkee, to see the principal of my alma mater. I thought of trying my luck there for a job in my own province and to request the principal to recommend my application for the post of assistant engineer to the chief engineer of irrigation department of the NWFP. My old principal, Lt. Col. E.H. Atkinson, had since been recalled to Britain on the outbreak of the First World War and appointed chief engineer of the first army corps which initially landed in France. The current principal was Mr W. Gunnell Wood, CSI, a retired chief engineer of the PWD (B&R branch) of UP, India. Though he did not know me, the principal saw me in his bungalow while he was dressing to leave on a shooting trip to the countryside with a number of friends. He put my application in the pocket of his *shikar* jacket.

At long last I arrived home. A week later I received a telegram from the secretary irrigation, NWFP, offering me an appointment to the post of assistant engineer in the Indian Imperial Engineering Services (later Indian Service of Engineers [ISE]). The principal had evidently not dropped or lost my application during his *shikar* trip and the palmist's prediction had also come true.

8 Mardan, 1917

Mr Smith, a sub-divisional officer of the Lower Swat Canal Sub-division, with whom I was attached for practical training in irrigation practice, was a man of about forty with a youthful appearance, which went rather oddly with his greying temples. He had a flair for oriental languages, having acquired a smattering of Arabic, Urdu, Punjabi, and Pashto, during the course of his service in Egypt and India. Mr Smith was married to a lady older than himself. She, however, suffered from somewhat impaired sight and hearing but was a perfect gentlewoman. Mr Smith fiercely guarded her from possible unwelcome attentions by his British acquaintances, particularly military officers. He had given strict standing orders to his domestic servants not to allow any *beltwala* (military officer) to enter his house in his absence. I accompanied Mr Smith during his tours of inspection of the canal system and the canal irrigated areas, and also learned the administrative routine in his office. I had received only a month's training when Mr Smith proceeded on leave and I received orders to take charge of the Lower Swat Canal Sub-division from him.

A few days later, Sir Thomas Ward, inspector general of irrigation in India, arrived on an official tour of the NWFP, and in due course also visited Mardan. I paid my respects to him. On learning that I was new to the irrigation service and was holding charge of an important sub-division, he gave me a long lecture on irrigation practice, from distribution of water among cultivators and the system of measurement of crops and collection of water rates, through maintenance of outlets, distributaries, canals, and head works to the design and construction of irrigation works. He further asked me to revise my college courses on the subject of irrigation and suggested some further advanced reading. I felt gratified at the attention the head of the irrigation department in India gave me and, needless to say, took his valuable instructions to heart.

The next day, Sir Thomas Ward left for Swabi by car. I rode to the submerged causeway of the Kalpani stream road-crossing in the Hoti Bazaar quarter of Mardan to see Sir Thomas Ward's car safely through. On my return through the crowded bazaar, my horse, a borrowed beast of unknown temperament, took fright on seeing a number of gaudily clad Hindu women. It was a vicious beast of enormous proportions, and I discovered later that it was also blind in the left eye. Moreover, it had fattened without exercise, on green alfalfa fodder during the past few months. The frightened animal took the bit in its mouth, rendering me powerless to control it and bolted through the crowded bazaar. Failing to notice a telegraph post on its blind side it collided against it and, knocking me down, and bolted back home. When consciousness returned, I found myself lying in the military hospital at Mardan, with every limb in my body aching. Luckily, no bones were broken. Sir Thomas Ward and Mr F.A. Burkitt (executive engineer), my boss, sent me messages of sympathy.

The Lower Swat Canal was planned as a political measure chiefly for the pacification of the Pathans of the Peshawar district. Surveys for the canal started in 1874, and construction was completed in 1891. The canal takes off from the Swat River below the point where it emerges from the gorge into the Peshawar plain. It runs parallel to the peripheral hills, cutting across a number of deep hill torrents, by spectacular cross-drainage works, to irrigate an area of over 100,000 acres. The landowners, as well as the local tenants, were Yusafzai Pathans who raised wheat crop on winter rains. It could not be determined whether they would agree to use canal irrigation for raising the summer crop of corn, sugarcane, and orchards, which involved a lot of labour and the use of expensive manures. Any apprehensions about the development of the canal areas were, however, soon dispelled. Industrious Mohmand tribesmen from the neighbouring tribal area settled in large numbers in the canal irrigated areas. In a few years, irrigation on the Lower Swat Canal reached 135 per cent, including double cropping at 35 per cent of the area.

The Mohmand tribesmen occupied a large slice of the tribal area along the north-western periphery of the hills surrounding the Peshawar plain. A large section of the tribe also spilled over into Afghanistan across the Durand Line. Consequently, Afghan propaganda had affected the tribesmen on both sides of the border. The First World War was at a crucial stage in 1917. The Mohmand tribesmen of the tribal belt were technically at war with the British Indian government. The bulk of the Indian Army, both the Indian and British elements, had been dispatched to the various war fronts in Europe and the Middle East, leaving only skeleton British troops behind in India, who were constantly on the move in the country to give the impression that the British troops were still in effective occupation of India. The cantonments in India were protected by troops from the loyal Indian states.

The Abazai front against the hostile Mohmand tribesmen was manned by a contingent of troops from the Hyderabad Deccan state. A permanent barrage was under construction on the Swat River at the head of the Lower

Swat Canal to replace the existing temporary diversion work. As sub-divisional officer of the Lower Swat Canal, I was technically in charge of the construction work. The hostile Mohmand tribesmen took pot-shots at the labourers working on the barrage from their distant fortified positions on the far bank of the river. The defending Hyderabad troops, however, appeared disgruntled with their task. 'Look,' an officer of the Hyderabad contingent confided in me, 'In Hyderabad we had to do only the morning parade and afterwards we ate *pulao* for lunch and rested for the day. Here we are up against serious warfare and on top of it everyone here carries either a rifle or a pistol!'

As mentioned earlier, the Lower Swat Canal tract was fully developed, and my main task as the sub-divisional officer was to deal with the numerous petitions from the farmers concerning equitable distribution of canal supplies. The conditions were ideal for my subordinate revenue staff to reap rich graft from the all too willing farmers. But since it was unthinkable to offer a bribe to an 'officer', the Khans, who were the past masters in the art of diplomacy, usually checkmated my scruples by providing lavish hospitality to me whenever I visited their village *hujras* (drawing room) on official business. One old Khan, during the course of the conversation, addressed me as 'Your Authority,' in the same sense as one might use 'Your Excellency' to an exalted personage. He probably used that title while addressing all officials of various government departments. Another old diplomat used the quaint and mouth-watering phrase of 'lamb's meat in your mouth', in place of the usual phrases like 'You know', or, 'You see'.

One day, I had a visitor of quite a different category. He was a Khan from the important village of Utmanzai. During our conversation he said that the sole purpose of his visit was to make my acquaintance. He had learnt about my appointment in the irrigation department and, furthermore, that I was a fellow Pathan from the Hazara district and was interested in the welfare of the Pathan people. I questioned him about his own activities. He informed me that he had constructed a *mandi* in the Utmanzai bazaar, in competition with the Hindus of the village, who held a monopoly over that business. A *mandi* was a clearing house for the exchange of agricultural produce from the neighbouring rural areas to the wholesale dealers in the town. He was also running a number of primary schools for Pathan children from private donations. My visitor was Abdul Ghaffar Khan,[1] later to become the leader of the famous Red Shirts Movement.

In November 1917, I was transferred to the Maira sub-division of the Upper Swat Canal. My new boss, Mr W.B. Harvey (executive engineer), invited me to accompany him on his tour of inspection of my prospective sub-division. We drove together on a Tonga along a dirt road for the first 7 miles of our journey to the village of Garhi Daulatzai, where our horses had been sent ahead, for riding the rest of the journey to my future headquarters at the Jagannath rest house on the Maira branch of the Upper Swat Canal. Mr Harvey was a tall, lanky officer, of a reserved temperament. We had exchanged only a few monosyllables during our tonga journey at the end of which we silently mounted our horses and started on our 10-mile ride. When we emerged from Daulatzai village, Mr Harvey, without warning, put his horse to a fast canter along the narrow village track. I automatically followed suit. After about half a mile of fast riding we halted before a large watercourse barring our way. Mr Harvey casually looked back and found me close behind him. It dawned on me that Mr Harvey was testing my riding skills. Apparently satisfied, he opened up, and started questioning me about my technical qualifications which, apparently, were also satisfactory.

My firm but courteous manner also seemed to please him. We were soon riding leisurely together on cordial terms. Riding a few miles further we reached the large village of Yar Hussain. While riding along the dirty winding lanes of the village, through which our path led, Mr Harvey was greeted by a villager, who invited us for a cup of tea. As it was considered impolite to refuse the hospitality of a Pathan, Mr Harvey graciously accepted the offer. Our host, who was of humble means, brought out a ramshackle cot on which we sat in the lane waiting for our tea. After a long wait, the readymade tea was brought out in an old blue enamelled kettle along with two old enamel cups, one without a handle. We drank our tea and, thanking the host, resumed our journey; by the time we reached the Jagannath rest house, we were on pleasant terms. Mr Harvey had evidently approved of me. He returned to Mardan, leaving orders for me to take charge of the Maira sub-division forthwith.

Mr Harvey was a thorough gentleman and a capable engineer. With Mr Stoddard, he was the co-inventor of the Stoddard-Harvey module outlet for field water courses which was a considerable improvement over the existing structures. An improved version of the Stoddard-Harvey outlet was later universally adopted by the NWFP, Punjab, and Sindh irrigation systems, in the form of Mr Crump's Adjustable Proportional Module outlet. Mr Harvey was scrupulously honest about public funds at his disposal. He would not even use an office pencil for his private use. His wife, on the other hand, was sometimes impatient with his meticulously honest ways. She was bitterly critical of the government for not paying him adequately for his talents and hard work, and often complained that she could hardly make ends meet with his meagre salary.

Mr Harvey, though strict in punishing any corrupt official working under him, was quite soft at heart. On one occasion, he was at the point of dismissing a corrupt subordinate when the culprit brought out a bullet he had received in his body when a hostile Mohmand tribesman fired at him while he was working on the construction of the canal barrage at Abazai. On seeing the bullet, Mr Harvey turned scarlet, and let off the man with a warning. During our leisurely tours in my sub-division, Mr Harvey and I sometimes discussed world affairs. I recall that he had no great respect for the famed military conquests of Alexander the Great. Judged by the scale of the prolonged bloody battles fought at Somme, Ypres, and Verdun between the Allies and the Germans against the occupation of France during the First World War, Alexander's achievement, i.e. the conquest of the vast Persian Empire in a single battle was, in his judgement, not very impressive. Mr Harvey proceeded home to England on a six-month furlough and handed over charge of the division to Mr Smith from whom I had received practical training. Returning from leave to join duty in the NWFP, Mr Harvey met a tragic end while travelling from Bombay to Peshawar by the Bombay Mail. He was looking out, leaning against the door of the railway carriage when it suddenly opened outwards, throwing Mr Harvey out of the compartment from the fast-moving train.

It was the spring of 1918 when Mr F.W. Carne, superintending engineer and secretary for irrigation, NWFP, arrived on an official tour of my sub-division, accompanied by my boss, Mr Smith (executive engineer). They arrived at my headquarters hungry and thirsty after a long and tiring ride on horseback. I had prepared a sumptuous Pathan-style lunch for them after which Mr Carne asked me whether I was sometimes depressed living in a lonely rest house, miles away from the bustle of city life. From the windows of the dining room of the rest house, there was a fine view of the country around, facing the purple Buner hills in the north. 'Sir, have a look at the country around,' I replied impulsively, 'the green Maira (heath) is studded with red poppies, and purple and yellow flowers; it is sheer delight to ride out for canal inspection tours in the country. Besides, remembering my recent service far away from home in Assam and Bihar, I am thankful to be living with my family, at a spot only a few hours away from my home at Abbottabad. What more could a young man desire?' Mr Carne, who was on the verge of retirement, listened to me silently with a faraway wistful look in his eyes.

On 13 September 1918, the death of a young man from influenza was reported at Mardan. I was at the time living alone at my headquarters at Jagannath rest house. The rest house, fortunately, was situated at a lonely spot; the nearest village from it was 3 miles away at Mardan. From that date onwards, people began to die like flies from the influenza epidemic in the villages all around. I received daily reports of deaths from my *baildars* who came for work on the canals from different villages. The death toll was appalling. Thirty to fifty deaths per day were reported from each village. The remaining days of September slowly dragged on in an agony of apprehension. I took daily doses of the standard influenza mixture, knowing full well that its efficacy was negligible against the epidemic.

On 1 October, small clouds gathered in the sky and a small shower occurred. The epidemic was reported to have reached even the remotest hamlets situated high up in the Himalayan mountains, and deaths due to influenza were reported to have exceeded the death toll on all the fronts during the four years of the First World War. During my entire tenure in Mardan district I did not have any permanent residence but, along with my wife and young son, constantly moved from one rest house to the other, situated at vantage points along the canal system. There was a canal rest house at the head of the bifurcation at the tail end of the Upper Swat Canal and two rest houses each, on the Maira and the Indus branch canals. That area comprised my sub-divisional charge and the rest houses were our residences. We travelled very frequently during the period of the influenza epidemic and were lucky to have dodged the dreaded disease.

Outlaws and Unrest

Almost the whole of NWFP had remained in a disturbed state during the period of the First World War. The roads were unsafe for travel after sundown. Conditions in the Peshawar vale were the worst and remained so even after the Armistice. A band of armed outlaws headed by an Afridi tribesman named Aitbar Shah was still operating within the hump enclosed by the Maira and the Indus branches, looting the villagers and travelling across the area with impunity. One fine winter morning, I left alone on horseback from Jagannath rest house for Toru rest house, along a direct village track, cutting across the distributary channels. As I reached the top of a rising, I suddenly saw, a hundred yards ahead, Aitbar Shah's band of about twenty outlaws sitting in a row against an embankment, with their rifles placed on the ground in front of them. The footpath ahead led to within fifty paces of the seated outlaws.

As I suddenly burst into view, all eyes were turned on me. It would be cowardly and, in any case, futile to turn back. I was an easy target for them; so I decided to keep moving, slowly, in their direction, keeping my eyes down but covertly looking at them to watch any movement of a hand towards the rifle. But there was no such move. I remained under the silent scrutiny of the outlaws as long as I was visible, for about 200 yards on the high ground. As soon as I got down the sharp depression on the other side and out of the outlaws' view, I put my horse into a gallop till I reached my destination. I returned to my headquarters at Jagannath rest house in late afternoon, with an armed mounted escort of Khasadars which I had requisitioned by phone. The next day, one of my canal *baildars* brought a message for me from Aitbar Shah, the leader of the band of outlaws, to the effect that I need not have looked so scared while riding in front of his band the previous day as they had no intention of harming me 'because I carried a good reputation as a just officer'. I stood high in the eyes of his bandits! Despite this, I reported the incident to my superior officers and the civil authorities and asked for special measures for my protection during official tours. However, the authorities assured me that, according to intelligence reports, Aitibar Shah's band had no intention of harming me!

In February 1919, I attended, for the first time, the annual session of the Punjab Engineering Congress at Lahore as a fully-fledged member. Membership of the engineering congress was open exclusively to 'first-class' officers of the irrigation, buildings and roads, railways, and electricity departments of the Punjab and NWFP. The total membership was about 200 engineers, of which about 70 per cent were Europeans, 25 per cent Hindus, and 5 per cent Muslims. In the evening, after the first day's session, I went out for a stroll on Lahore's Mall road starting from Anarkali street and going towards the Charing Cross. As I was walking along the footpath past the Bank of Bengal building, I saw a large number of people approaching from the Charing Cross direction. They came running silently in a disorderly fashion and appeared to be in great panic. On enquiry, a hurrying passer-by shouted, 'There has been firing,' and sped away. Presently, the crowd passed, leaving the Mall road deserted and silent. On their heels came a cordon of armed police covering the entire width of the Mall road from footpath to footpath. At the junction of the Mall and the road leading to Anarkali street, a policeman shouted: 'They have taken the Anarkali route' and, completely ignoring me, the line of armed policemen followed the crowd.

British Brutality

On the previous day, Mr Gandhi had been taken off a railway train and arrested at Palwal station while travelling to Lahore. He had come to India from South Africa to lead the Indian National Congress in their struggle for self-government. A large deputation of Hindus, consisting mostly of young college students, had gone to personally petition the [lieutenant] governor of Punjab for Mr Gandhi's release. They had gathered at the Mall road gate of the Government House and, on their refusal to disperse, were fired upon. A second firing on the same fleeing crowd took place later at the far end of the Anarkali street.

On the morning of the second day's session of the engineering congress, the atmosphere in the university hall, where the meetings were being held, was electric. The congress president of the year, a veteran British chief engineer of the Punjab Irrigation Department, referred to the demonstrators, who had been fired upon the previous day, as 'insurgents'. Numerous airplanes flew low over the university hall apparently to protect the European participants of the congress. Before the commencement of the morning session, Sir Herbert John Maynard, the financial commissioner of the Punjab government, walked in. Stepping up to the dais, he addressed the members of the congress in somewhat the following words:

> Gentlemen, I have been instructed by Sir Michael O'Dwyer, the [Lieutenant] Governor of Punjab, to deliver the following message to [the] members of the engineering congress: 'After four and a half years of the bloodiest war in history, Great Britain has emerged the greatest military power in the world. The insurgents who are creating disturbances in the city are no better than mosquitoes. Now a mosquito at its worst is more than a nuisance, but it can be crushed between a finger and the thumb. Sir Michael desires that the Indian members of the Engineering Congress in particular should take this message of His Excellency to their homes.

On that day, another firing incident took place on a crowd in the Gawalmandi street of Lahore.

On 13 April 1919, the dreadful Jallianwala Bagh massacre took place at Amritsar. A large crowd of Indians had assembled in a small square in the town.[2] The said square was enclosed on all sides by two-storey houses, with only one lane providing access to the ground. General Reginald Dyer arrived on the scene with a contingent of troops and, without issuing any warning, ordered his men to start firing point blank on the trapped crowd. It was estimated that between 1200 and 1800 people died in the ghastly tragedy, after a few minutes of firing. Then followed the infamous Crawling Order,[3] everyone living in the lane leading to the square, and in the square itself, was ordered by the British troops occupying the lane to crawl through the lane while going in and out.

A wave of indignation swept through the country as soon as the news of the tragedy became known. The government was forced to hold an inquiry into the incident.⁴ During cross-examination General Dyer, confessed that the firing stopped only after the ammunition had been exhausted. The inquiry brought out in all its nakedness, the ghastly features of the tragedy and the brutality of the British rulers.

In May 1919, Afghanistan declared war on the British Indian government. Martial law was declared in Peshawar district. Since after the Jallianwala Bagh tragedy, the loyalty of the Indian troops was doubtful, British troops were ordered to advance through the Khyber Pass. After meeting tough resistance from the Afridi tribe inhabiting the Pass, they occupied the town of Dakka at the far end of the Pass in Afghan territory. But strangely enough, the British troops did not advance beyond Dakka town. The rumour was that a cholera epidemic had broken out among them. Another rumour said that the British troops were tired after fighting in the First World War and wanted to return home. The activities of the British forces were thereafter confined to occasional air raids on the Afghan territory.

Mr Smith, executive engineer and my immediate boss, usually brought me news of the progress of the war with Afghanistan during his frequent official tours of my sub-division. During the early stages of the war with Afghanistan, Mr Smith arrived one morning in high spirits. 'Have you heard the latest about the war,' he said. 'No Sir, I am unfortunately cut off from news here.' 'Well,' he said, 'Would you believe it, our RAF [Royal Air Force] boys from Risalpur left for Kabul yesterday morning after breakfast, they bombed the Amir's palace in Kabul, and returned to Risalpur before lunch!' 'Marvellous show!' I exclaimed, and added, 'When are we going to occupy Kabul?'. 'Oh well, in about a week's time from now, I think,' he said. But my tactless query had made Mr Smith suspicious about my loyalty, because negotiations for peace with Afghanistan were already underway.

A few days later, while we were riding along the Lahore Minor, a small irrigation channel in my sub-division, named after the village situated on its bank and alongside extensive ruins of the ancient Lahore village of the Buddhist period close by, Mr Smith suddenly said, 'Mr Rahman you are a knowledgeable person. How long do you think the Buddhists ruled India?' 'For over a thousand years,' I replied. 'From the fourth century BC to the seventh century AD.' 'And how long did you fellows rule here?' 'For about 800 years, from 1000 to 1800 AD.' And then came the bombshell: 'And for how long do you think we shall rule India?' The question was so sudden, it left me speechless. I felt completely at a loss for words. I realised that, in spite of the fact that Britain had emerged victor in the titanic struggle during the First World War, India, for the first time, was defying the British rule in India, through the Non-Cooperation Movement started by M.K. Gandhi who had been released from prison. The movement promised to be the thin end of the wedge against the British power in India. 'Speak up, Mr Rahman why are you quiet?' I suddenly realised that my silence would be misinterpreted as disloyalty. 'Sir,' I said, 'Yours is the greatest empire in history. Indeed, you are the greatest military power in the world today (quoting Sir Michael O'Dwyer). Your rule is just and impartial. If you continue to rule in India with justice and impartiality as in the past, there is no reason why you shouldn't rule India for a thousand years!'

My reply had evidently not satisfied Mr Smith. 'Mr Rahman,' he snapped, 'you seem to be fond of talking in terms of "ifs" and "buts",' while dilating his nostrils, a habit with him whenever he was angry. I had a premonition of coming danger. Sure enough, only two days later, I received a demi-official letter from him to the effect that since the Mohmand Pathan tribesmen were fighting against the British Indian government in the current Afghan War and since I too was a Mohmand Pathan, would I explain why I think I should not be treated as an enemy. Naturally, I got alarmed because of the existence of martial law in the district.

Early the next morning, I got my horse saddled and rushed to Mardan, announcing myself at Mr Smith's bungalow while the family was at breakfast. He politely invited me to join. During breakfast, I broached the subject of his demi-official (DO) letter. 'Yes, what about it?' he said somewhat casually. 'Sir, it is well known,' I said, 'that I am from the Mohmand clan. However, my forefathers came to India with the army of Ahmad Shah Abdali and we are settled in the Hazara district of the NWFP. So, I could not possibly have any connection with the present rebel tribesmen of my clan.' 'Oh well,' Mr Smith murmured, 'your clarification clears the matter. The incident may be considered as closed.' I was greatly relieved by his reassurance. Nevertheless, I remained apprehensive because Mr Smith was already perfectly aware of the facts stated by me during the breakfast; I realised that his DO letter had been in the nature of a threat.

In July 1919, a peace treaty was signed with Afghanistan, which secured that country's independence from the British Empire. The British troops never advanced beyond Dakka village, although Mr Smith had boasted that they would occupy Kabul within a week. Moreover, the British forces had shown a strange weakness in tactics and strategy during the Afghan War. They were ordered to retreat in haste 40 miles from the fortified Tal Fort to Kohat, as soon as Nadir Khan, the Afghan commander-in-chief, showed up suddenly on the top of the hill overlooking Tal village, and fired a number of shells at the village from a field gun, from that vantage point. They evacuated the whole of North and South Waziristan, abandoning large quantities of arms and ammunition, after being misled by false intelligence that Afghan forces were poised for the invasion of Waziristan in large numbers.

Mr Smith's Reprisal

The news of the debacle of the British forces in the Afghan War, which became common knowledge, further embittered Mr Smith; his behaviour towards me grew cool and formal. Matters took a serious turn in August 1919. Returning to duty after a short leave in Abbottabad, via the Jahangira–Swabi road, I was perplexed to notice that both the Jahangira minor canal and the branch minor, the tail minors of the Indus branch canal in my charge, were bone dry, although all the canals were expected to be running at full supply, during August, for irrigating the extensive corn crops which required watering after every eight days. Following the canal road along the Jahangira minor I reached Jalbai canal rest house to investigate.

Here my subordinate staff met me with gloomy faces and informed me that both the Maira and the Indus branches, in fact my entire sub-division, had to be closed because their channels were choked with heavy deposits of silt which was brought down from the adjoining high ground by the recent exceptionally heavy rains. Unfortunately, 'catch water' drains, and cross-drainage works had not yet been constructed along certain reaches on those channels, which allowed free access to muddy rainwater into the channels from the adjoining sloping high ground. Thus, although the disaster was due to no fault of mine, the weekly supply of water to the standing corn crop was in jeopardy. Considering the magnitude of the task of clearing the silt from the choked-up channels, within a week, I shuddered at the idea of wholesale destruction of the Kharif crops in my sub-division.

I was informed further that Mr Smith was camping in my sub-division and that he was staying in the Lahore rest house and was anxiously awaiting my return from leave. I started immediately for the Lahore rest house, 10 miles away putting my horse on the gallop all the way and met Mr Smith in the early afternoon. Mr Smith was almost incoherent with rage. He put the entire blame for the disaster on me and kept on repeating 'How are you going to clear your channels within a week and restore irrigation to the starving crops?' 'I will do my best, Sir,' was all I could say. He evidently did not believe me and appeared pleased at the imminent prospect of the ruin of my career. He shifted his camp to Jalbai rest house at the tail of the Indus branch the same evening. I also returned to Jalbai with him and we both put up in Jalbai.

That night, I sent an urgent message to the Khan of the neighbouring Jalbai village, a friend of mine, to gather as many able-bodied men as he could, above the age of fourteen, equipped with shovels for silt clearance of the Indus branch, on liberal daily wages. The next morning after an early breakfast, Mr Smith and I were sitting in the fortified tower of the rest house, from which there was an extensive view of the surrounding countryside. The large village of Jalbai was spread out on low sloping ground among tall corn fields a mile below us. Precisely at sunrise the beat of drums was heard from the village and shortly afterwards the spearhead of a large crowd of villagers emerged and started moving in our direction. At the head of the crowd were a couple of drummers, with three teenage boys clad in bright red clothes executing a dance in unison while walking briskly. The crowd swelled in numbers as it emerged from the village to more than 500. Mr Smith looked fascinated at the approaching crowd. 'Who are they?' he asked. 'Sir, they are the labourers for silt clearing of the Indus branch,' I replied. The crowd halted for a while right below our tower. The dancing boys executed a 'Khattak' dance to the tune of a flute and the lusty beat of drums and passed on to the Indus branch. 'Excuse me,' said I, 'I must go along with the labourers and put them to work.'

Later that day, Mr Smith inspected the dry Indus minor canal. A small brick masonry fall on the minor was found damaged due to excess supply which must have run in the minor during the late heavy rains. 'What do you propose to do about repairs to this fall?' asked Mr Smith. 'Sir, I will take up repairs to the fall after attending to the immediate task of restoring supplies in my canals,' I said. 'So, you will allow the fall to be further damaged,' he shouted. Exasperated, and losing my temper, I replied sharply: 'Sir, in the interest of the urgent task of irrigation I would not care if this small fall is further damaged.' 'Oh, so you do not care for the protection of government property,' snorted Mr Smith, 'This seems a fit case for reporting to the higher authorities.'

That day we returned to the Lahore rest house. The next morning, in response to my urgent messages to the local Khans, a large crowd of village labourers turned out from Lahore and Jalbai villages and was spread out along the upper reaches of the Indus branch for silt clearing. The process was repeated at the Jagannath rest house on the third day to clear the Maira branch, where the damage had been comparatively slight. Thus, enough waterways were cleared in both the branches, restoring sufficient supplies in their minor branches in good time to save the corn crop.

Mr Smith decided to prolong the duration of his tour of my sub-division to a month. His attitude remained persistently hostile. It was evident that he was out to find fault with my work. In due course, a copy of a DO letter from the superintending engineer was passed on to me by Mr Smith in which

I was ordered to submit an explanation regarding my disregard for the safety of government property, as reported by my superior. There was another charge against me: namely, the failure on my part to fix a temporary wooden outlet on a minor, which had been ordered by the executive engineer the previous year on the request of a landowner. Quite upset with the way I had been treated, I let myself go. In my letter of explanation, addressed directly to the superintending engineer, I defended my remarks about disregarding immediate repairs to the small masonry fall, on the minor, in the interest of emergency irrigation to save the corn crop. Regarding my alleged failure to fix a temporary outlet, I explained that Mr Smith had given verbal orders at the site for fixing a dozen temporary outlets on various minors of the Indus branch, eleven of which had been duly fixed. I regretted to state that I had forgotten about the twelfth outlet. The executive engineer himself had failed to confirm later in writing his verbal orders, given at the site.

Pressing home the charge by quoting facts and figures, I claimed that Mr Smith's policy of granting temporary outlets indiscriminately had actually resulted in a decrease in the seasonal irrigation figures on the Indus branch, due to the failure of supplies to reach the tail reaches of the minors where the firm loam soil offered better chances for development. I further recommended that the tails of the minors of the Indus branch canal should be extended beyond the Jahangira–Swabi road into the potentially fertile non-irrigated lands situated between the road and the right bank of the Indus River.

The following decision of the superintending engineer was conveyed to me by Mr Smith under a DO cover: 'Mr Abdur Rahman Khan should realise that whatever his personal opinions, he is an officer subordinate to the executive engineer, and is duty bound to obey the orders of his superior officer. He is hereby reprimanded and is warned to strictly follow the instructions and orders of his superiors in the future.' I had thus saved my job but at the expense of damage to my future career. Official orders were simultaneously received for the extension of the Indus branch minors beyond the Jahangira–Swabi road, as I had suggested in my explanation.

Mr Smith continued to camp in my sub-division for a full month, trying to pick further holes in my work. He made confidential enquiries about the mode of transport of my family from one rest house to another during my camp shifting. He suspected that I might have been living most of the time at my headquarters while submitting bogus travelling allowance bills. He must have been disappointed to learn, however, that I owned two horses, one of which was used by my wife while shifting camp. The small cot of my infant son, Mohammad Aslam Khan (later chief engineer), was carried on the back of a bullock. We thus moved alone, escorted by our own armed sub-divisional 'treasure guards', because Atibar Shah's gang of outlaws was still operating within the limits of my sub-division. At that stage, Mr Smith and I ceased to be on speaking terms. We exchanged only strictly official correspondence and that too on slips of paper. He established himself with his family at the Lahore rest house while, to save him inconvenience, I occupied the 'subordinate' rest house within the same compound.

The source of water supply at all the rest houses in my sub-division were deep open wells with corrugated lining about 5 feet in diameter, which in most cases was buckled in the sandy substrata. Since the groundwater in most cases was over 100 feet deep, it was not practicable to draw water from the distorted wells during the season of annual canal closure for repairs. Consequently, it was customary to obtain drinking water supply from wells in neighbouring villages. Since the canals were closed for the silt cleaning work in progress, I obtained my drinking water supply from the well in Jalsai rest house 12 miles away. Mr Smith was, however, obtaining his supply independently from a neighbouring village 3 miles away. It came to my notice that cholera was raging in the village from which Mr Smith obtained his drinking water. Getting alarmed about the safety of his family, I wrote an urgent warning on a slip of paper and sent it to him through his personal servant. In the note, I suggested that I would personally arrange to procure drinking water for his use from Jalsai rest house. His servant, who could understand English, reported that my note was read by both Mr and Mrs Smith, and the lady had reprimanded her husband for harassing 'that young man who is so anxious about our safety.' Mr Smith replied promptly, thanking me for the valuable warning. The next day he called me to the rest house and, after an exchange of cordial greetings, left with the family for his headquarters at Mardan. He was reported to have remarked to his British colleagues in the department that I had greatly improved in my work recently due to his guidance during his long stay in my sub-division.

The development of irrigation on the newly constructed Upper Swat Canal was presenting unexpected difficulties. The costly project had been undertaken on the basis of expectations of its speedy development, as was the case of the neighbouring Lower Swat Canal. But conditions proved different in the areas of the new canal. Due to more copious annual rainfall in the Upper Swat Canal areas, it was possible to raise Rabi (winter) crops of wheat, without much labour. On the other hand, Kharif (summer) crops of corn, sugarcane, orchards, etc. required a lot of labour in hoeing, weeding, and putting manure for fertiliser. The new areas being far from villages, the farmers were reluctant to take up cultivation of Kharif crops that far away from their homes. Another drawback was that the farmers' holdings in the new areas were scattered in different locations, so that they only cultivated the areas nearer their homes. The canal officers and others were accordingly encouraged by the government to submit proposals for the speedy development of irrigation on the Upper Swat Canal. In due course, after careful study of local conditions, I submitted three proposals for consideration of the government: first, new colonies for tenants should be constructed in the new canal areas; second, the work of consolidation of agricultural holdings should be expedited; and third, work of *killa-badi* involving the survey and division of land into square blocks of one acre each, along the lines running north-south and east-west, and the construction of water courses along these dividing lines, be carried out.

My proposals were examined by the civil authorities. Mr Olaf Caroe,[5] assistant commissioner of Mardan, who was the concerned officer, reported that he fully agreed with the proposal for the consolidation of holdings through rectangulation surveys. He, however, did not recommend the construction of new colonies for tenants in the canal areas, which would antagonise the local Khans or landowners, which was not desirable for political reasons. The work of consolidation of holdings and rectangulation was taken up later, but it took years for new colonies to be built through private enterprise on the new canal areas. In the meantime, owing to slow progress, the Upper Swat Canal was declared an 'unproductive' undertaking. However, at the end of the year, I received double the usual amount of annual increment to my salary, to Mr Smith's astonishment, who had not given me any special recommendation. Such was the double-edged justice dispensed by the British in India. On the one hand, my professional career had been damaged in order to maintain the prestige of a British officer while, on the other, I was rewarded for good work. In the spring of 1920, the plague epidemic struck the surrounding villages. I, therefore, applied for three months' leave, which was sanctioned, and so I relinquished charge of the Maira Sub-division of the Upper Swat Canal.

9 Yusufzai Territory

The Hamzakot Sub-division, which I took over on return from leave, was adjacent to my former charge, the Maira Sub-division, and was situated upstream on the main channel of the Upper Swat Canal. Cutting across a limestone spur through a number of short, unlined tunnels, the main line of the Upper Swat Canal ran parallel to the Swat and Buner hills, at a lateral distance averaging 3 miles. It crossed several large hill torrents through magnificent cross drainage works, notably a steel tube siphon, two large aqueducts, and a super passage. Numerous distributaries and minors branched off crosswise from the main line, irrigating rich lands on which many villages depended.

The inhabitants were Pathans of the Yusufzai tribe; both men and women were sturdy and good looking. Many had extremely fair complexions with grey eyes and golden hair, presumably due to the intermixing of Greek blood during the long occupation of the country by the Greco-Bactrians. Pathan boys and girls were of a spritely and cheerful nature, fearless of strangers, and frank and friendly in approach. Both the young girls and youths decked their ears with wild flowers in spring and were fond of humming love songs in Pushto. The Pathans of the area were brave to a fault, hospitable to strangers, and quick to take offence. The main crops sown in the area were wheat, mustard, and alfalfa in winter, and tobacco, sugarcane, and corn in summer. With the advent of the canal, citrus fruit orchards were also introduced. Yusufzai Pathans of the Peshawar district and some surrounding tribal areas claimed to be of Jewish origin. The name of the tribe Yusufzai—literally sons of Joseph—is itself suggestive.

The entire area covered by my sub-division had numerous ruins of the Buddhist period. The tract, evidently, had been the heartland of the Gandhara civilisation. There was a shrine within the pine-crowned top of the Karamar hill, rising to the sheer height of 2000 feet, venerated as if it was that of a Muslim saint, but it was evidently an old Buddhist shrine. At the foot of the hill was a megalith consisting of a circle of vertical stones; it was a pre-historic relic. Nearby was the large modern village of Shahbaz Garhi, with seventeen edicts of the Buddhist king, Ashoka, engraved on a couple of huge rock boulders close by. Shahbaz Garhi was said to be a large walled city during the Buddhist period. From that city, daily provisions were carried up to the top of the Karamar hill, for the monks living there. It was said that the donkeys were so familiar with the track leading up to the shrine that they carried their daily burden and returned unescorted.

There was a large Buddhist stupa and a colony of cells for the monks at Jamal Garhi near the village of Palo Dheri. A small Buddhist village was perched high up on the hill slopes at a site which was hidden in a narrow valley to which access was only possible through a gorge. The stone walls of the village were intact, but the roofs were missing. At the foot of the valley, there was a small spring which had since dried up. The principal path leading up to the village was worn through by innumerable steps. The village site was littered with small and large, broken and damaged, statues of Buddha. On the top of a low hill, a few miles from the Shewa rest house, were the extensive ruins of the Buddhist village of Rani Ghat. The ruins of the rani's palace were located at a higher level on an adjoining peak. That site was also littered with numerous damaged statues of Buddha.

There were four fortified rest houses located at an average distance of 10 miles from each other, built on the Upper Swat Canal, in my sub-division. My headquarters were located at Shewa rest house in the middle. Shewa village lay 3 miles away from the rest house, at the foot of the Karamar hill. Shortly after taking over my new sub-division, I rode out to Shewa village, and called on Mohammad Umar Khan, the Khan of Shewa. He was an old man of about seventy years of age, but looked handsome, with a grey beard and a fair complexion. Wearing his silk turban from Mashhad and gold embroidered *kulla* (peaked cap around which the turban is wound) he reminded me of the paintings of a Mughal emperor. The Khan maintained a stable of about a dozen horses, with one noble animal for his personal use and the rest, somewhat nondescript animals, for the use of guests and retainers. He kept his small bungalow-type guest house meticulously clean and had a fastidious liking for good cuisine.

I was soon on friendly terms with the Khan. Shortly after our acquaintance, one day he asked me to write a farewell letter in English (he did not know English), to Mr Olaf Caroe, assistant commissioner of Mardan, on his transfer to another station. Among other things the Khan disclosed that on hearing about Mr Caroe's transfer he had felt as if a near relative of his had passed away. I was deeply touched by the Khan's feelings for Mr Caroe. After I had finished the message to Mr Caroe, he requested me to write another letter for Mr Wiley (later Sir John Wiley, governor of Bihar and Orissa) the incoming assistant commissioner. To my astonishment the Khan asked me to write that on hearing about Mr Wiley's posting to Mardan, he (the Khan) had felt like 'He was dying out of sheer joy.' In response to my somewhat amused and questioning looks, the Khan confided to me that this was the way to deal with the British rulers!

To my great relief Mr Smith was transferred to the Punjab, and Mr W.G. Yeaman took over from him. Mr Yeaman remained at his headquarters in Mardan for a month, communicating with his sub-divisional officers through mail, and then commenced a prolonged tour of his division. After a week's tour of the neighbouring Maira Sub-division, he paid his official visit to my sub-division. Noticing some ongoing earthwork repairs on the main line of the Upper Swat Canal, Mr Yeaman enquired about the rate at which the earthwork was being executed. The information was duly supplied. 'How is it,' he remarked, 'that, in spite of the fact that your channels are larger than those of your downstream neighbour the SDO Maira, your rates for earthwork are lower than his.' 'Sir,' I replied, 'Pawinda labour comes from Afghanistan in large numbers for carrying out earthwork on our canals, so we can bargain for the rates which are considerably lower than the sanctioned schedule of rates.' 'I see,' said he, impressed by my reply. After a week's stay in my sub-division, Mr Yeaman returned to Mardan not only satisfied with my work, but we parted on cordial terms. From that time onwards, Mr Yeaman became my friend and paid frequent visits to my sub-division.

I recall one occasion when, on a cold winter morning, Mr Yeaman, clad in a fur-lined light Burberry overcoat along with myself, and Aziz ur Rahman (overseer), left Hamzakot rest house for a visit to the Kashmir Smast cave. We rode along the bank of the Upper Swat Canal for 3 miles up to the foot of the Palo Dheri spur of the lofty Pajja hill and started a vertical climb of 2500 feet up the steep limestone slope, reaching breathless, but refreshed by the exhilarating air on the top of the spur. We walked a couple of hundred yards along the gently rising hump towards the point where the spur met the loftier slopes of the Pajja hill. At the junction point we were rewarded with the appetising sight of roasted fowl, a heap of peeled medium-boiled eggs, and the traditional *parathas* (freshly made flat bread fried in clarified butter) prepared at the site by the retainers of the Khan of Palo Dheri, who was there to receive us in person. After a hearty breakfast, we walked another couple of hundred yards up the slopes of the Pajja hill to the Kashmir Smast cave.

The mouth of the cave, about 30 feet in diameter in pure limestone rock, was an impressive sight. Inside the cave, which was about a hundred yards in length, there were a number of mini Buddhist shrines. The cave narrowed as one proceeded further inside along a gently rising slope, till it ended in a beautiful vertical tunnel about 50 feet high through which soft bluish light filtered into the cave. There was an air of freshness in the atmosphere, as we emerged from the cave into golden sunlight. From our high positions there was an enchanting view of the Peshawar vale which was green and yellow with the wheat and mustard crops, with the pine-crowned Karamar hill rising from the plain in middle foreground, and the amphitheatre of hills surrounding the vast Peshawar valley in the distance.

This was the period of rising political aspirations among the Indians and their growing assertiveness in the form of M.K. Gandhi's Non-Cooperation Movement (launched 1920). Initially, the movement only asked the Indians to resign from their government posts and give up their titles, and the students not to attend government schools and colleges. It was a mild affair, and only a few Hindus and probably no Muslim followed Mr Gandhi's directive. The Muslims were reluctant to join the movement, fearing that it would place the Muslim minority in the hands of the Hindus who were in the majority. However, the infamous Treaty of Sevres which sought to dismember even the remnants of the former Turkish Empire caused considerable agitation among the Muslims. During the First World War, Arabia, including Palestine and Iraq, had already been occupied by the British, while France had grabbed Syria. By the Treaty of Sevres, the British took Mehmed VI the Turkish Sultan and the last Khalifa of Islam, under their 'protection', and permitted the Greeks to conquer the plateau of Anatolia, the heartland of the Turks, by force, expecting that the Greeks would easily accomplish this task because the Turkish Army had been defeated and stood disbanded.

Visit of Edward, Prince of Wales

The Indian Muslims deeply resented the role of the British and started the Khilafat Movement (1919–22), accusing the British of committing sacrilege by imprisoning the Turkish Khalifa of Islam. The movement was especially active in the NWFP, where thousands of Muslims sold their properties at a nominal price and migrated to Afghanistan with the intention of fighting against the British from the base of a free Muslim country. Caravans of Muslims, particularly from the Hazara and Peshawar regions crossed the border and entered Jalalabad in Afghanistan as exiles. The British government in India was bewildered and did not know how to deal with this quasi-religious movement.

Mr Gandhi, who had been very conscious of the indifference of the Muslims towards his Non-Cooperation Movement, now saw a golden opportunity of making common cause with them against the British by lending his support to the Khilafat Movement. The Muslims were naturally grateful to Hindus for their sympathy and gladly made common cause with the Hindus and the Indian National Congress in the struggle to free India from the British rule. The British thus faced a united India against them. The British government sent Edwin Samuel Montagu, secretary of state for India, with the package that came to be known as the Montagu-Chelmsford Reforms (1918) according to which Indians could aspire to high 'cabinet' posts as members of the councils of viceroys and the governors. A limited number of Indians was also admitted into the officer rank of the Indian army. However, the selections for those 'high posts' were made by nomination from within the well-known and loyal families.

In 1921, the British government sent Edward, Prince of Wales, later King Edward VIII, and heir to the British throne, on a goodwill mission to India. He arrived in Peshawar, the capital of NWFP, when the country was seething with discontent over the atrocious provisions of the Treaty of Sevres and also the inadequacy of the Montagu-Chelmsford reforms. It was arranged that the prince should see the famous Qissa Khwani Bazaar of Peshawar. However, the citizens of Peshawar observed complete *hartal* (strike) when the prince visited the city, and all the shops of the historic Qissa Khwani Bazaar were closed. Despite elaborate precautions which had been taken by the police, huge placards appeared in the Qissa Khwani Bazaar bearing the words: 'Tell Father, we are Unhappy'. Thousands of tenants of Nawab Mohammad Akbar Khan of Hoti were brought by the government to Peshawar from Mardan and made to stand in front of the closed shops. The prince could not, however, have failed to notice the public discontent with the British rule in India. Following his visit to Peshawar, the sleek, creamy-white special train of the Prince of Wales, arrived at Mardan railway station, on a sunny morning, and stopped a few hundred yards ahead of the small railway station building on the open railway track where the Nowshera–Mardan road ran parallel with the railway track. A red carpet had been laid from the railway track up to the road, where a large Rolls Royce awaited the prince. The door of the prince's compartment was exactly in front of the red carpet when the train stopped.

Two rows of Indian officers of the civilian departments, in the category of 'Class I, or First-Class officers' stood on both sides of the red carpet, throughout the short length, from the railway boundary fence to the Nowshera–Mardan road, each bearing a basket of rose petals to shower on the prince, as he passed by. Thousands of Pathan men, women and children lined the Nowshera–Mardan road to catch a glimpse of the prince. The tall and strong Sir John Maffey, chief commissioner of the NWFP, and the shorter but equally impressive Nawab of Hoti, had been waiting for the arrival of the prince like two nervous schoolboys outside an examination hall. The prince's reception was their joint show, and one suspected that most of the Pathan spectators were the nawab's tenants.

I stood second on the left row of the Indian officers holding a fancy basket of rose petals. The first Indian officer in my row was the elderly and respected, Sahibzada Fazal-e-Rahman (extra assistant commissioner), and a cousin of Sir Sahibzada Abdul Qayum (founder of Islamia College, Peshawar). The prince emerged from the special train clad in a riding suit consisting of Jodhpur breaches and a sports jacket of creamy-pink light wool material; he wore a bowler's hat. He was of medium height, slim, and of pleasing countenance with not a trace of snobbery about him. He walked along the red carpet with a slight limp—the result of a recent riding accident. The prince was said to be extremely fond of polo and had, in fact, come to Mardan to play polo with the famous Guides Cavalry regiment that afternoon. After that, he was to visit the historic Malakand Fort where, during the 1895 war with the Yusufzai tribes, Winston Churchill had acted as a war correspondent.

As soon as the prince arrived near the row of the Indian officers, Sahibzada Fazal-e-Rahman, at the head of the left row, grabbed a fistful of rose petals from the basket he was carrying and, as if taking a deliberate aim at the prince's face, threw the rose petals forcefully at him. The prince was, of course, unaware of the contents of the stuff in Sahibzada's hand. Fearing that perhaps a projectile was being thrown at him, he abruptly halted and stepped back, holding his left hand before his face to parry the blow. But when the 'blow' turned out to be a cascade of petals, the prince laughed and continued walking. The rest of us, taking notice of the Sahibzada's gaffe, showered the rose petals gently at the prince as he passed by us. When the prince arrived at the end of the red carpet and was about to get into the royal car, a rustic Pathan woman from the crowd flanking the road suddenly stepped forward and threw a dirty white shirt stained with patches of dark dried blood at the prince's feet and, in impassioned tones in the Pushto language, demanded justice for the brutal murder of her son by his enemies. The blood-stained shirt was her son's, which he had worn at the time he was murdered.

Everybody was stunned by this extraordinary incident. The prince appeared to be at a loss and looked silently at Sir John Maffey and the Nawab of Hoti. Both the gentlemen themselves were stupefied. Then the Nawab of Hoti pushed the protesting woman back into the crowd, assuring her that the prince had heard her story and full justice would be done to punish her son's murderers. The prince played polo in the afternoon and was reported to have immensely enjoyed the game. 'Mardan is heaven in India,' was the prince's verdict. The next day he visited the Malakand Fort before returning to Peshawar. On the day following the prince's return from Malakand, the chief commissioner gave a large garden party in his honour at the Government House, Peshawar. The prince, dressed in Mufti, moved about freely in the crowd. My own impression during the garden party was that the prince had the personality and bearing of an inconspicuous young Englishman without any royal airs about him.

Confronting Armed Marauders

One sunny winter morning, I was cycling along the smooth, motorable patrol road which ran along the Upper Swat Canal's main line in the direction of its tail. The canal was running full, its placid waters reflecting the blue and purple hills marking the boundary between the tribal territory and the administered district, running parallel with the canal a few miles away. As I proceeded downstream, the tribal territory of Buner descended from the crest of the hills into the plain, and progressively came nearer and nearer to the canal. A few hundred yards ahead lay a super passage over the canal where a dry stream with a shingle-strewn bed crossed the canal. At that point, the boundary of the tribal territory in the plain was only a few hundred yards from the canal. On looking around, I noticed that no one was in sight except a young rustic Pathan carrying a long spear, who stood watching my progress a few hundred yards away, within or near the boundary of the tribal territory. He suddenly started running in my direction. I calculated that he would reach the canal at the super passage, almost at the same time as I would reach it, and where I would have to dismount from my bicycle in order to cross the bed of the stream. I guessed his motive but, as I carried a loaded pistol in my pocket, I continued my progress towards the super passage.

As I had calculated, both of us reached the super passage at about the same time. I nonchalantly got down from my bicycle and started walking towards him, while he stood in the middle of the stream, barring my way. However, as soon as he noticed the bandolier of pistol bullets slung across my shoulder and guessed that I had a loaded pistol in my pocket, he greeted me respectfully. 'Khan, may you never be tired,' he said. 'You bloody bandit,' I shouted at him, 'You came in haste to rob me, but seeing that I am armed, you choose to be polite. Run away you rascal, before I put a bullet in your chest!' And the fellow ran back to the tribal territory in as much haste as he had come.

Near the upper tail of the Upper Swat Canal, a distributary, known as the Pahur branch, took off; it was aligned for a few miles with the tribal territory before re-entering the administered Peshawar district, via an unlined tunnel through a spur of hills. One afternoon, accompanied by my overseer and a canal *baildar*, I was inspecting the part of the Pahur branch which lay within the tribal territory when an armed bandit suddenly emerged from the shelter of a tall maize field close to where we stood on the canal bank. He was a picturesque ruffian, armed to the teeth. He was carrying two well stocked bandoliers of rifle and pistol bullets along with a rifle. A pistol and two daggers were also slung around his sides. I realised with inward alarm that I was unarmed, having failed to take the precaution of carrying a firearm, as the site was not far from a canal rest house, where I was camping. There was, therefore, nothing for me to do but to put up a bold face.

The bandit was evidently an absconder from justice who had taken shelter in the tribal territory after committing a crime in the administered area. Unfortunately, by treaty, the British Indian government had no jurisdiction within the tribal territory, so murderers could take refuge in it without fear of arrest. Such criminals and murderers were termed *mafrurs* (absconders from the law) who usually visited their villages at night, and lived with their families, and left for the sanctuary of the tribal territory before dawn. The *mafrur* in question, realising that I was a canal officer, instead of harming me, chose to use me as a means to convey a message. 'Tell the *thanedar* (inspector of Kalu Khan Police Station),' he said, 'not to harass me while I visit my family at night. If he does not stop pursuing me, I will put a bullet in him from the shelter of a maize field.' I promised to convey his message to the *thanedar*, which I did. The police inspector smiled and informed me he was well prepared for such an eventuality!

In those days (1921–2) the countryside of Peshawar district was in a highly disturbed state. Aitibar Shah's gang of bandits was still operating undisturbed in the Maira area within the district. There were frequent raids from the Buner tribal territory adjoining Mardan and Swabi tehsil of the Peshawar district. The roads were unsafe for traffic after dark. There had even been a number of daylight holdups on the Mardan–Swabi road. The Khilafat Movement was in full swing among the Muslims. There was also the aftermath of the recent Afghan War. The British authorities appeared apathetic about the law and order situation. Apparently, they did not wish to antagonise the Pathan population of the district by adopting stringent measures. The tribal territory of Buner ran parallel to the length of the Upper Swat Canal at a lateral distance of only about 3 miles. All rest houses in my sub-division were, therefore, most prone to raids from the Bunerwals. As a precautionary measure, twelve *chowkidars* armed with smooth bore Martini-Henry rifles with plenty of ammunition, were posted in each of my rest houses. In addition, I had my treasury guard, including one daffadar (army sergeant), and three Burkandazes (armed mercenaries from disbanded armies of Indian states), as well as armed personnel who accompanied me on my tours. Their function was to guard the 'treasure chest' which a canal sub-divisional officer carried for making petty cash payments, and disbursing monthly salaries to the numerous subordinate establishments, including the overseers, working under him.

One winter evening, while camping at Hamzakot rest house, my treasure guard brought a sum of Rs 5000 in cash (in those days 'cash' was generally in circulation in silver rupees) from the Mardan sub-treasury on a bullock-cart, as was the usual custom, for disbursement of salaries. After about 2 a.m., the sharp sound of a rifle shot woke me up. At the same time, the daffadar came running and gently tapped at my bedroom door, and informed me that the rifle shot had come from outside, and that the bullet had struck against the stone masonry of the wall of the rest house; he further said that the rest house was surrounded by raiders from Buner tribal territory. All the rest houses on the Upper Swat Canal were fortified structures. In one corner of the rectangular fort-like structure was the rest house for officers, constructed on a 10-foot-high plinth with earth filling underneath. In the corner of the rest house was a 20 by 20-foot bastion, further protected by a parapet wall equipped with loop holes for firing.

On the diagonally opposite corner, was the subordinates' rest house, a double storey structure, equipped with a similar bastion for defence purpose. The rest of the rectangular fort was occupied by rows of rooms opening into a spacious courtyard. The outer walls of the roof of these rooms were also provided with a 5-foot-high stone parapet wall, with loop holes for shooting. A deep well within the courtyard provided drinking water for the residents. A large gate, armoured with thick iron plates, provided entry to the fort. The gate could be effectively closed in times of danger. All the woodwork of the doors and windows of the officers' rest house were of expensive teak wood, with exquisite antique statues of Buddha of the Gandhara period built into the walls of the central sitting-cum-dining rooms. Two bedrooms with attached bathrooms, and a small office room, flanked the central corridor.

It was a pitch-dark cold night. The shot appeared to have come from the east side of the fort from the shelter of the large stock of timber lying outside of the rest house compound. Soon, however, shots started coming from all directions. It was evident that we were surrounded. I stationed our twelve-armed *chowkidars* and two men from our treasure guard on all four sides of the fort, under the general command of the daffadar. At first, I felt a peculiar feeling of anxiety on realising that we were under fire. The feeling, however, soon dissipated when I realised we were fairly safe behind the walls of the fortified fort, and the attackers' shots were either going astray, or striking the walls of the rest house without doing any damage to the solid stone structure. Sporadic firing went on from both the sides for three or four hours until early dawn. I had given orders not to shoot indiscriminately in the pitch dark, but to save ammunition, and to fire only to respond where the flash of gun fire had come from, and no damage had been caused to our men so far. The sound of firing for so many hours must have travelled far in the still night. As soon as there was a glimmer of light of early dawn, we heard from a distance a din of crowds of people approaching the fort from three directions. The Khans of the neighbouring villages of Nawi Killi, Charguli, and Hamzakot had, at last, sent *cheghas* (posse) of men of their respective villages to our rescue. In answer to their shouts we shouted back that in the uncertain light it was not possible for us to distinguish friend from foe. We, therefore, advised them to take up positions around the rest house, outside the range of our rifle fire.

In the meantime, on the approach of the rescue parties of the villagers, the raiders had noiselessly slipped away. As soon as there was enough light, I heaved a sigh of relief, and fell into the sound sleep of exhaustion. I woke up at 10 a.m. to find the Khans of the three villages waiting in my office in the rest house. The first thing we did was to take a walk around the outer walls of the rest house. There, we found numerous footprints and some *chappals* made of *mazri* grass (*Nannorrhops ritchiana* (Griff) Aitch.) which the raiders had abandoned in haste.

I telegraphed a report of the incident to Mr Yeaman, my immediate boss, and to the assistant commissioner at Mardan commending the services of the three Khans. In due course, *sanads* of appreciation were received by the Khans from the chief commissioner of the NWFP for their services to the government. In addition, the Khan of Nawi Killi was granted a license to purchase a rifle, loading thirteen cartridges at a time, from the government's *malkhana* (warehouse) at a concessional rate. He had specially requested me to put in that recommendation. That very day, I shifted my camp 8 miles downstream to Shewa rest house, where the Buner hills were farther away, and felt reasonably safe. In the night, however, I received an urgent telegraphic message from the inspector in charge of the Hamzakot police station, which was situated close to the foot of the Buner hills. The gist of the message was that during the previous night's raid on the Hamzakot rest house, one raider had died and two were seriously wounded. Consequently, a retaliatory raid by the Bunerwals was imminent, and I should remain vigilant. Unfortunately, at the Shewa rest house we had only eight armed *chowkidars* and had two enclosed compounds to defend. However, we kept shouting throughout the night to give the impression of a large numbers of defenders. Luckily, no retaliatory raid took place during that night.

The next morning, I shifted my camp to Shahbaz Garhi rest house which was situated in the close vicinity of large and populous villages and which, moreover, was only 7 miles from Mardan. There was no telegraph office in Shahbaz Garhi village, so I took my telegraph clerk from Shewa rest house with me, along with a portable telegraph set, which he had connected to the telegraph wires running in the vicinity of the rest house. I was thus able to keep in touch not only with all the canal rest houses in my sub-division, but also with Mardan, the district headquarters.

I told my wife to prepare a feast of thanksgiving to celebrate our deliverance from the raiders. While we were partaking of the feast at night, I received an urgent message from my Overseer at Hamzakot rest house, stating that a *lashkar* of about 150 raiders from Buner was hiding in the neighbouring Karamar hill close to Shahbaz Garhi rest house, and was reported to be ready to carry out a retaliatory attack on us. As it was not safe to move out of the rest house at night, I telegraphed the situation to Mardan. The civil authorities were prompt in taking action. I was informed by telegraph that fifty mounted constabulary *sawars* were being kept ready to rush to my aid on receipt of telegraphic information from me. Furthermore, a telegraph clerk would be in attendance throughout the night, at the Mardan Telegraph Office, to receive any message from me. We had only six armed *chowkidars* at Shahbaz Garhi rest house but kept shouting throughout the night to keep up a bold front. Sounds of the beat of drums to gather men for fighting the raiders were being heard from villages, near and far, bordering the Karamar hill. Scared by our shouts, perhaps, the raiders did not attack our rest houses but raided the neighbouring Bala Garhi village instead, where they looted the houses of two wealthy Hindus, and shot dead a young Muslim villager who had gone to the rescue of his Hindu neighbours. In the morning, we saw the melancholy spectacle of the bier of the young Muslim victim being taken to the graveyard for burial.

The raid on the Bala Garhi village had occurred during the early hours of the morning. The shouts of the villagers of Bala Garhi repelling the raiders were clearly heard in the Shahbaz Garhi rest house. An armed guard with loaded rifle was guarding our sub-divisional treasure chest throughout the night. On hearing the shouting in question, he got excited and his rifle went off unintentionally. Unfortunately for the poor fellow, the thumb of his left hand, the size of a medium walnut, happened to be blocking the end of the rifle barrel at the time, through which the bullet passed making a clean hole. We rushed down expecting to see an accident but could not help smiling at the sight of the stupid giant holding his thumb tightly in his armpit. The wound was bandaged, and I telegraphed Mardan in the morning for a doctor. The doctor, a Hindu gentleman, duly arrived and on examining the wound declared it to be a bullet wound. 'You must be mistaken doctor,' I said. 'The poor fellow has got his thumb stuck on an iron spike.' The doctor looked dubious, but he dressed the wound properly, applying the well-known ointment which had been developed for treating wounded soldiers during the First World War.

That day, I moved to Mardan and gave notice that I would not return to my headquarters on the Upper Swat Canal, unless adequate measures were taken by the government for my protection. The government was sympathetic. A contingent of fifty mounted *sawars* of the Frontier Constabulary of the Afridi tribe were allotted and dispatched to each of the two rest houses in my sub-division, namely Hamzakot rest house where the raid had taken place and at Guwahati rest house. The constabularies were stationed permanently at those rest houses, living in tents until permanent barracks were built for them outside the rest house compounds. Thus, I went back to my sub-division with a sense of security.

A few months later, Mr Wiley (ICS), assistant commissioner of Mardan, while on a tour of inspection, happened to be staying with me at Hamzakot. He informed me in a half-serious, half-joking vein that I would no longer be attacked by raiders from the Buner tribal territory. 'You will soon hear sound of gunfire in the Buner tribal area, "across yonder hills",' he added somewhat cryptically. Sure enough, a few months later a great battle took place between Miangul Abdul Wadud, the grandson of the famous Akhund of Swat, Abdul Ghaffur, and the Nawab of Amb, for the possession of the beautiful upper Swat River valley and the adjoining tribal territory of Buner. There was a confused account as to which party had the upper hand in the conflict, but both the territories in question were conquered by Miangul of Swat, evidently with the blessings of the Indian government. Miangul assumed the title of the Wali of Swat. Under his ruthless but strong administration, not only did complete peace reign in the Buner territory but tribal raids in the adjoining British administered territory also ceased. The Swat valley itself had been a highly turbulent area. The Indian government decided to pacify both the valley and Buner through Miangul who already enjoyed considerable prestige in the valley.

10 Dera Ismail Khan (D.I. Khan)

A City in Peril

Two of my former Roorkee colleagues, Minhajuddin and Khalil, had also been posted in the NWFP Irrigation Department. In 1921, Minhaj completed his nine years of service in the department and was consequently promoted to the rank of executive engineer. But he was forthwith transferred to the Punjab because, for political reasons, it was not expedient to post an Indian to the high post of an executive engineer in the NWFP. Only two years later, in 1923, Faqir Mohammad Khan, a Pathan, a qualified engineer from Roorkee, was appointed officiating executive engineer in a neighbouring division. He was to officiate as executive engineer for only three months and then had to revert to his post of assistant engineer.

Faqir Mohammad Khan and I were good friends. On one occasion, we both happened to be on tour in Nowshera town, where he took me to meet his friend Dr Abdul Jabbar Khan, who had recently resigned from his post in the Indian Medical Service, in response to the Khilafat Movement. Dr Abdul Jabbar Khan, who later became famous as Dr Khan Sahib, was running a small private clinic in Nowshera town at that time. He was the elder brother of Abdul Ghaffar Khan, the Red Shirt leader. Both the brothers subsequently threw in their lot with the Indian National Congress. The career of Abdul Ghaffar Khan, as the Frontier Gandhi, is well known. Both of them suffered imprisonment in British jails and were also later interned, for some years, in Pakistan. Subsequently, Dr Khan Sahib accepted Pakistan and was appointed chief minister of West Pakistan. While working in that capacity, he was assassinated by a Punjabi petty revenue official, because of an alleged grievance against Khan Sahib's government. When I met him at Nowshera for the first time in 1923, Dr Khan Sahib was tenderly treating the inflamed eyes of a dirty baby, carried by a poor rustic Pathan woman. During our conversation he mentioned his immense pride in the Pathan race.

In the summer of 1923, Faqir and I were vacationing at Abbottabad on short leave. One day he called at my house in Abbottabad and gave the unexpected news of the appointment of Kirpa Ram as the sub-divisional officer in charge of an important construction work, comprising a stone protective embankment on the right bank of the Indus River at Dera Ismail

Khan [henceforth D.I. Khan], to safeguard the town from rapid erosion by the mighty river. The D.I. Khan Sub-division, with Kirpa Ram in charge, was under Faqir Mohammad Khan, as the executive engineer. 'It is a great shame,' said Faqir Mohammad Khan, with some bitterness 'that an Imperial Service engineer with six years of canal experience [meaning myself] should have been ignored and that fellow, Kirpa Ram, has got away with this important position.' 'But how can I get the job, when orders for the posting of Kirpa Ram have already been issued?' I asked. 'I have a stratagem,' he said. 'We go to Nathiagali together; you seek an interview with Mr R. Cannel, secretary for irrigation, and present your case and while you are doing so, I shall walk in. As I am Kirpa Ram's boss, Mr Cannel will naturally ask my opinion, and I will strongly back you.'

Our stratagem succeeded. Telegraphic orders were issued by Mr Cannel, appointing me to this important construction position, and Kirpa Ram was transferred elsewhere. 'Mind,' said my friend Faqir Mohammad Khan, as we emerged from Mr Cannel's bungalow, 'you will retain the job for only three months and shall hand it over to me when I am reverted to the post of assistant engineer, three months hence.' 'Agreed,' I replied.

So, in August 1923, I and Haider Ali (sub-overseer) an experienced subordinate engineer, were transferred to D.I. Khan town, to tame the mighty Indus, which was then in high flood due to summer monsoon. The main current of the river was hugging the right high bank and, after short intervals, big chunks of the bank collapsed and disappeared in the swirling current all along the length facing the D.I. Khan town. The river was perilously approaching the town, the nearest point of the riverbank at that time being only about a mile from the main city.

D.I. Khan town was founded by a Baloch chieftain in the fourteenth century, giving the town his own name. He marked the outline of the town, with its two main streets running at right angles to each other in the form of a cross and located its site 2 miles to the west of the then right high bank of the river. During the course of centuries, the west bank of the river shifted steadily westwards, completely overwhelming the Baloch chief's town. The present town was founded on the same pattern by a Pathan chieftain during the eighteenth century. It was about 4 miles to the west of the river's west bank. Within the next two centuries, the riverbank had shifted to within a mile of the present town. The westward shift of the Indus River, and that of some other great rivers of the subcontinent, has been attributed by some scientists to the west-east rotation of the earth. The city of D.I. Khan was now in peril. A future flood of large magnitude could wash away the city. There was growing anxiety amongst the public and the government. Swift action was called for, to protect the city from encroachment by the mighty Indus River.

My initial task, during the summer flood season of 1923, was to measure the depth of scour of the main flood current hugging the right (west) bank, and then to establish a relationship with the river's velocity and maximum scour depth below the surface of the flood current. That data was required for designing the dimension of the sloping stone embankment along the west bank of the river. While we were at work on the riverbank, the citizens of D.I. Khan town would visit the river to watch the progress of erosion of the bank. The Hindu citizens in particular, who formed the influential and most vocal section of the city's population, felt dissatisfied that a Muslim had replaced a Hindu to carry out this critical assignment. 'Will these two persons who have been sent by the PWD be able to tame and harness the mighty Indus River?' they asked, within our hearing.

The contract for the construction of the protective embankment had been signed by the secretary for irrigation and the contractor at Nathiagali, the hill station (elevation 8000 feet) and the summer headquarters of the NWFP government. The major portion of the stone for the embankment was to be quarried from Bilot hills, situated 40 miles upstream along the river, and transported to D.I. Khan by boats, at a rate specified in the contract document. After completing our task of observation and recording of scour depths and surface velocities of the flood current of the river, I went to Bilot, 40 miles upstream from D.I. Khan with the contractor to locate the quarries for stone for the embankment. However, we found that the Bilot hills were composed entirely of earthy shales, and there was no stone quarry at or near Bilot for miles around. On the other hand, we located limestone in abundance 20 miles further upstream, at the Chashma spur of the Mali Khel hills. The contractor, however, flatly refused to start work on the Chashma quarries unless his transportation rate was increased.

Since it was beyond my authority to increase the contractor's rates, I wrote an urgent demi-official letter by name to Faqir Mohammad Khan (executive engineer), explaining the situation, and requesting him to urgently visit the site and settle new rates with the contractor. I also requested him to supply me the working plans of the design of the proposed embankment to enable me to start the work. Unfortunately, there was no reply from him. I sent him repeated reminders every few days but still there was no response. In the meantime, the contractor, an influential local Hindu of D.I. Khan town, positively refused to start supplying stone from the new Chashma quarries until enhanced rates were settled with him.

There was another source of supply of stone, i.e. the limestone quarries in the Sheikh Badin hills situated 20 miles away on the D.I. Khan–Bannu road. It was proposed to lay the railway track along the raised earth berm of the D.I. Khan–Bannu road, so as not to interfere with the road traffic. The requisite length of rails with iron sleepers and fish bolts, and forty tip wagons and two engines of a narrow gauge, Duncanville railway, had been supplied by the Punjab Irrigation Department for transport of stone from that source. I spent three months of inaction in laying the railway track. After the lapse of three months, not a single stone had arrived at the site of work at D.I. Khan. The influential citizens, mainly Hindus, of D.I. Khan town, sent telegrams to the government complaining about the lack of progress on the work and demanding a change of the construction personnel. Thus, a storm had suddenly burst over my head.

To my dismay, I learnt that a new executive engineer, Mr A.N.M. Robertson, an able and experienced engineer from the Punjab Irrigation Department, had replaced Faqir Mohammad Khan, who had been transferred to the Punjab. Both Mr Robertson and Mr R. Cannel (secretary for irrigation), unexpectedly arrived in D.I. Khan in response to the protests of the notables. The next day, a kind of a tribunal consisting of the above two officers and the deputy commissioner of D.I. Khan, Mr J.G. Acheson (1889–1973),[1] was held in the Circuit House to decide my fate. It was openly said that I should be removed and replaced by Kirpa Ram on the construction work. I waited outside the Circuit House, to hear my sentence.

At last Mr Robertson came out looking pretty grim. He told me that it was not as bad as I was expecting. The tribunal had carefully gone into the case, and they had taken note of my repeated requests for a decision by the executive engineer, and the lack of response on his part. The tribunal had, therefore, put the main blame for lack of progress on the work on the executive engineer. The tribunal had also blamed me for showing a lack of initiative in failing to start the transport of stone from the Chashma quarries by boat, on my own responsibility, by promising an enhanced rate to the contractor, to be decided later through negotiation. It was, therefore, decided to retain me in my position in the department and that I would also remain in charge of the construction work. I was, however, issued an official warning for having shown lack of initiative; my annual increment was stopped for one year, to be restored with retrospective effect, if I showed good progress the following year.

Mr Robertson very soon sent me working drawings of the proposed protective embankment. He had designed them himself as a Guide Bank for harnessing the flow of the river and protecting the city from its incursions, under the heaviest flood conditions. The total length of the embankment was over two-and-a-quarter miles, about 12,000 feet. The first 1000-foot length from the downstream end was straight, being placed in the bed of a small creek of the river which dried up in winter. The remaining length of the embankment facing upstream gently curved away from the riverbank, leaving a substantial piece of land between the riverbank as it existed at the close of the flood season of 1923 and the proposed alignment. This was to allow for any scouring by flood waters during the flood season of 1924.

The embankment consisted of a sloping earth bank armoured by a 2-foot thick dry stone, resting on a 12-inch thick base consisting of small stone pieces of up to 2 inches in size. The stone pitching was hammer-dressed and lay smoothly in layers. The foot of the embankment rested on a stone apron 40 feet wide and between 3.5 feet and 6 feet in thickness. The specification for stone in the apron required that each individual stone should weigh 70 lb or more or be at least 9 inches thick. It was assumed that as the main flood current would flow parallel to the embankment, the stone in the apron, resting in the river bed, would settle down in the scour created by the flood, at approximately the same slope as the embankment, and would support the stone pitching and keep it in place. To meet the contingency of the main flood current, striking the embankment at a skew angle and thereby causing deeper than normal scouring at the point of contact, provision was made in the estimate for keeping a permanent reserve of stone on top of the embankment to meet any emergency. It was further proposed to lay the Duncanville railway track permanently on top of the embankment, in order to rush reserve stone supplies on tip wagons to the threatened spot. The height of the embankment was kept 4 feet above the highest estimated flood level of the Indus River at D.I. Khan. The proposed width of the top of the earth embankment was 40 feet to accommodate a metalled road on top in addition to the Duncanville railway track.

Since it was not practicable to complete the entire construction of the embankment before the flood season of 1924, it was decided not to carry out any work along the 1000-foot straight length of the embankment along the bed of the dry creek lying downstream, but to start construction from a point 1000 feet upstream. This was necessary in order to counter the thrust of the main channel of the river, which was expected to hug the bank against the newly reinforced part but leave the creek below the 1000-foot mark high and dry. The remaining work was to be completed the same year after the flood season.

I laid the curved part of the embankment on the ground personally by theodolite. An enhanced rate for the transport of stone from the Chashma quarry was fixed and work had simultaneously started at a brisk pace on the embankment at D.I. Khan, at the Chashma quarries, and at the Sheikh Badin quarries on the Bannu road. Five hundred country boats from the long reach of the Indus River, between Kalabagh and Sukkur, were commandeered at generous rates for the transportation of stone 60 miles by river from the quarries to the site of work at D.I. Khan. They made a return trip in a fortnight and worked throughout the year bringing in about 200,000 cubic feet of stone per month. The daily supply of stone by the Duncanville railway from the Sheikh Badin quarries was 3200 cubic feet. The Pathans of the Khattak tribe and Odes, a Hindu tribe, who were experts, carried out the earthwork. The pitching work was done by Kashmiris who were experts in that field. I insisted that the work of stone pitching should be done in hammer dressed masonry and laid in regular courses, on the rates allowed for ordinary pitching. All together about 5000 labourers of all categories were at work at D.I. Khan and at the stone quarries of Chashma and Sheikh Badin, including the boatmen and the Duncanville railway staff. This mammoth mobilisation of men, material, and transport is reminiscent of the colossal projects of ancient Egypt.

At that stage, Sir Norman Bolton, the chief commissioner of the NWFP, inspected the construction work of the protective embankment at D.I. Khan, and expressed his satisfaction at the progress of work. He assured the citizens, who had assembled at the work site to greet him, that the progress on the construction was satisfactory, and that the PWD staff appointed for the work (meaning myself and three overseers working under me) were adequate. The last statement was meant, in particular, for the Hindu citizens, who had protested about lack of progress in the past. Thus, their hopes of replacing me by Kirpa Ram, were extinguished.

During the early stages of the summer flood season of 1924, the position of the river channels was the same as that at the end of the flood season of 1923. The main current was scouring the right high bank opposite the curved part of the protective embankment, as expected, where we had completed construction of a sizeable portion of the work, including the dry-stone apron. Lower down, however, the main current swerved to the left towards the middle of the wide channel, leaving a small creek flowing parallel to the bank.

Thus, after satisfying myself that all was well at the site of the construction work at D.I. Khan, I went to Chashma, 60 miles upstream, on horseback, to inspect the Chashma quarries. While I was returning from Chashma, and had reached Bilot at the head of the Paharpur Canal, 40 miles from D.I. Khan, I received an urgent telephone message from Mr J.G. Acheson (deputy commissioner, D.I. Khan), urging me to return immediately, as the main current of the river had suddenly changed its course and was now attacking the lowermost 1000-foot length of the unprotected bank, with full force.

An Epic Struggle with the Mighty Indus

The quickest means available for getting back was by boat. So, I hired a light boat manned by two boatmen, with a free board of only 18 inches, and started down the Bilot Creek. The junction of the creek with the main stream a few miles lower down was an awe-inspiring sight. Flood waters were rushing down in a mile-wide channel with the speed of a mill-race, emitting a thunderous roar. We were swiftly borne down along the current and could see the lines of trees along the far-flung banks racing upstream. A few miles downstream, all the flooded branches of the river joined together, forming a single channel 8 miles wide, carrying perhaps a million cubic feet per second of flood discharge. We felt as if we were riding on air, while on the placid surface of a vast lake.

Our blissful state, however, was to be replaced soon by stark terror! A few miles further down, we noticed faint streaks on the downstream horizon which soon took the form of a number of islands which again divided the current into smaller channels. We further noticed that our boat was being irresistibly drawn towards a narrow channel between two islands. On getting closer, I noticed with alarm that the current was heaving up in rollers on entering the gap. In a panic, I shouted to the boatmen to steer the boat to an adjoining placid channel, or to the far away banks, but it was too late! The oars were useless against the force of the current. We dropped the 40-foot long iron anchor chain to dampen the speed of the boat. It failed to touch the bottom. The rudder, too, was useless. The boat turned around, presenting a broadside to the rapidly approaching rollers.

I was fervently praying to God to deliver us from the imminent threat. Suddenly, we were shot up on the first roller, which must have been 20 feet high, and down the valley between the next one. With our eyes fixed on the succeeding rollers, we hung to the sides of the boat, which was being tossed about from crest to valley, and up again, twisting and turning helplessly between the rollers. Realising, however, that we were still afloat, and the rollers were diminishing in height, we took courage. I silently poured forth my heartfelt thanks to the Almighty for our deliverance. The rollers continued to toss us around for the next 5 miles. We had to go through a similar harrowing experience in two more reaches lower down the river, before we touched the bank at D.I. Khan, where Mr Acheson, was anxiously awaiting my return. Mr Robertson, my boss, arrived by air the next day.

And now began an epic struggle to save D.I. Khan from the raging flood torrents of the Indus, by stopping, or retarding, its relentless erosion of the unprotected length of the high bank, the former creek, which it had suddenly invaded overnight. The struggle lasted for a fortnight, at all hours of the day and night. The European cemetery lay a hundred yards away from the bank, and behind the cemetery lay the wide expanse of the cantonment bungalows and roads. By mutual consent, Mr Robertson, executive engineer and my superior, agreed to work under my general supervision. I assigned him the day duty of twelve hours and took over the night duty myself. A large number of labourers were employed to throw cascades of stones in the jaws of the eroded chasms which carved up the bank at short intervals and at unexpected points, but to no avail. We were steadily losing ground. We found out, however, that very large pieces of 'stone balls' about 5 to 6 feet in diameter securely tied within one-eighth inch gauge wire mesh, and rolled down in the jaws of the scour, did have a dampening effect on the progress of the scour. So, this strategy was adopted.

Unfortunately, the flotilla of boats had just left on their slow upstream journey to Chashma, after discharging their stone cargo, which had been used up in the upstream parts of the embankment under construction. So, the only available source of stone to meet the emergency was the daily supply of 3200 cubic feet of stone from the railway. Another possible source was to shift the stone from the upper reaches of the embankment which were not under attack. But that was slow work as only donkeys could be used for the purpose. Then, as though by inspiration, I thought of the brick kilns around the town and placed unlimited orders for the quick supply of bricks for constructing huge crates for dampening the erosion of the bank. That measure somewhat eased the situation.

And then began the titanic struggle against the fury of the mighty river's flow, as it was now attacking the bank at numerous points simultaneously, day and night. Our job was to roll the huge crates of enmeshed stones and bricks into the jaws of the scour wherever it occurred. That was hectic work requiring all concerned to bring forth every ounce of effort and vigilance to the task. Mr Robertson worked like a 'mate' (labour supervisor) during the daytime. His voice got hoarse due to constant shouting. I did not mind the all-night duty, but since I was primarily responsible for the construction work and also for preserving the alignment of the embankment, as far as possible, anxiety often drove me to the scene of action on the river, even during the daytime. After a fortnight of constant and unrelenting struggle against the torrents, we had slowly gained ground by only 50 yards. On the morning of the sixteenth day I was too exhausted to go back home for sleep. As the sun arose in the eastern horizon, illuminating the wide expanse of the majestic river, I could hardly believe my eyes at what I saw. The main current of the river was roaring about a mile away from the beleaguered bank. Only a gentle current was passing close to it. The miracle had taken place during the latter part of the dark night, abruptly changing the course of the torrent.

Mr Robertson had been absent from the site of the work since the day before. I got anxious about him and made enquiries. He was stricken down with typhoid fever and was under treatment in the Cantonment Military Hospital. I paid frequent visits to the hospital to enquire about his welfare. After three weeks he came out of the hospital in a weak and emaciated condition. The embankment and D.I. Khan town itself had been saved! As the summer flood season passed and the river settled down in its winter channels, we lost no time in completing the embankment in the downstream 1000 feet reach which had been the scene of our hectic struggle during summer.

Contractor's Deceit

During the early winter of 1924–5, the contractor intimated as usual that the flotilla of boats had brought down a large quantity of stone from the Chashma quarries which had been stacked on the riverbank. I was requested to inspect the stone stocks and to give permission for their use in the stone apron. On inspection, I found that more than 25 per cent of the stone stacks—20 feet wide, 3 feet high, and about a furlong in length—consisted of small pieces, which were below 65 lb weight for individual stones, or less than 9 inches in measurement, required under specifications, to prevent the stones in the apron from being carried away by the flood current. I prohibited the dumping of stone from the stacks into the apron until the pieces that were below the specification had been separated and removed.

The next morning, I received a short note on a slip of paper from Col. Charles Edward Bruce, the deputy commissioner of D.I. Khan, who had recently taken over on the transfer of Mr Acheson. The note said, 'SDO come and see me.' I hastened to the deputy commissioner's office at 10 a.m. sharp and sent in my card. Promptly, a chair was brought out for me in the veranda of the office by the peon, implying that I must wait outside until called in. I waited outside for hours. People went in and out of the deputy commissioner's room, callers, as well as his clerks with piles of official files. I reminded the peon two or three times to intimate my presence to the deputy commissioner. The peon told me in somewhat embarrassed tones that the deputy commissioner was perfectly aware of my presence, and would call me in, as and when he was pleased to do so. At 1 p.m., closing time of the office, I was called in. Without offering me a chair to sit down the Col. Bruce said in curt tones: 'SDO, I have received a complaint from the contractor of the protective embankment that you have demanded illegal gratification from him.'

'Sir,' said I, boiling with indignation, 'the contractor's allegation is false!' He regarded me angrily for a moment. 'Have you prohibited the contractor from dumping the stone supply recently brought by the boats, into the apron?' 'Yes, that is so,' I replied. Col. Bruce added sharply, 'Why have you done so?' 'Because a large percentage of the stone brought by the contractor is below specification,' I replied in anger. 'Are you the judge of the specification of the stone?' he asked. 'Yes sir, I am,' I replied. 'How is that?' he asked incredulously. 'Because I have written instructions to follow the specifications for the work,' I replied. 'Where are those specifications?' 'They are in my office.' 'Can you show them to me?' 'Yes, any time you please.'

Somewhat nonplussed at my bold show of indignation, but still obviously disbelieving me, Col. Bruce said, 'Very well, come to my office at 6 a.m. sharp tomorrow morning. I shall inspect the stone stacks myself. Bring your specifications along.' The next morning, punctually at 6 a.m., I entered Col. Bruce's bungalow, one room of which he used as his office. A car was waiting outside, and the contractor was standing beside it. Col. Bruce was pacing back and forth in the veranda. 'Has the SDO come,' he had asked his staff, a number of times before 6 a.m., somewhat irritated. Col. Bruce drove the car himself, inviting the contractor to sit in the front seat beside himself, and motioned me to sit in the back seat with the peon. Silently, he drove to the site of the work. There was a long line of neatly stacked stone. Impressed, the colonel said. 'SDO, what is wrong with these stone stacks?' In reply, I silently handed over to him my copy of typed specifications for the work. 'Now sir, you will notice, that the specifications say that each individual stone for the apron must weigh a minimum of 65 lb, and at least 9 inches in measurement. Now kindly look at those stacks and judge for yourself what percentage of stones meet this specification.'

The colonel studied the specifications and then looked at the stone stacks. He was bewildered. 'Rai Bahadur,' he addressed the contractor, 'The SDO is right.' Saying this, he silently drove back to his office, with the contractor and his peon, leaving me at the site of work! In a strong tone, I wrote a long demi-official letter to my immediate boss, Mr Robertson (executive engineer), protesting the insulting treatment that I had received at the hands of Col. Bruce. Furthermore, since my honour and integrity had been challenged, I requested for immediate transfer to some other station.

In reply, I received a blind copy of a demi-official letter which was sent by Mr Robertson to Col. Bruce. Mr Robertson had written that Mr Rahman was not the type of SDO of the Military Engineering Services who was generally a promoted subordinate but was a member of the Imperial Service of Engineers and, as such, had the same status as Mr Ambrose Dundas[2] (ICS), who was working under him (Col. Bruce) as assistant commissioner at D.I. Khan. Mr Robertson further wrote that Mr Rahman was entitled to the same courtesy as he (Col. Bruce) extended to his assistant commissioner. Lastly, for Col. Bruce's information, he had added that the government had full confidence in the integrity, and efficiency of Mr Rahman as an engineer.

A few days later I received a demi-official letter from Col. Bruce, in remarkably civil language. 'Would you make it convenient to see me in my office?' Col. Bruce met me graciously in his office, courteously offering me a chair. He said he was very glad to learn from Mr Robertson that I was putting in good work in the construction of the protective embankment, and that he had ordered the contractor to remove the undersized stones from the stone stacks at the site of work. And he, Col. Bruce, would be grateful if I could find time to take over the honorary task of advising his civil staff on the problems of Rod Kohi (flood) irrigation from hill torrents in the D.I. Khan district. I accepted the honorary duty, and we parted as good friends. In 1930, when I was conferred the title of 'Khan Sahib' by the Government of India for meritorious services, Col. Bruce, then holding the high office of agent to the governor general in Balochistan, at Quetta, was the first to send me a telegram of congratulations.

The construction of the protective embankment posed no problems or crisis after the memorable flood season of 1924. During the summer of 1925, the intervening strip of high ground between the riverbank and the upper curved portion of the embankment had vanished due to river erosion. The river had invaded the protective embankment itself, but we were fully prepared for the eventuality. The affected reach of the embankment had been completed in advance. By October 1926, the construction work on the entire embankment had been completed. On the completion of the construction of the protective embankment, I submitted my 'completion report' on the work and showed a saving of Rs 100,000 in the expenditure. Both Mr Fletcher, the new executive engineer who had replaced Mr Robertson and Mr Walker were so impressed that they both paid a visit together to D.I. Khan to personally verify that the saving in the final estimate was justified. I explained the reason for the saving as follows:

My instructions were to excavate a 40 feet wide trench to take the stone for the apron, down to the spring level. The spring level, however, rose by a foot or 15 inches during summer. But, since the top level of the stone apron was to be kept at a specified level, this meant that less quantity of stone would be placed in the apron during the summer months of 1923–6. That circumstance accounted for the saving in the estimate. Mr Walker was astonished to hear my explanation. He said, 'Rahman, your honest services will not be forgotten!' After the completion of the construction work, I formally handed over the embankment to Mr Dundas, assistant commissioner of D.I. Khan, along with the stacks of reserve stone on top of the embankment, with written instructions for annual maintenance and for dealing with emergencies. I was honoured with a civic reception by the elders of D.I. Khan, Municipality. The embankment continues to stand firm until today and is remembered by the local citizens as 'Mian Sahib's Bund', after the honorific title of 'Mian' or 'Syed' by which I had become popularly known in the district.

The most important irrigation work under the civil officials was the Paharpur Canal. This canal had originally been constructed by the NWFP PWD as an inundation canal during 1905–7. Its 'head' was located at Bilot village, on the Bilot Creek of the Indus River, 40 miles upstream of D.I. Khan town. It ran parallel with the right (west) bank of the Indus River with its tail a few miles downstream of D.I. Khan town. It irrigated about 50,000 acres of land, lying partly in the low bed of the river, technically known as *Katchhi* lands, in the head reach, and partly high *Da'man* lands in the tail reach, all within the narrow area enclosed between the canal and the west bank of the river.

The flow in the canal started around mid-May when the river levels rose high enough and continued to flow during the summer months up to the middle of September, when the river levels fell below the bed of the canal. The full supply discharge of the canal was about 600 cubic feet per second. At the very outset, however, the canal had proven to be a failure. Its banks were badly breached at numerous points by cross drainage from hill torrents during summer, even during the very season it was opened to flow. Consequently, irrigation from the canal was badly interrupted and the revenue from canal water rates was not adequate to maintain the canal in reasonable repair, let alone pay interest on the capital outlay. The canal was, therefore, declared unproductive. In 1920, it was handed over to the civil department to be maintained by the cultivators themselves under the guidance of civil officials.

During the winter of 1924, as honorary advisor to the civil department, I inspected the Paharpur Canal at its head, at Bilot at the time of the annual silt-clearing of the channel by the cultivators. The canal was, of course, dry. Its bed was about 8 feet higher than that of the Bilot Creek of the Indus River, which fed it in summer. The Bilot Creek was only about 100 feet wide and I noticed that a modest discharge, of about 500 cubic feet per second, was passing in the creek at that time. My enquiries revealed that about the same volume of discharge was always present in the creek during winter. I had a sudden inspiration, if the channel of the Bilot Creek was completely dammed up by an earth embankment about 12 feet high, the winter flow could be headed up and diverted into the Paharpur Canal. In other words, the Paharpur Canal could be turned into a perennial channel. I promptly issued instructions for damming up the creek by an earth embankment, as soon as the work of silt clearing of the bed of the Paharpur Canal was completed. My instructions were, equally, promptly carried out by the willing and intelligent cultivators.

The Paharpur Canal was thus converted 'overnight' into a perennial canal! It flowed throughout the winter of 1924–5, providing a sizeable amount of irrigation water for the Rabi (winter) wheat crop without interruption from the disastrous cross drainage summer floods. The winter rains due to the western disturbances were usually not of sufficient intensity to cause damaging floods in the hill torrents. Towards the end of May, when levels in the Indus River (and consequently the Bilot Creek) rose high enough to cause spontaneous flow in the Paharpur Canal, the earth embankment was washed away but was promptly rebuilt by the cultivators at the end of the summer in late September for the next winter crop. These idyllic conditions on the Paharpur Canal lasted for many years. The canal areas flourished and land prices rose.

Some years later, in the autumn of 1931, I came up with the idea of building a permanent weir across the Bilot Creek while it was still in flow in order to avoid the temporary period of the canal's closure during the period when the summer levels in the Indus River went down and the Bilot Creek was dammed up. Laying the foundation of the weir on the sandy bed of the creek within over 10 feet deep water, however, posed a technical challenge. This was solved by constructing 40 by 40 feet mattresses of Pilchhi (*Tamarix dioica*) bush, which was plentiful in the dry islands of the Indus River, and floating them down the creek to coincide with the proposed alignments of the weir. The mattresses were then sunk by loading them with heavy stones brought by country boats. Thus, the entire foundation of the weir was covered with the thick Pilchhi mattress. I had invited Mr Arthur Oram (superintending engineer), to witness the process of our sinking the mattress on the sandy bed of the creek. He appeared to be impressed. The stone weir, with an overflow spillway, was soon completed, and from that time on the Paharpur Canal became a truly perennial canal. Shortly thereafter, I constructed a motorable dirt track on top of the left embankment of the 40-mile long canal making it much more convenient to reach Bilot, than on horseback.

The Controversial Kalabagh Right Bank Canal

The construction of a dam on Indus River at Kalabagh is presently a subject of considerable controversy among Pakistan's provinces resulting in an impasse on the implementation of this vital project. Ironically, many years before the proposal of the Kalabagh Dam was considered and even before the construction of the barrage on the Indus River at Kalabagh (started in 1939 and completed in 1946), the Government of NWFP was planning the construction of a right bank canal on the Indus at Kalabagh, which would irrigate millions of acres in the southern districts. Had this proposal been revived in the initial inter-provincial discussions on Kalabagh Dam, where the principal party objecting to its construction was the Government of NWFP, a favourable trade-off might have been negotiated and construction on the dam allowed to proceed.

In early 1925, Mr R. Cannel (secretary for irrigation, NWFP), was transferred to the Punjab and replaced by Mr Samuel Walker. In 1925, Mr Walker sent orders for the preparation of an album of photographs of the 7-mile hilly reach of the Indus River, comprising the Kundal–Kafir Kot–Chashma–Umar Khel salient of the Mali Khel hills where the right bank of the Indus River hugged the hillside. A stand camera taking glass slides of cabinet-sized photographs was supplied for the purpose. I took the photographs myself and sent the album to Mr Walker. I was astonished to receive an 'honorarium' of Rs 50 from the government for my photographs. Shortly after taking over charge of the drainage and surveys sub-division at Mardan, I was ordered to start surveys for a right bank canal from the Indus River, starting from Kalabagh with a capacity of 10,000 cubic feet per second. The proposed canal would irrigate an area of about two-and-a-half million acres of land lying along the right (west) bank of the Indus River in the D.I. Khan, Dera Ghazi Khan, and Sukkur districts. The proposed canal was to be aligned on the hill slopes, in the 7-mile long hill section of the Kundal–Kafir Kot–Chashma–Umar Khel salient of the Mali Khel hills. This explained Mr Walker's orders for a photographic album of the right bank of the river in the hilly reach in question. Emerging from the Kalabagh gorge, the Indus River fans out in a diverging channel, and dissipates its surplus energy in scouring out a deep channel and deposits the scoured-out material in the form of a curved bank at the downstream end of the fan. The river divides itself into a number of channels as it passes over the submerged ridge of the scoured-out material. The old railway bridge has been built over the ridge in question.

A party of eight handpicked overseers, experienced in survey work, was placed under my charge to assist me in the survey. Our instructions were to align the 85-mile long main line of the canal down to Saidowali, where it emerged into the plain of the D.I. Khan district, at a longitudinal slope of 1 in 10,000, leaving permanent stone pillars at 1000 feet intervals. I took up the work of alignment of the canal with theodolite, deputing two of the most experienced overseers for double levelling of the alignment and the remaining three pairs for taking cross sections across the alignment. A camel loaded with stone pillars for the alignment always accompanied the survey party as the work progressed. About forty survey *khalassis* were recruited for the survey party. A small Swiss Cottage tent for myself and the requisite number of smaller *chholdari* tents for the staff and surveying instruments were provided by the department. At my request, the deputy commissioner, Mianwali district, instructed the headmen of the villages in the Isa Khel tehsil to render all help to the survey party, en route.

We started work from a benchmark on a west masonry platform of a bathing *ghat*[3] on the edge of the river at Kalabagh town situated upstream of the shingle bar and ran the alignment of the proposed canal parallel to the west bank of the river, at a safe distance away from it. I took the lead with the theodolite traverse of the alignment, the rest of the survey party trailed behind me, along a long line of waving red-and-white survey flags. A British executive engineer of the discharge division of the Punjab Irrigation Department happened to be visiting Kalabagh on an official tour and on catching sight of the line of our survey flags along the opposite bank of the river, crossed over in a small boat to investigate. He was directed by my surveyors to meet me at the far end of the line. 'Good morning,' said he, 'may I enquire what kind of activity is going on here?' Feeling somewhat nettled, I replied that we were surveying for a right bank canal which would take off at Kalabagh to irrigate an area of about two-and-a-half million acres lying along the west bank of the river. 'But where is that area?' he said, somewhat bewildered looking incredulously at the narrow, hill girt area of the Isa Khel tehsil, which was visible from our high point of observation. About 30 miles lower down the encircling hills met the Indus River, forming the Mali Khel hill section and thus completely enclosing the lands of the Isa Khel tehsil. 'The area to be irrigated lies much beyond those hills,' I pointed out and added, 'About half-a-million acres lie in the D.I. Khan district.' 'To what department do you belong?' he asked. 'I am an assistant engineer in the NWFP, Public Works Department, and am carrying out the orders of Mr Walker, secretary for irrigation, for the Government of NWFP.' 'But don't you know the entire winter supply of the Indus River has been earmarked for the proposed Thal Canal which will take off from the opposite (east) bank of the river?' 'I know nothing about that,' I said. Saying that he must report the matter to his government he left in a huff. We, however, paid no attention to the episode, and continued our work.

As the survey work progressed, we camped at various villages situated near the west bank of the river, in Isa Khel tehsil of Mianwali district, making our purchases of milk, chickens, and eggs from the villagers, through the courtesy of the village headmen. About 8 miles from Kalabagh, we reached the town of Isa Khel where we replenished our supplies of vegetables and purchased mutton. I also called on the local notables of the town. A few miles beyond Isa Khel was the Kurram River crossing; the river at that point is close to its junction with the Indus. It divides into a number of channels and has a width of about half a mile between its high banks. It was considered necessary to prepare a contoured plan of the river for some distance both up and downstream of the alignment of the crossing that I had proposed, in order to exercise options of the exact crossing, later on.

A few miles further downstream, we reached Kundal village, from which point the Indus riverbank hugs the hill side, in a length of 7 miles down to Umar Khel village. This part of the Mali Khel hills was familiar to me as the source of stone quarries for the construction of the protective embankment at D.I. Khan. In that hilly reach, we usually pitched our camp on higher ground in a small side valley. Luckily, there was always a strip of dry sandy bed, 50 to 100 yards wide, between the west bank of the river's winter channel and the toe of the hill, along which we laid our alignment. From the traverse alignment it was necessary to take cross-sections of the hillside at 100 feet intervals, a total of 350 cross-sections in all, as the actual alignment of the canal that would fall along the hill slopes.

Taking cross-sections of the hill slopes by successive shifting of the position of the levelling instruments was extremely slow and painstaking work. Anxious about meeting our deadline, I made an urgent request for the services of a dozen extra surveyors to assist me for taking cross-sections of the hill slopes. I immediately received an assurance from the executive engineer that arrangements were being made to provide the extra staff.

However, luckily through an inspiration, I saw a way of accomplishing the task without the help of the extra staff. By mathematically calculating the hill cross-sections using only a theodolite and the 12-foot measuring staff. In this way, time consuming physical measurements were considerably reduced. I wrote a hasty demi-official letter to my boss, explaining the procedure I had adopted and suggested that there was no further need to send the extra survey staff. My demi-official letter was forwarded in original by the executive engineer, to Mr Walker, who sent his approval to my proposals. From that moment, progress on taking the hilly cross-sections was rapid, as almost the whole of the remaining alignment for the proposed mammoth canal down to Saidowali ran in the hilly terrain.

It was pleasant work, doing surveys in winter in the rose-coloured limestone hill range of Mali Khel running along the west bank of the mighty Indus River. The width of the river in that area between the high banks was about 10 miles. From our vantage points in our elevated camps we had splendid views of the wide expanse of the river with its network of silvery winter channels. The tree-lined far bank of the river was also to be seen miles away. In the foreground was the almost stark Mali Khel range of hills, covered with tufts of grass and bramble. The dry strip of the riverbed at the toe of the hillside had occasional patches of a young, green wheat crop. The hillside was teaming with grey partridge and Sissies (a smaller variety of grey partridge) which greeted the rising sun with melodious song and came forth from their refuge in the bushes on the hillside to feed on the patches of sprouting wheat crop in the riverbed. Each morning at sunrise I traversed the hillside on foot to the spearhead of the alignment, to resume the theodolite work, lugging my trusty W.W. Greener shotgun. I was generally lucky to shoot a brace of partridges en route, which I sent back to the camp to be prepared and served at the communal lunch for the survey party. Similarly, on my way back to camp in late afternoon under the dying rays of evening sun, I frequently bagged a couple of birds or more, for our dinner. Now and then, we purchased a fat-tailed sheep from the village en route and enjoyed a feast of tikka, roasted leg of lamb, and *pulao*. The field survey work of alignment of the 85-mile long main lines of the proposed Kalabagh right bank canal took three months to complete. I went to Peshawar to work on the plotting in Mr Walker's office.

In the meantime, the executive engineer of the discharge division of the Punjab Irrigation Department had promptly reported our activities to the Punjab government. The latter had taken up the matter with the Government of India and had protested about the attempt on the part of the NWFP government to claim the winter supply of the Indus River at

Kalabagh for their proposed right bank canal. Whereas, the whole supply in question had been earmarked according to a decision by the Government of India for the proposed Thal Canal which would take off from the left bank of the river and would exclusively irrigate the area lying in the Punjab. Mr Walker was, consequently, called to Simla and was reportedly told off and he had to give up his proposal for a right bank canal from Kalabagh. On his return from Simla, Mr Walker gave me confidential instructions to report on the unfavourable features of the proposal. This was evidently a 'face saving' device.

Reluctantly and with some remorse, I stressed the constraints of the project in my report. I mentioned the prohibitive cost of aligning a canal of 10,000 cusecs capacity on the hilly slopes for a length of 12 miles. Besides, there was the formidable Kurram River crossing and numerous other cross-drainage works involved. Above all, the proposal envisaged construction of an 85-mile long Main Line Canal, the greater part of which would have to be lined. My report was approved by Mr Walker and that ended the matter.

11 Waziristan, 1927

The Untamed Borderland

In August 1927, a few months after the completion of the Kalabagh right bank canal survey, I received the surprising and welcome notification from the NWFP administration that I had been promoted to the rank of executive engineer in the Indian Service of Engineers (ISE), and was appointed executive engineer of the proposed Waziristan Survey Division. I was, furthermore, asked to report to the irrigation secretariat at Peshawar to receive instructions about my duties from Mr Walker (secretary for irrigation). I hurried to Peshawar and was instantly taken to Mr Walker's room, who was not present there at the time. A voluminous official file of papers was lying open on his table. Unable to restrain my curiosity, I started reading a report on Waziristan written by a political officer. The report stated that since the complete evacuation of Waziristan during the Third Afghan War of 1919, the South Waziristan area had not been visited by any British officer. The area was in a disturbed state and it was considered unsafe for a British executive engineer to be posted there for survey work. That much information was enough for me, as I quickly resumed my seat.

Shortly thereafter, Mr Walker arrived. He congratulated me on my promotion and I duly thanked him. Then he started giving me verbal instructions about my new duties, the gist of which were as follows: The Gomal hill torrent with a huge catchment area of about 15,000 square miles drained the major portion of South Waziristan. The torrent had its source in Afghanistan, and it entered Waziristan at a place called Domandah, where it cut across a range of low hills forming the boundary between Afghanistan and Waziristan. At that point, the torrent received its first tributary, the Kunder, from the south principal which drained portions of Afghanistan and Balochistan. About 10 miles lower down, the Gomal torrent entered into a narrow gorge at Gul Kachh, which seemed a splendid reservoir site on the upstream side of the gorge. He (Mr Walker) had personally made the discovery of the possibility of a dam and large reservoir on the Gomal torrent at Gul Kachh from the topographical survey maps and wanted me to carry out surveys at that site in order to test its feasibility. (Note: The dam was inaugurated in 2013 and was constructed with the assistance of USAID).

About 10 miles lower down, the Gomal torrent received its second tributary, the Wana Toi, which drained the Wana plain of South Waziristan. About 12 miles further downstream, the largest tributary of the Gomal torrent, the Zhob, joined the Gomal from the south, above the well-known gorge of Khajuri Kachh. The Zhob torrent drained the extensive Zhob valley of Balochistan, and had a larger catchment area, and greater discharge, than that of the Gomal at this point. About 20 miles lower down, the combined Gomal torrent emerged into the plain of the D.I. Khan district of the NWFP. A large amount of Rod Kohi (flood) irrigation was carried on from the Gomal torrent in its 50 miles stretch of the D.I. Khan plain, before its junction with the Indus River.

Schemes for the storage of flood waters of the general torrent by the construction of a dam at the Khajuri Kachh gorge below the junction of the Zhob and Gomal torrents had been prepared by the Punjab Irrigation Department from time to time, but were turned down because of excessive silt carried by the Gomal torrent which would restrict the life of the reservoir to a few years. Mr Walker's plan was to construct delay reservoirs, by constructing regular dams but with open sluices for the delayed but speedy passage of floods temporarily stored in the reservoirs in order to considerably lower the peak discharges of the floods due to temporary storage, as well as to prolong the life of the reservoir by reducing silting by delaying the passage of the floodwaters. For this purpose, I was asked to carry out surveys for dams and reservoirs at Gul Kachh and Khajuri Kachh sites on the Gomal torrent and at Brunj and Badin Zai sites on the Zhob River, situated a few miles downstream and upstream, respectively, of Fort Sandeman station in the Zhob valley. Initially, however, the survey of the Gul Kachh site was to be undertaken.

Mr Walker borrowed the idea of reducing the flood peaks through 'check dams' and 'delay reservoirs' from those constructed on the Miami River (USA) and its branches for reducing the disastrous peaks of the combined floods which flooded the city of Miami, causing heavy damage to property. Mr Walker gave me the volumes of the Miami Conservancy Project to study the techniques of designing and constructing these reservoirs.

The same survey party, which had worked with me on the Kalabagh right bank canal survey, was placed at my disposal for the work in Waziristan. The senior most overseer among them, Mr Ahmad Bakhsh, was promoted to the rank of sub-divisional officer to assist me in the supervision of the work. We initially started survey work on the Tank Zam torrent, basing our camp in the Jandola Fort which was built on a plateau behind the low range of hills forming the boundary between D.I. Khan district and Waziristan. There were five possible dam and reservoir sites on the Tank Zam torrent. Two of them lay in the territory of the Bhittani tribe and the other three in the Mahsud tribal territory. All five could be surveyed from our base in Jandola.

The Tank Zam torrent is situated in Waziristan. It takes off from the snow-clad range of Pir Panjal and Sheridan mountains near the Afghan border and runs parallel to the Gomal torrent in a west-east direction and emerges into the D.I. Khan plain from the gorge at the mouth of the Zam about 30 miles to the north of the Gomal gorge at Murtaza. Its catchment area is contiguous to the Gomal torrent lying on the north. Tank Zam is a much smaller stream compared to Gomal, having a catchment area about one-tenth of the Gomal catchment, but is comparable to its sister torrent in perennial and flood discharge because of greater annual rainfall received in its catchment.

Issuing out of the mouth of the Zam in the D.I. Khan plain, the Tank Zam torrent divides itself into five branches and irrigates a considerable area in D.I. Khan district like the Gomal torrent. Its summer floods, likewise, do a lot of damage to roads, railways and villages, and ravage good farmland. It also breaches the banks of the Paharpur Canal. The control of the Tank Zam torrent and its main tributary, the Shahpur torrent, through 'delay reservoirs', was also a part of our survey programme in Waziristan. We were instructed to initially start work on the Tank Zam torrent as arrangements for the safety of the survey party, during its work at different sites, could be controlled from Jandola fort. Moreover, arrangements for the survey party to work in the remote Gul Kachh area in the upper reaches of the Gomal River, not far from the Afghan border, required careful evaluation, as no defences in the form of fortifications existed in that area.

Jandola Fort is situated in the Bhittani tribe's territory and was the headquarters of the South Waziristan Scouts. It was a square structure, with high walls and with bastions on all four corners. Access to the fort was through a heavily armoured gate. The sides of the fort formed the back walls of rows of rooms with verandas opening into a courtyard. The rooms were occupied by a large contingent of South Waziristan Scouts. The outer walls of the fort had 5-foot high parapets with holes for shooting against invaders, in case of an attack. The night watch on the fort consisted of four armed sentries, one on the roof of each tower; a sentry patrolled the roof in the middle of each of the four walls of the fort. Sentries changed duty every two hours. In addition, a mobile force of about half a dozen sentries was constantly on the move on the roof, behind the parapets. They kept a vigilant eye on the eight stationary sentries. Furthermore, about a dozen scouts, fully armed and awake, were always at hand in the courtyard below, in case of emergencies. No lights were permitted outside the fort at night. The night watch worked in perfect silence. This was the routine observed in all the defence forts in Waziristan.

The South Waziristan Scouts, both mounted and foot soldiers, were recruited from the various Pakhtun tribes residing in the tribal areas outside Waziristan. They had a reputation of intrepid courage and bravery due to which they were made to control the reputedly bravest and most troublesome of the independent tribes, the Mahsuds. The South Waziristan Scouts were under the command of the British officers of the Indian Army who were posted to Waziristan for a number of years. These officers were usually junior in rank and were bachelors. Many of them, who were in financial distress, were afforded an opportunity to recover their debts through a heavy 'Waziristan Allowance' coupled with heavy travelling allowance bills, in addition to their salaries. The force was commanded, at the time, by Captain Johnson who was stationed at Jandola, with two lieutenants, namely Leeper and Fitzpatrick. The mess and living quarters of the British officers was a small single-storey cottage built in the middle of the courtyard of the fort. I and my survey party were lodged in a couple of corner rooms that formed the guest house of the fort facing the courtyard. Survey *khalassis* were provided from amongst the Bhittani and Mahsud tribesmen from outlying areas.

Other forts of the South Waziristan Scouts were located at Kotkai and Sararogha on the Razmak road and at Sarwakai on the Wana road—all three in Mahsud territory. In addition, there was a post at Tanai in the Spin plain in Wazir territory. At Wana, the centre of South Waziristan, there was a large fort which could accommodate a brigade of the Indian Army. In those days, there were only two black-top roads in Waziristan; one was a part of the outer circular road through Waziristan connecting Tank, Manzai, Jandola, and Sararogha with Razmak cantonment. The other black-top road forked from Jandola to Sarwakai Post. Beyond Sarwakai, a dirt road led to Tanai Post and Wana. Razmak cantonment was connected on the other side, with Bannu via Razani and Mir Ali Khel posts by a black-top road running through North Waziristan in Wazir territory, and completed the outer circular road through Waziristan.

The British Indian government dealt with the Waziristan tribes through two British political agents, one for the South and the other for North Waziristan; each assisted by a British assistant political agent. The winter headquarters of these officers were at Tank and Bannu, with summer headquarters at the Razmak cantonment, where a considerable force of the Indian Army was stationed. The actual day-to-day dealing with the tribesmen was carried out by Indian political officers, one for each of the south and north political agencies. These officers were assisted by a political tehsildar, and other petty establishment who were stationed in these posts. The Indian officials in Waziristan were all Muslims and mostly Pakhtun. Their job was to supervise the work of *badragas* (levies) recruited from among the local tribesmen for guarding the roads and for carrying out political negotiations with the tribes under the general supervision of the British political agents. A lot of money was given annually to the tribesmen in the form of wages for road protection *badragas* and allowances to tribesmen. That expenditure was exempt from audit from government departments and there was, reportedly, a lot of overt corruption going on among the ranks of the political department.

The main task of the scouts was to keep the communications to Razmak cantonment safe and open to traffic. It was performed by their mounted armed patrols under the command of British officers. Another important task was to ensure safe passage to caravans of Pawindahs from Afghanistan through their hazardous 50-mile trek through the Gomal valley, along the Gomal River, while passing through the country, from marauding Mahsud tribesmen in the lower reaches of the Gomal River. The Pawindah tribesmen consisted of the Sulaimankhel, Nasar (or Nasir, Nassiri, Nasiri), and Kharoti clans that seasonally migrated, bag and baggage, along with their families, tents, sheep, and goats. They brought camel loads of dry fruit and other exportable goods from Afghanistan, such as the karakul furs and rugs, at the beginning of the cold season in October. While passing through Afghanistan, the Pawindah caravans were fully armed and moved slowly, protected by moving Pawindah pickets. But when they crossed into Waziristan, the intrepid Mahsuds and Wazirs considered them their legitimate prey, and it was the duty of the scouts to protect the Pawindah caravans during their perilous journey through Waziristan. On arriving in the administrative district of D.I. Khan, the Pawindah tribesmen deposited their arms at the Murtaza Post before entering into Indian territory. The arms were returned to them on their return journey to Afghanistan the next spring, in April. The Pawindahs deposited their goods bought in Afghanistan in the town's Pawindah Sarai, where it was disposed of in bulk and purchased by Hindu moneylenders of the Bhatia caste. The Bhatias not only provided the Pawindahs with cash against their purchase of the Afghan goods but lent them extra money on favourable terms of interest.

Leaving their tents, families, and animals in the grazing grounds in the foothills of the NWFP, or further afield in the lush plains of the Punjab, the Pawindahs went generally to Calcutta, traversing the entire length of the Indian subcontinent. Thousands of Pawindahs would be seen crowding the

Hindu cloth merchants' shops in the 'Barra Bazaar' of Calcutta, purchasing cheap cloth for sale among the pliant peasantry of Bengal, on credit. When the Bengali peasants had gathered the three months' winter rice crop, the Pawindahs returned to the villages to collect their dues. They measured out the cloth on credit to any villager who wanted it on easy terms, demanding interest or delayed payment of about Rs 7 for the cloth worth Rs 5, a profit of 40 per cent, in three months.

They returned to their families in the Punjab and the NWFP in March/April and made preparations for their return to Afghanistan in caravans which were sometimes 40 miles long, seen plodding along on the D.I. Khan–Tank road. Before their departure for Afghanistan, the Pawindahs purchased foodstuff such as flour, sugar, tea, *gur* (jaggery), matches, salt, and kerosene oil—again through the agency of their friends, the Bhatias of D.I. Khan. They collected their weapons from the Murtaza Post and ran the gauntlet of the Gomal Pass through Waziristan, before entering the cold highlands of Afghanistan. In their mother country, the Pawindahs were mercilessly fleeced of much of their earnings by the Afghan officials. This cycle of seasonal migration of the Pawindahs of central and southern Afghanistan via the Gomal Pass and Ghilzais of northern Afghanistan via the Khyber Pass into India and back has continued through centuries. As a result, the Hindu Bhatias of D.I. Khan and the Muslim fruit merchants of Peshawar waxed rich on the earnings of the nomads of Afghanistan.

There appeared to be perfect understanding and mutual trust between the Pawindahs and the Bhatias of D.I. Khan and the wholesale cloth merchants of Barra Bazaar in Calcutta. No case of fraud or cheating was reported. The Pawindahs, on the other hand, ruthlessly collected their dues from the docile Bengal peasantry. A widely circulating story concerned a Bengali Muslim debtor of a Pawindah who had in the meantime died. His Pawindah creditor reportedly continued to beat his grave with his staff till the relatives of the deceased implored him to stop and paid off the debt.

Hazardous Treks for Exploring Dam Sites

We started work on the Tank Zam torrent from our base in Jandola in mid-August 1927. It took us about three months, till around the middle of November, to complete our work on the five dam and reservoir sites in that area. About half a dozen Bhittani *badragas* (armed tribal guards) accompanied the party while it worked in the Bhittani territory. However, when we later started work on the sites in the Mahsud territory, only one armed Mahsud lad, about seventeen years of age, accompanied our survey party as our guard. It turned out that the Mahsuds enjoyed great prestige amongst the rest of the tribes in Waziristan, namely the Bhittani, Wazir, and Sulaimankhel of the Dotani sub-clan who occupied the Gul Kachh area which was contiguous to Afghanistan. So much so, that no one from the rival tribes would dare to wrangle with even a single Mahsud lad for fear of incurring the anger of the entire tribe. The Mahsuds occupied a comparatively smaller area of South Waziristan, but they were constantly exerting pressure on the neighbouring tribes of Wazirs and Bhittanis, particularly the latter, by frequently encroaching on the lands of the weaker Bhittani tribesmen in that area. One day I jokingly asked our guard, the Mahsud lad, why his tribe was hankering after the Bhittani lands which were merely stony wastes. The lad smiled, put his right hand on his rifle, and said, 'We could graze our sheep and collect firewood from the bushes in this area!'

The site of the Shahnur Tangi Dam presented an awesome spectacle. The Shahnur torrent, the principal tributary of the Gomal, had cut through a narrow gorge only 40 feet wide at the bottom, with sheer, almost vertical rocks, of mudstone rising a few hundred feet on both sides. The site was situated in the territory of the Jalal Khel Mahsuds, noted for their reckless bravery. Before the construction of the Jandola–Sarwakai road along the left side of the gorge, the tract connecting Jandola and Sarwakai forts led along the left side of the gorge. On one occasion, in 1924 or 1925, three or four Jalal Khel Mahsuds hid themselves in the rock crevices at the narrowest point of the gorge and halted a brigade of the Indian Army. The brigade was marching along the bed of the gorge to provide relief at the Sarwakai Fort which was surrounded by a *lashkar* of hostile Mahsuds. The unerring shots from the rifles of the hiding Mahsuds are said to have kept the brigade at bay for a week. The defendants of the gorge sneaked away one night under the cover of darkness. The well-known zigzag method of retreat of the Mahsuds evoked admiration of strategists of the Indian Army operating in Waziristan. The retreating Mahsuds would start shooting from cover from a particular spot thus attracting enemy fire to that point. In the meantime, they would quietly slip away to another position in a zigzag direction and start shooting from the new position, diverting enemy fire to that position. And so on, until safety!

The main dam site, located in Gul Kachh which was at the heart of Mahsud territory, was vulnerable to attack and ambush. At last, the government arrived at a decision on the number and kind of military cover to be provided for the protection of the survey party during the survey of the Gul Kachh site. Gul Kachh fort—abandoned during the Afghan War of 1919, and believed to be in ruin—was situated on the bank of the Gomal River in its upper reaches in a wide and desolate basin, far from any habitation in a remote isolated region. The fort completely lacked all communication means. As the fort had not been re-occupied or even visited since 1919, it was a matter of speculation as to what kind of resistance the survey party might expect. It was decided that a force of 600 infantry soldiers and 50 horse soldiers of the South Waziristan Scouts under the command of Captain Johnson, assisted by Lieutenants Leeper and Fitzpatrick, with four machine guns, would accompany the survey party. In addition, a system of heliostat signals was established on peaks of some high hills en route, to maintain contact with the Gul Kachh force and the Sarwakai Fort, at a distance of some 30 miles. Homing pigeons (*Columba livia*) were also taken along to carry back news in case of any emergency. The Wana Brigade of the Indian Army was also moved from Razmak cantonment to Wana in South Waziristan, a distance of about 30 miles from Gul Kachh. The redeployment of the Wana Brigade was to provide backup to the comparatively small scouts force accompanying the survey party.

We received orders to proceed to Kachh. The march from Sarwakai to Gul Kachh and back would take six days. Eight days were allowed to the survey party for carrying out a quick reconnaissance survey of the dam and reservoir site at Gul Kachh, comprising an area of about 40 square miles, as estimated from the government's topographical survey plans. A couple of small monoplanes flew daily over the Gul Kachh area, dropping mail and other essential supplies in the camp and carrying back news of the Gul Kachh force. In addition to all the above military arrangements, political support for the survey campaign was secured through recruitment of about forty mounted tribal representatives of the Mahsuds, Wazirs, and Dotani Sulaimankhel tribes, through whose territories the force would pass. Those tribal Maliks were attached to me, as the head of the survey party. They also supplied mounts for me and other members of the party. We recruited forty sturdy young men from among the Marwat tribe residing within the administered districts of D.I. Khan and Bannu as survey *khalassis*. Tents, *chholdaris* (small tents), and other camp equipment were supplied by the chief commissioner's camping department at Peshawar. That unique gesture, by itself, showed the keenness of the government to ensure success of the Gul Kachh survey operations.

We recruited thirty camels from the Mahsud tribes, through the political department, to carry our personal belongings, rations, and surveying equipment. The reason for recruiting the Mahsud camel men was that the party had to pass through territories of the tribes which were hostile to each other. While a Wazir or a Sulaimankhel would hesitate to enter the territory of another tribe, a Mahsud would go anywhere and without fear! The scouts force recruited 100 camels from the same source. Both the survey party and the escorting scouts force assembled in the Sarwakai Fort. One early morning, our caravan left Sarwakai on the first leg to our long trek to Gul Kachh. The caravan of 130 camels was 2 miles long. Our progress was slow. That day's target was only about 10 miles, in the spin plain, in the vicinity of the scouts' Tanai Post—the last fortified post in that area. The planned camping site at the end of the march that day was at the foot of a solitary outcrop of rock rising in the middle of the plain. It was about 200 yards in diameter, and about 50 yards high in the centre. The rock was known as the Babar Ghundai after one of the pre-invasions probing excursions of the Mughal Emperor Babar in India, through that route.

The bulk of the infantry scouts marched in the front of the caravan, constituting its front guard. The cavalry disposed themselves as the flank and the rear guard. Whenever the column entered a gorge, batches of infantry from the front guard swarmed up the flanking hills and formed moving pickets. They moved along the ridges of the hills on both sides, parallel to the long moving column. Many a time, they disturbed a solitary deer or a pair of them who came crashing down the slope and, passing in front of the column in the valley, clambered up the opposite slope. At such times, the column stopped, allowing the deer to pass. Shooting was prohibited, lest it should alert the enemy, for the march was conducted in complete silence except for the sound of shuffling of the camels' and the pedestrians' feet and the sound of horses' hooves. I, along with my survey party and the tribal Maliks, rode at the rear of the column. Our survey *khalassis* also marched on foot at the rear.

At a narrow bend in a gorge, one of the long poles of my personal tent, supplied from the chief commissioner's camp equipage, got stuck in the vertical rock of the gorge. It was extricated with much trouble. Lt. Leeper, who was supervising the operation, lost his temper and looked at our survey party with exasperation. Both the young British lieutenants kept riding back and forth along the flanks of the moving column to ensure that everyone was in line. Noticing a pair of our survey *khalassis* carrying a crate of aerated water bottles, and a basket of fruit for me and my SDO, Lt. Leeper shouted at the two bewildered men in Pashto. 'Hey there, this is not Qissa Khwani Bazaar of Peshawar!' I observed with surprise that all the three British officers had donned the uniforms of common scout soldiers and had thereby rendered themselves indistinguishable from fair Pathan soldiers, many of whom had almost matching complexions. They wore the smart, standard uniforms of the scouts—baggy trousers with long shirt of the coarse grey cotton militia cloth, with the shirt kept in place by a tight fitting leather belt; a grey turban on the head wound around khaki cotton peaked cap; and sturdy leather *chappals*, smartly polished.

I was the only person in the column wearing a solo hat, a khaki linen jacket, breeches, and leather leggings over Jodhpur English shoes. When we set up the camp on arriving at the Babar Ghundai in the early afternoon, I was further dismayed to note that the three British officers were to occupy the inconspicuous *chholdari* tents of the common scout soldiers, while my spacious tent stood in the middle of the camp, an obvious target for any hostile force. The British offices wore the scout soldiers' uniform throughout the duration of the expedition. The site for the camp was selected close to the base of the Babar Ghundai rock outcrop. The camp was laid out according to a pre-arranged plan. A square with its sides measuring about 100 yards was marked on the ground. At each of the four corners 10-foot diameter lines for circular bastions were marked. A shallow trench was speedily dug around the perimeter of the square and the circulars bastions, and the excavated material thrown in front of the trench and dressed in the shape of breastwork. Large pieces of round boulders were collected from the neighbouring ground and laid at short intervals on top of the breastwork. Two 3-foot wide paths, laid crosswise, divided the square into four equal parts. *Chholdari* tents were pitched close to each other in all the four sections of the square. The entire task of erecting the camp took about two hours. As soon as the camp was ready, the trenches were occupied by a row of crouching scouts with rifles pointing outwards. The four machine guns were installed in each of the four corner bastions. A fortified picket was also established on the top of the neighbouring Babar Ghundai. Guards were changed in the trenches at two-hour intervals. Evening meals were quickly prepared and all lights were extinguished early in the night. The camp remained in complete silence and darkness throughout the night. This routine was observed throughout the expedition.

Very early the next morning, long before dawn, the camp was astir; a morning breakfast, and lunch were hastily prepared and, at sunrise, the column was on the march. It was a fine winter morning. Our path lay through the level Spin plain in the Wazir territory along a small stream running southwards in the middle of the plain, which was bone dry in its upper reaches, but developed a measure of clear flow, as we proceeded downstream along it. The Spin plain was about 10 square miles, surrounded on the west and east by bare low hills of a frowning black colour. On the north side, the plain was screened by a lofty range of hills which divided it from the Wana plain. The soil of the Spin plain was brown loam, free from the dark surface boulders, which is a common feature of the stony plateau of South

Waziristan. About 2-foot high grass with white blossoms covered almost the entire width of the plain, giving it its characteristic name 'Spin' (white).

About 8 miles lower down the Spin Nullah, now a perennial stream, we entered a 2-mile long gorge. At the upper entrance of the gorge there were ruins of a small post. Captain Johnson, commanding the column, informed me that the post in question was built during the time of the Mughal Emperor Jahangir, along the Gomal Pass route. I marvelled at the knowledge of local history of the British scouts officer. On clearing the gorge, the wide Gomal River valley suddenly burst into view, with the Toi Khulla Post 2 miles away. This was the Sulaimankhel territory. The Wazir country was left behind in the Spin plain. That realisation caused considerable relaxation of tension and anxiety. We camped for the day near the Toi Khulla Post situated on the high left bank of the Gomal River. The post was occupied by a few Khasadars of the Dotani sub-clan of the Sulaimankhel tribe that was settled within South Waziristan. The next morning, we started on the last leg of our trek to Gul Kachh. After crossing the shallow waters of the Wana Toi stream, which joined the Gomal River near Toi Khulla Post, our track led away from the riverbank, behind the low range of hills flanking it. A few miles further on, the track swerved back towards the riverbank, gently rising to a line of high ground. On reaching the top of the high ground, the magnificent Gul Kachh basin suddenly burst into view. The reality was grander even than what Mr Walker had visualised from the Government Trigonometric Survey (GTS) plan. The width of the basin between the high ground on which we stood, and the far high bank was estimated to be not less than 10 miles. The blue ribbon of the winter channel of the river between the existing low banks flowed in the middle of the basin along the lowest drainage line, about 5 miles away.

The Gul Kachh Post, situated near the far bank of the river, looked tiny and deserted in the distance. The space between the far-flung high banks, and the present low banks of the river in the middle of the basin had a series of silt terraces like a descending staircase and also some solitary silt mounds. They marked the former bed levels of the river, as it had gradually cut down the bed of the Gul Kachh gorge, across the range of low hills, which barred its passage. The blue ribbon of the river could be seen entering the upstream mouth of the Gul Kachh gorge. On the upstream side, the high banks of the river stretched away as far as the Warsak hills forming the boundary between Afghanistan and Waziristan. Beyond the far bank of the river, the flat Girdao plain of Balochistan stretched down to the horizon. Gul Kachh appeared to be a magnificent dam and reservoir site!

In the near foreground, just below the ridge of the high ground where we stood, were what appeared to be extensive ruins of a city covering a large area of the basin. The deserted city appeared to have towers, roofless houses, lanes, and open spaces. 'This is the city of the wicked,' informed our Dotani Maliks. And the high ground on which we stood was appropriately known as the *Kanzur Ware Narai*, or the 'Wicked City Pass'. Apparently, it was the old Jewish traditional doomed city of Lot [Prophet Lut (AS)]. The Sulaimankhel [the tribe of Prophet Sulaiman/Solomon (AS)] claimed to have Jewish origins. Actually, what appeared to be a city were the highly cut up remnants of former heavy silt deposits in the higher sheltered parts of the basin. The Gul Kachh basin was apparently a lake, prior to the erosion of the Gul Kachh gorge, by the Gomal River.

A hurried march down the basin brought us to the Gomal River. Its blue waters were fordable. Crossing the river, we pitched camp near the far bank, close to the ruined Gul Kachh Post. The extensive basin had fine loam soil and was cultivable. However, except for bushes and tall grass, there was no sign of cultivation. The basin was absolutely deserted. That evening, I informed the scouts that the next day the survey party would start work in the Gul Kachh gorge, the upstream entrance of which was 4 miles away. The next morning, when we arrived at the gorge, we found groups of scouts already occupying the ridges of the hills flanking the gorge for over a mile under the command of Lt. Fitzpatrick.

I selected the narrowest section of the gorge as the tentative dam site. Taking a survey *khalassi* along with a flag, I swarmed up the right slope of the hill flanking the proposed dam site and told the SDO to climb up the opposite slope. We fixed the flags on the hill slopes at an elevation of about 400 feet from the bed of the gorge, for taking a cross section of the gorge at that point. Lt. Fitzpatrick was watching our rapid progress up the hill slopes while comfortably seated in the nullah bed below. On his return to the camp that evening, he narrated the incident to Captain Johnson, adding that the members of the survey party had worked all the time like *jinns*. That incident considerably raised our prestige with the scouts who, hitherto, we suspected had regarded us with a kind of disdain.

After taking the longitudinal profile and cross sections of the gorge, we quickly emerged into the vast reservoir site of the basin. There we laid a levelled base-line running parallel to the course of the river and took cross-sections of the basin at short intervals. In taking cross-sections of the flanking high silt mounds we adopted the theodolite-tachometer method which I had developed during similar work on the survey of the Kalabagh right bank canal. The survey job at Gul Kachh was thus completed within the specified period of eight days. One evening, returning from work, we found the Gul Kachh basin bustling with life. A very large caravan of Pawindahs of the Sulaimankhel tribe from Afghanistan, on their way to India in the course of their annual winter migration, had set up their temporary camp in the lower Gul Kachh basin. Thousands of black tents had sprung up on the opposite bank of the Gomal River close to our fortified camps. Herds of camels, horses, sheep, and goats, jealously guarded by fierce Afghan dogs, were placidly grazing on the tall grass in the basin. A group of Pawindah men, 'amongst the finest specimens of humanity' (according to a Norwegian lady author who visited Afghanistan) were strutting about, fully armed. They were sturdy and picturesque, clad in embroidered waistcoats, white voluminous turbans, long pleated shirts, and baggy trousers. Gul Kachh basin was a part of Waziristan and, therefore, of India, but it was under the occupation of the Bohtani sub-tribe of the Sulaimankhel. The Pawindahs, therefore, considered Gul Kachh as part of their own homeland.

Rows of comely Pawindah women lined the bank of the Gomal River, filling their water buckets from the limpid stream. Their fair faces, arms, and hands were ruddy from exposure to the weather and the hard task of loading and unloading camels, and helping to set up the tents and performing other household chores. Some of the young women were strikingly beautiful. The Pawindah women offered chickens and eggs for sale, which were eagerly purchased by the young men from my survey staff, who bargained merrily with the young Pawindah girls, and even cracked jokes with them. On hearing about this, Captain Johnson quietly advised me to caution my staff against taking liberties with Pawindah women. 'We are, at the moment, surrounded by 5000 armed Pawindah warriors.' he said. 'Should they take offence and attack us, we shall be wiped out before the Wana Brigade can move out its barracks!'

One dark night, the Pawindah Maliks arranged a dance show of young Pawindah lads for our entertainment. A huge bonfire was lit near our camp. Large crowds of Pawindah men and women gathered around the bonfire in a wide circle, the men occupying one half of the circumference, the women the other half; young girls sat in the first row with the older women standing behind them, many rows deep. Chairs were placed for the British scouts officers and for me and my SDO, inside the circle. A dozen or so sturdy and comely Pawindah youths, in their late teens, clad in embroidered vests, white shirts, and baggy trousers, with haughty looks and glittering kohl-lined eyes, entered the arena and started a rhythmic dance in a circle around the bonfire. They bent forward, stood up and moved around in frenzy with increasing tempo to the beat of the drums, their bobbed tresses moving forwards, backwards and sideways in unison with the tempo of the drumbeat. It was a thrilling show enacted in the remote wilderness, with the fire illuminating the shining faces of the youthful dancers, and its glow reflected from the pretty faces of the young Pawindah maidens sitting in the front row and watching with wistful eyes.

One morning, Captain Johnson invited me to visit the Afghan border which was only about 10 miles away from our camp to the west. We would have lunch there and also examine the boundary pillar no.1 of the famous Durand Line which formed the boundary between India and Afghanistan. The party consisted of Captain Johnson, Lieutenants Leeper and Fitzpatrick, escorted by the entire contingent of mounted scouts, and myself. In addition, about half a dozen of the Dotani Maliks formed my personal bodyguard. They were, in fact, politically responsible for the safety of our entire force at Gul Kachh. The scouts officers were mounted on their indiscernible ponies, while I and my Dotani bodyguard had local Dotani horses.

Crossing the wide bed of the Gomal River near our camp, we climbed up the north high bank of the river and, gaining the high ground of the hill bordering the Zermlan plain, we rode along the north bank all the way to the Afghan border, which at that point comprised a low range of limestone hills. The Gomal River burst through the thin line of hills into Waziristan across a gap of about 100 yards, through which the flat loess plain of the Afghan highlands was visible up to the skyline. The Kunden tributary of the Gomal also joined its parent river from the south at that point. The pillar no.1 of the Durand Line was located in the Waziristan plain on the west bank of the river about 100 yards away from the foot of the hills. At the site of the pillar, there was now only a 4-foot square depressed foundation. The pillar itself had disappeared.

We took our lunch while seated on a narrow limestone ledge at the foot of the hill, well within the Afghan territory. After lunch, a number of group photographs were taken at the lunch site, against the picturesque background of the limestone cliff. The group comprised, besides the British scouts officers and me, a number of non-commissioned Pathan officers of the scouts force. In one of the photographs taken by Captain Johnson, I was invited to sit in the middle of the group. After the photographs were taken, Captain Johnson proposed to re-build the pillar no.1 of the Durand Line, and ordered the scouts to collect large boulders for the purpose. Instead of re-building the pillar on the original foundation, however, he selected a new site about 100 feet further inside the Afghan territory. Noticing my enquiring looks, he smilingly remarked: 'Britannia never retreats, it always forges ahead!' True to the discipline of the force, the scouts obediently erected the pillar at the new site. A photograph of the scouts force and the British scouts officers was taken by Captain Johnson around the new boundary pillar. I was not part of that group and stood aloof along with my Dotani bodyguard. The Dotani Maliks, however, audibly murmured their displeasure at Captain Johnson's illegal act and vowed they would remove the new pillar. They had my full sympathy. We, Indian Muslims, took pride in King Amanullah Khan's free Afghanistan.

12 Across the Frontier into Balochistan

A Long Walk to Fort Sandeman

As the day of our departure from Gul Kachh was drawing near, I was given my next task, i.e. to survey the dam and reservoir sites on the Zhob River in Balochistan, in the vicinity of Fort Sandeman. Zhob River, the principal tributary of the Gomal, joined it from the south, at Khajuri Kachh, about 25 miles downstream of Gul Kachh. Captain Johnson strongly recommended that I should take a 24-mile long diagonal short cut through the Girdan plain which would take me to the Zhob Levy post of Mir Ali Khel on Zhob River. From there, I could reach Fort Sandeman by a leisurely trek up the Zhob valley, a total distance of only about 70 miles from Gul Kachh. Otherwise, the distance by the alternative route by rail and road via D.I. Khan, Sukkur, and Quetta would be 700 miles. I agreed, and so a day before our departure a contingent of 100 Zhob Levies of the Afridi tribe, under the command of a Pakhtun non-commissioned officer, arrived at Gul Kachh to escort my survey party to Mir Ali Khel in the Zhob valley.

The thirty Mahsud camels that were hired at Jandola to carry our personal luggage and surveying instruments to Gul Kachh now departed, while the scouts force also returned their camels. The Zhob Levies brought along 100 bullocks to carry their own as well as our luggage to Mir Ali Khel. On the morning of the day of departure, the troops of the South Waziristan Scouts and the Zhob Levies arose long before dawn. The camps were struck and the baggage was loaded on the camels and bullocks before sunrise. Both the forces held a joint parade in that eerie wilderness (the Pawindahs had since left) and saluted each other, according to the British military custom. Farewell bugles were sounded, and the two forces separated and departed for their respective destinations. The Zhob Levies had not brought any riding ponies for me and my survey staff. Their Pathan commander alone had a mount. The reason given was that all their ponies were in service at Fort Sandeman because of the state visit of Lord Reading, the viceroy of India, to Balochistan. The levies commander and I rode his horse by turns, the rest of my survey staff walked on foot. Nevertheless, we left for Balochistan in high spirits conscious of having successfully accomplished an arduous task.

The Gomal River formed the boundary between the NWFP and Balochistan. Ascending the south high bank of the river we entered the vast, flat Girdan plain of Balochistan. The plain had light brown, good quality loess soil, but was bone dry. At midday, having travelled half the way, about 12 miles, we stopped for a hasty lunch, with the Afridi commander urging us to hurry in order to place as much distance between Waziristan and ourselves as possible. 'The Wazirs and the Mahsuds are ruthless marauders,' he said. Somewhat unheedingly, we consumed most of our drinking water supply with our lunch and, in consequence, suffered from thirst during the remaining half of the journey. In the last few miles our path lay along a narrow, dry valley flanked by frowning jet-black hills, known as the Girdan Tangi. We trudged wearily under the sinister shadow of the black Tangi. As the sun was nearing the western horizon, we suddenly burst into sunshine and the Zhob valley. The limpid bluish-green waters of the Zhob River flowed placidly in the boulder bed of a channel about 50 yards wide and about knee deep. We rushed into the Zhob channel and quenched our burning thirsts with its water, despite its strong, oily smell, and saline taste. The Mir Ali Post built on high ground on the far bank was a welcome sight. At Mir Ali Khel, we exchanged camels for bullocks for carrying our baggage.

The route to Fort Sandeman lay along the east bank of the Zhob River along the Zhob gorge. The Zhob valley was considered comparatively safe, so from Mir Ali Khel onwards only half a dozen foot soldiers of the Zhob Levies accompanied us up to Fort Sandeman. Riding horses being still unavailable, we had to trek the rest of the journey on foot. On the third day of our march from Mir Ali Khel, a fine wintry day, the Zhob gorge opened out into the plain of Fort Sandeman at Viala, 11 miles from our destination. The hills on the east side of the gorge receded into the distance. At that point, the Kapeep torrent met the Zhob from the east in a broad valley. At noon, our foot-sore and weary caravan reached the outskirts of Fort Sandeman, in a straggling line about half a mile long, carrying our casualties (one case of pneumonia, two of malaria) on camels.

The Balochistan administration had received instructions from the Government of India, to meet the executive engineer, Waziristan Survey Division, NWFP PWD, and his party at Fort Sandeman, with due courtesy. So, we were agreeably surprised to meet half a dozen officials of the Balochistan administration outside Fort Sandeman. They were headed by an assistant political agent, a tall Pathan with rosy cheeks and a pronounced paunch, wearing a fur coat and karakul cap. The officer was evidently not impressed with the sorry-looking caravan from the NWFP. He informed me in somewhat lofty tones that the political agent at Fort Sandeman was expecting me for lunch. 'But what about my party?' I asked. 'They will also be taken care of,' he said. Piqued by his patronising tone, I curtly said: 'Look, I need an ambulance immediately to transfer my casualties to the hospital and I must first personally see to the proper lodging and feeding arrangements for my party before I can meet the political agent.' 'But it is past one o'clock and lunch time, and the political agent is waiting for you,' he said, somewhat taken aback. 'The lunch can wait,' I said. It was 2:30 p.m. when I ascended the winding road along the hill, leading to the castle built on its summit—the famous Fort Sandeman, the official residence of the political agent. It was an imposing stone structure built in the style of an English country house, originally for the residence of Captain Sandeman. Fort Sandeman town was built around the solitary hill crop crowned by the castle.

Khan Bahadur Sharbat Khan, the political agent, Fort Sandeman, an Afridi Pathan from the Kohat district of the NWFP was patiently waiting for my arrival. On learning the cause of my delay, he fully endorsed my action. His family being away, he was living in the castle with only his young and handsome son, whom he introduced to me. The young man had recently finished his medical studies in the UK and had been inducted into the prestigious Indian Medical Service (IMS). He retired as a colonel from the IMS and, later on, after the creation of Pakistan, served as vice chancellor of the University of Peshawar.

I was impressed by the luxurious decorations and furniture in the castle. The house was literally a museum of priceless Persian carpets. Sharbat Khan informed me that only a few days ago, Lord Reading had been his guest and the house was exactly in the same condition in which it was during the visit of the viceroy of India. I remained Sharbat Khan's guest for a couple of days in order to give rest to the survey party, which quickly revived under the care of our hosts. With some reluctance, Sharbat Khan gave me leave to depart. His gracious hospitality and solicitous concern for me and my party were quite overwhelming. He also seemed pleased to be in the company of a fellow Pathan officer, with whom he could converse openly and we became very good friends.

From Fort Sandeman, we moved to the Badin Zai Dam site 10 miles upstream of Fort Sandeman and set up our camp in the rest house. By this time, my survey staff had acquired sufficient skill in the work and so I had ample time to shoot black partridge in the foothills. From Badin Zai, we moved to the Brunj Dam site 12 miles downstream of Fort Sandeman. At Viala, the deep part of the Zhob River teemed with small fish of the Rohu and

Mahseer varieties. They were evidently ignorant of baited lines. We caught thirty or forty small fish in as many minutes using kneaded flour as bait.

Having finished our job in Balochistan we returned by easy stages down the Zhob valley to Khajuri Kachh Post, at the junction of Zhob and Gomal River in Waziristan. During the last stage of the Zhob River, between Mughal Kot and Khajuri Kachh posts, we were escorted by only half a dozen Mahsud *badragas* because the area was considered hazardous for the Zhob Levies. I asked one of our Mahsud escorts where he got the excellent rifle that he was carrying. With pride, he said that it was a part of the loot of the arms which the British army had abandoned in their hasty retreat from the Wana Fort in Waziristan, during the Afghan War of 1919. From Khajuri Kachh we took a shortcut of 18 miles to Sarwakai Post and so back home to our headquarters at D.I. Khan.

On my return to D.I. Khan, I found official orders waiting for me; according to the orders my division, The Waziristan Survey Division, had been upgraded to Waziristan-cum-Balochistan Survey Circle [hereinafter referred to as Circle]. Mr Arthur Oram, ISE, from the Punjab irrigation, was made in charge of the Circle as superintending engineer. He had already arrived at D.I. Khan, the headquarters, and taken over charge of his new duties. I was assigned new responsibilities of field investigations and surveys of dam and reservoir sites in Waziristan and Balochistan and establishing a system of stream gauges for measuring the discharge and runoff data of the streams in this area. Mr Oram's duties were the design of dams, for which he was assisted by office staff.

I accompanied Mr Oram during his next inspection tour of the various dam and reservoir sites on the Tank Zam River in Waziristan, whose survey we had earlier carried out. We made our temporary headquarters in the Jandola rest house from where we could conveniently inspect the various sites. Major Hey, political agent of South Waziristan, usually accompanied us. One day at the Jandola Fort, a party of Mahsud tribesmen of the Abdur Rahman Khel sub-clan of the Jabal Khels brought some problems relating to the irrigation of their lands, to the political agent. They occupied a strip of territory along the Shahur torrent, a tributary of the Tank Zam. The fields in question were situated at a place called Chagmalai about 5 miles from the Jandola Fort alongside the branch road that led to the Sarwakai Post. We promptly went to inspect the site. The party included Major Hey, his Pathan assistant political officer, and political tehsildar (a Pathan subordinate official, responsible for day-to-day dealings with the tribe). We drove in our own cars, while the Mahsuds and the scouts followed in the scouts' bus. Since the problem was related to irrigation, Major Hey referred the matter to Mr Oram, who promptly passed it on to me.

The Mahsud Malik explained the problem to me in Pashto, drawing imaginary lines in the air with his curved knife, a miniature scimitar, to illustrate his points, repeatedly bringing the knife perilously close to my nose in the process. I, however, maintained an unconcerned attitude, as if it were the most natural thing in the world to hear the details of a problem explained with the help of a curved knife pointed at one's face. The obvious solution was to construct a stone and brushwood temporary diversion weir across a neighbouring dry hill torrent, and divert its flood waters during the scanty summer monsoon season, on to the fields. I explained the solution to the knife-brandishing Mahsud Malik. Impressed by my unconcerned attitude, and observing that I was treated on equal terms by Major Hey and Mr Oram, while the assistant political officer and the tehsildar were respectfully keeping their distance, the Mahsud Malik asked Major Hey who I was. The Major again referred the matter to Mr Oram. The latter, at a loss to explain my status, had a bright idea. He said that I was a 'Major'. The tribesmen were duly impressed. The Mahsud and Wazir tribesman evidently respected only the political and army officers, in Waziristan.

On a cold January morning in 1929, at D.I. Khan, I donned my warmest clothing and, along with Mr Oram who was clad likewise, climbed into the front seats of a lorry that was to take us to Sarwakai Post in Waziristan, 80 miles away. The SDO, our servants, along with our cooking utensils and extra heavy beddings, were crowded in the rear of the lorry. A few miles beyond Sarwakai, we were to get down from the lorry and make a round trip of the Spin plain on horseback, accompanied by half a dozen Wazir Maliks who owned the plain, for investigating the possibilities of irrigating the fertile plain, which was lying absolutely barren under a perpetual covering of tall grass with white blossoms. The trip was being undertaken in response to the request of Major Williams, the political agent of South Waziristan. At the end of the tour of the plain, we were to pass the night in the scouts' camp which had been temporarily set up for our benefit at the head of the plain under the foot of the Karab Kot peak. Major Williams would be waiting for us at the camp.

As our lorry started on our long journey, we noticed with apprehension a fairly strong breeze blowing from the west—the dreaded Gomal. In winter, a strong cold breeze, arising in the cold uplands of Afghanistan and Waziristan, sometimes blows for days on end, through the funnel-like passes in the hills, affecting large parts of NWFP. The Gomal wind was so strong that its effect was felt as far as the Indus River 50 miles away from the

Murtaza gorge, forming the mouth of the Gomal River. So much so, that the breeze helped the navigation of the country boats upstream along the river by hoisting the sails. I was familiar with the Spin plain, having traversed it twice during the course of my annual trek to Gul Kachh. In days gone by, a karez[1] from the Wana Toi stream cut across the low hills, in the north-west, and irrigated the Spin plain. The karez, about 4 or 5 miles in length and mostly in the form of horizontal tunnel had long since choked up and was out of commission.

On the other hand, a small local drainage, starting as a dry depression at the head of the plain, took definite shape as it flowed southwards towards its confluence with the Gomal River, at Toi Khullah. In the lower reaches the Spin Nullah, it became perennial with a considerable permanent discharge of 10 or 12 cusecs, which was enough to irrigate the whole plain, if it were possible to tap it upstream. The local Wazirs thought that instead of going to the huge expense of putting the lengthy tunnel-karez from the Wana Toi stream, it would be cheaper to tap the perennial discharge of the Spin Nullah itself at some point upstream by constructing a new karez. Major Williams wanted our professional opinion on the proposal of the Wazir Maliks.

We arrived at Sarwakai Post at 10:30 a.m., making slow headway against the wind which was constantly increasing in intensity as we progressed forward towards our destination. At Sarwakai, the wind was blowing with the fury of a blizzard. A political official braved the storm to come out of the post to give us information about the whereabouts of the Wazir Maliks who were waiting for us somewhere ahead on the road. The political official could scarcely make himself heard from a distance of only 5 feet from the lorry. About 10 or so miles further on, the Wazir Maliks, along with their ponies, were hugging a high road cutting to take shelter against the driving wind. We had our lunch within the lorry and then emerged to start a nightmarish journey on horseback in a blinding windstorm. Only one decent pony had been sent by the scouts for Mr Oram's use. The SDO and I had to mount two sorry-looking nags that had been brought for us by the Wazir Maliks.

Our journey started from the upper part of the plain near the Babar Ghundai. We rode in the downstream direction for about 10 miles along the Spin Nullah, and then we crossed over. From there, the road wound 4 miles up to the western foothills of the plain, and then back towards the head of the plain till we reached the old tunnel karez at the north-west corner of the plain. From there we followed the derelict karez, a distance of 3 or 4 miles till its point of emergence into the plain. And finally, as it grew dark, we rode up to our camp. Our progress had been slow owing to the blinding dust filling our eyes and the blizzard chilling us to the bone. And all the time the tiresome Wazir Maliks had been unceremoniously jostling past us to explain their views to Mr Oram who was riding ahead. It was quite dark when we reached the scouts camp. We were dead tired, shivering with cold and longed for a hot cup of tea and our dinner. But we were amazed to find our kitchen tent cold with no fires burning, and no tea or dinner as there was no fuel wood in the camp.

My enterprising cook, however, managed to obtain some fuel wood from Mr Oram's and Major Williams's cooks and so brought me a kettle full of steaming hot tea. He also set about preparing chicken curry and *chapatis* for my dinner. The SDO and the assistant political officer who had overheard my talk with my cook from their tents, both coughed meaningfully. I invited them to have tea and dinner with me, which they gratefully accepted. The assistant political officer vowed to punish his subordinate first thing in the next morning, who had failed to supply fuel wood and had found refuge in the high picket overlooking the camp. After dinner we crept into our heavy beddings fully dressed, but were shivering with cold all the same. Heavy rain had come down in the meantime adding to the blizzard, which was still raging with unabated fury. I do not remember when I feel asleep. When I awoke the next morning the weather was calm, the golden sun was peeping above the eastern horizon, in a turquoise blue sky. And the whole scene outside had changed. The mountains and the plain were covered with a mantle of snow. The mortar-shaped Karab Kot peak had a hoary covering of snow, looking like the bald head of an old man with snow white sidelocks. Mr Oram and Major Williams emerged from their tents. Mr Oram explained, to the Major, our joint conclusion, reached the previous day during our tour of the spin plain, that the perennial flow in the Spin Nullah in its lower reaches was due to seepage from the extensive alluvial flat of the spin plain; that there was no underground stream in the upper reaches of the plain near Babar Ghundai to be tapped; and the only practicable means of irrigating the plain was to put the tunnel karez from the Wana Toi stream in order.

Mr Oram and Major Williams left for Wana Post and I and the SDO prepared to return to D.I. Khan. But the assistant political officer asked us to see a *tamasha* (show) before our departure. He called the political *muharrir* (the culprit who was responsible for the breakdown in our fuel wood supply the previous evening) from the perch of the picket. Despite his protests, the poor fellow was stripped naked above his waist, and fistfuls of snow were rubbed on his head, face, and torso. And so we returned to D.I. Khan quite amused by this unconventional punishment.

A few days later, I received instructions from Mr Oram to accompany him back to Fort Sandeman. I joined him to discuss our travel arrangements with Major Hey for our journey along the Gomal valley, from Murtaza to Khajuri Kachh at the junction of Zhob and Gomal. That part of the journey was along a hazardous route which lay in South Waziristan. As was quite customary among British officers, Mr Oram, with me in tow, walked unannounced into Major Hey's drawing room. The major was at the time scolding his Indian bearer for some dereliction in duty. He casually nodded to Mr Oram but took no notice of me. Mr Oram, in undertones, apologised for his host's behaviour. 'Don't mind him,' he said. 'He is angry with his bearer.' After having breakfast with Major Hey, during which he thawed somewhat, we left by car for Murtaza Post, where the Gomal River emerges from the Waziristan gorge into the vast D.I. Khan plain.

There, an escort of a contingent of mounted South Waziristan Scouts was waiting for us along with two thoroughbred horses, one for Mr Oram and the other for me. We pushed along in the Gomal valley. Our route was flanked by stark low hills of conglomerate and fantastically coloured bright red and green shales. We stopped for the night in the small Niti Kachh Post, 10 miles from Murtaza, which was perched on a high spot on the bank of a dry boulder-strewn side nullah, flanked by bare shale cliffs. The next morning, we descended from the post into the nullah bed for resuming our next leg of the arduous 15-mile journey to the Khajuri Kachh Post. From there we were to follow the Zhob valley up to our destination, Fort Sandeman. A Wazir caravan was coming down the nullah bed, from the Khajuri Kachh side, bound for Murtaza, and the plains of D.I. Khan. Before mounting our horses, we let the caravan pass. A beautiful young Wazir girl formed the vanguard, walking about 50 paces ahead of the caravan. The girl was breath-takingly beautiful. Mr Oram looked at her and fell into a deep trance, which lasted until the whole caravan passed silently before us.

The old road running deep down the bank of the winding Gomal gorge between Murtaza and Khajuri Kachh, having broken down at numerous places, followed the circuitous route along boulder-strewn and trackless dry streambeds flanked by bare and almost vertical cliffs of shale leading up to the *Gor-lora-Liar Narai*—literally the ox-entrails pass (so named because of the existence of black and white forms in the grey limestone rock of the pass). The pass overlooked the Khajuri Kachh Post. There was a steep drop of a couple of miles down to the post. The Gomal gorge, the historic Khajuri Kachh dam site, also cut across the Gos-Lara range, a couple of miles to the west of the pass. About midway between Khajuri Kachh and Nita Kachh, was the Dozakh Tangy (hell gorge) of the Gomal River, which was reputed to have hot springs, with the water charged with sulphur. We made many unsuccessful sorties from our route, down to the Gomal gorge to locate the fabled Dozakh Tangy, without success. At sundown, we reached the Khajuri Kachh Post, utterly exhausted by the full day's rough ride and, to our pleasant surprise, a sumptuous feast of lamb *pulao*, barbecued *tikkas* and potato curry, with hot *naans* and *halva* for dessert, was waiting for us. Our hosts were the South Waziristan Scouts who had killed a large fat-tailed sheep in Mr Oram's honour. Both Mr Oram and I ate with relish and went to bed before sundown in the only surviving room of the roofless post, which had been destroyed by the Mahsud tribesmen during the retreat of the British forces from Waziristan, during the Afghan War of 1919.

The next morning, we started early for the Mughal Kot Post on the Zhob River, 15 miles from the Khajuri Kachh Post. During that stage we were also escorted by the South Waziristan Scouts because that part of the country was dangerously near Mahsud territory, and only the scouts force was considered competent enough to cope with them in case of an ambush. From Mughal Kot onwards, as far as Fort Sandeman, we were escorted by a mounted escort of the Zhob Levies. The next day we left Mughal Kot for the Mir Ali Khel Post. It was now mid-winter, January of 1928. The surface of the Zhob stream had a thin coating of transparent ice under which the water of the stream flowed swiftly. Our feet grew numb with cold while riding. So, we got down from our horses at intervals and jumped about on the road to restore circulation in our semi-frozen feet.

On arrival at Fort Sandeman, Mr Oram and I immediately called on the political agent of Fort Sandeman at the castle. Khan Bahadur Sharbat Khan had since been transferred and a British army major of the Indian Political Service (IPS) had only recently replaced him. Mr Oram had to stay in a small rest house and was served with meals by the cook of the rest house, because the wife of the British political agent declared that, as they had just moved, she could not take in a guest. I, however, fared far better. I was promptly taken charge of by Jaffar Khan, the assistant political agent, a bright young officer of the IPS, a scion of Afghan royal family. Jaffar Khan lodged me in the luxurious guest room of his bungalow, which was situated on a minor hilly outcrop in the vicinity of the castle, and entertained me lavishly. He sent me a fur overcoat as protection against the freezing cold of Fort Sandeman. In view of our recent arduous journey through the Gomal and Zhob valleys, immediately following the six months' survey work in Waziristan, I was tired and requested for my return via the longer but more comfortable bus and railway route through Sibi, Sukkur, and Multan. Mr Oram, who was staying longer at Fort Sandeman, readily acceded to my request.

The bus route lay from Fort Sandeman to Loralai and then via Sanjawai to the rail-head at Harnai station situated in the valley of the Nari River. Jaffar Khan had phoned Amir Ahmad Khan, the assistant political agent at Loralai station to receive me, and to arrange for my road transport onwards to Harnai railway station. Amir Ahmad Khan was a Qizilbash Pathan of the Peshawar district. He welcomed me cordially into his house. After dinner, we remained absorbed in conversation far into the night, for he was anxious to hear news of his home district, until his wife tapped at the connecting door reminding her husband that it was past bedtime. Amir Ahmad Khan strongly urged me to see Quetta town, the headquarter of the Balochistan administration and the only large town in the province, before returning. He had his own house in Quetta. It is a pity I did not avail of his generous offer to see Quetta town then in its heyday. The town was subsequently totally destroyed in the earthquake of 1935 with 25,000 fatal casualties.

13 Government of India's Inspection

In the late fall of 1928, I was instructed to take my survey party again to Gul Kachh for carrying out more detailed surveys of the dam and reservoir site at Gul Kachh which could serve as the basis for the design of a dam there. The previous year, on my return to D.I. Khan, I had confidentially reported on the feasibility of connecting the provinces of the NWFP and Balochistan through a road link, Tanas Toi–Khulla, Gul Kachh in South Waziristan and Gul Kachh–Mir Ali Khel in Balochistan, and a road bridge across the Gomal River at Gul Kachh. A couple of young British RE army officers accompanied us this time.

Our second march to Gul Kachh was precisely on the lines of the previous year's march consisting of the usual escort of mounted and foot scouts, machine guns, heliostat equipment, homing pigeons, and my Wazir and Dotani Khasadars. Like the previous year, we stopped en route at Babar Ghundai and Toi Khulla. Since the route was now familiar to us, we marched in high spirits, enjoying the adventure. The main difference was that I had discarded the high two-pole tent for myself for a small *chholdari* as used by the British military officers and the common scout soldiers. Another difference was that our new commandant was Captain Scotland, a somewhat burly officer with no noticeable waist, who heroically walked all the way to Gul Kachh to shed some pounds off his middle.

After completing our survey work at Gul Kachh, we marched back for the survey of the well-known Khajuri Kachh Dam and reservoir site on the Gomal River, below the confluence of the Zhob with the Gomal River. Our scouts escort at Gul Kachh had marched back to Sarwakai, leaving us in charge of only half a dozen Mahsud Khasadars. While in Mahsud territory, we were considered safe even with a small Mahsud escort. We now had ample time to examine the damage done to the Khajuri Kachh Post by tribesmen during the Afghan War of 1919. All the rooms of the post, but one, were roofless. The wood and ironwork of the doors, windows and rooves had been removed. On the walls of the ruins were inscribed in crude Persian the words, 'destroyed by the brave "Ghazis" of Islam'.

After completing our survey work at Khajuri Kachh we decided to return to D.I. Khan via the Sarwakai Post, situated northwards 18 miles from Khajuri Kachh across the east-west oriented Gomal valley. Sarwakai, moreover, was the nearest road-head from Khajuri Kachh. While a few miles from Khajuri Kachh, we were startled to hear sporadic rifle shots emanating from concealed positions on the uneven terrain from a distance of only about 200 yards to the west of the track we were following. Our Mahsud guide stopped and carefully surveyed the area in question. He concluded that a pitched battle was raging between a party of armed Pawindah migrants from Afghanistan, and the Mahsud raiders. On calmly assessing the situation, our Mahsud guide quickly led our mounted party down in the bed of a nearby dry hill torrent. We crawled quietly along the boulder strewn torrent bed for some miles under the comforting cover of the cliff which formed the west bank of the torrent until our Mahsud guide was satisfied that we were out of danger and the sound of rifle shots had ceased. We then emerged from the nullah bed and were led back on to the Sarwakai track which ran parallel to the torrent bed. So, we returned to Sarwakai Post, and back home.

During the fall of 1929, on the request of the NWFP government, the Government of India deputed Mr D.N. Wadia (superintendent geologist) to examine and report on the proposed dam and reservoir sites on the Tank Zam, Gomal, and Zhob rivers for which surveys had been carried out by our Circle. Mr Wadia was a Parsi from Bombay. He had the reputation of being one of the ablest geologists in India. The reputation was well merited, because his book *Geology of India* (1919) is considered a classic, and a standard work of reference.

The Parsi communities of Bombay and Karachi were among the most cultured and well-to-do communities in India. Despite the opposition of the Government of India, who were dead set against industrialisation in India, Tata,[1] a wealthy member of the Parsi community, had started the first steel mill in India.[2] A number of cotton textile mills had followed at Ahmedabad near Bombay, in competition with Lancashire in England. Every member of the Parsi community shared in this proud achievement. Mr Wadia, though of frail constitution, had nevertheless great dignity and dedication, which was due, no doubt, to his unusual proficiency in his profession. For instance, one day while we were riding in the Gomal valley, bound for the Gomal Tangi dam site at Khajuri Kachh, Mr Wadia, a poor horse rider, was going very slowly on an old Mahsud pony. 'Wadia, go a bit faster please,' said Mr Oram. 'This is a dangerous locality, and you are delaying the progress of the whole party.' 'But I prefer to go slow,' said Wadia, to the evident dismay of the mounted scouts' escort. Likewise, when we arrived at the Gomal Tangi dam site at Khajuri Kachh, Mr Wadia forthwith bent down to examine the rounded shingle and boulders lying in the riverbed. 'Look up Wadia,' said Mr Oram, pointing to the awe-inspiring limestone cliffs flanking the Gomal Tangi gorge. 'But I prefer to examine the shingle,' snapped Wadia.

The tools which Mr Wadia carried along during his tours comprised a hammer, an altimeter, a magnifying glass, a pair of binoculars, a note book, a 'one inch is equal to one mile' Government Trigonometric Survey (GTS Plan of the area in book form), and a supply of coloured pencils. As he went along, he marked on the GTS Plan, in colour, the successive geological periods of the rock formations, thus completing the surface geology of the area we passed through. I asked Mr Wadia many questions relating to geology during our joint tours of inspection of the dam sites on the Tank Zam, Gomal, and Zhob rivers. He generally replied, treating me as a layman. For instance, when I pointed out to him the brilliant scarlet and green shales in the Gomal valley near Nili Kachh Post, Mr Wadia said that the colour in a rock had little or no significance. He usually determined the period of a rock by examining its fossils with the help of his hammer and magnifying glass. Noticing a large vertical fault in the flanking limestone cliff of the Gomal Tangi gorge at Khajuri Kachh, Mr Wadia volunteered the information that the tremendous pressure generated against the sliding surfaces of the rock during the formation of a 'fault' sometimes transformed the base elements of the rock surfaces into gold! From Mr Wadia I learnt the sequence of rock formations across the 50-mile width of Waziristan, from the Siwalik conglomerates of the eastern hills forming the common boundary between Waziristan and the NWFP districts of D.I. Khan and Bannu, and the Jurassic limestones of the western hills forming the common boundary with Afghanistan. Mr Wadia's report on the dam sites in Waziristan was generally favourable.

The project was consequently submitted to the Government of India for allowing sanction for the construction of check dams and delay reservoirs at Gul Kachh and Khajuri Kachh. During the fall of 1929, Mr Harris, consulting engineer to the Government of India, came on tour to inspect the proposed dam sites on the Gomal River in Waziristan. Mr Walker (chief engineer), Mr Oram (superintending engineer), I, and Ahmad Bakhsh (sub-divisional officer), accompanied Mr Harris during his tour in Waziristan. The tour started from the downstream end of the Gomal gorge at Murtaza. We passed the first night at the Murtaza Post. The next morning after our luggage had been laden on the Mahsud camels for the journey up the Gomal valley to Khajuri Kachh site, the camel bearing Mr Walker's commode had to be brought back to the Murtaza Post for him to use the commode. That episode delayed our departure, and I suspect must have also irritated Mr Harris.

A few miles further on, a weak *mazri* rope with which the luggage was secured on the back of a camel suddenly snapped. Down came Mr Harris's stylish suitcases on the dry boulder-strewn path along which we were slowly moving. Mr Harris was visibly annoyed. He looked a bit comically incongruous in his fashionably cut 'sun-proof' cloth suit of orange-green colour, comprising a jacket and Jodhpur breeches. The spot where the minor mishap had occurred was a forbidding gorge with vertical cliffs of dun-coloured shales. And there we were stranded helplessly, on account of a broken rope of a Mahsud camel! At long last, another piece of rope was found, and the party moved forward. The Gomal Tangi dam site at Khajuri Kachh was a magnificent site. The width of the river at the gorge was only about a hundred feet. The limestone rock on both the flanks rose almost vertical to a height of 500 feet. Above that point, the cliff slanted away at a gentle slope to a height of about 1000 feet. Messrs Harris, Walker, and Oram were looking up at the gorge in silence. I felt a bit frustrated by the endless investigations of the dam sites and stepped forward and said, 'Sir, may I make a suggestion? We should build here a 500 feet high concrete gravity dam to impound the entire amount of flood runoff of the Gomal and Zhob rivers, combined.' Mr Harris looked at me with offended surprise. 'Not in my lifetime,' he snapped.

In due course, we reached the Gul Kachh dam site 20 miles upstream from Khajuri Kachh on the Gomal River. There, our party was joined by the political agents of South Waziristan and the Zhob Agency; the resident in Waziristan; and by Brigadier Evans, the chief engineer of the Northern Command of the Indian Army at Quetta. Everyone was treating Mr Harris with marked respect, and with flattering attention. We all mounted a 50-foot high hillock situated near the upper mouth of the Gul Kachh gorge from where we had a clear view of the magnificent Gul Kachh reservoir site, aided by the large-scale contoured site plan which I and the SDO spread before Mr Harris. All of us had climbed with ease the small hillock, affording a view of the reservoir site. Brigadier Evans, however said, 'Bring my charger.' A heavy built thoroughbred horse was brought, which carried him to the top of the hillock. Wherever there was even a small hill to climb, Brigadier Evans called for his 'charger'. He was in his late fifties and was wisely avoiding straining his heart.

Everyone was confident about Mr Harris sanctioning the construction of the Gul Kachh dam. Brigadier Evans offered to build a motorable road from the Balochistan side from Fort Sandeman to enable transport of material and construction equipment from the railhead at Fort Sandeman to the Gul Kachh site. The resident in Waziristan made a similar offer to construct a motorable road from Sarwakai Post to Gul Kachh in Waziristan. The prospective dam at Gul Kachh would serve the dual purpose of a dam and bridge linking the two provinces—NWFP and Balochistan. We hung on Mr Harris' verdict. To everyone's disappointment, Mr Harris' reply was an uncompromising 'no' to the proposed dam, on the grounds that excessive sediment was carried by the Gomal floods. In a panic, Mr Walker instructed me to speedily design an experimental dam at Gul Kachh for studying the behaviour of silt deposition in the Gul Kachh reservoir for obtaining Mr Harris's approval, before he returned to his headquarters at New Delhi. I hastily designed a 100-foot high rock-fill dam with an open sluice using the rock excavated from the spillway site in the body of the dam. The project cost the paltry sum of Rs 1 million. That small project was also vetoed by Mr Harris. Shortly after Mr Harris' return to Delhi, Mr Walker, chief engineer, the author of the Gul Kachh project, retired, a deeply disappointed man. Before his retirement, however, he recommended to the government to grant me the title of 'Khan Sahib' which, he confessed to Mr Oram, had been overdue, in view of my services as executive engineer, Waziristan Survey Division, and for successfully constructing the D.I. Khan Protective Embankment earlier.

So, I was conferred with the title of 'Khan Sahib' on 1 January 1930. Lord Irwin, viceroy of India, personally handed over the certificate and silver medal to me in a picturesque ceremony which was held at the Government House, Peshawar, in which various titles were distributed to recipients both Indian and European. My wife, on hearing of the event exclaimed: 'These ungrateful people. They should have conferred the title of "Nawab" on you.'

A directive was soon received from the Government of India to the effect that the NWFP administration should concentrate on improving the existing indigenous system of flood irrigation within the plain of the D.I. Khan district, instead of building storage or delay dams and reservoirs on the heavily silt-laden hill torrents of Waziristan. In 1950, while I was re-employed as a senior engineer in Balochistan overseeing its irrigation development, I noted that, after his tour of Waziristan in 1929, Mr Harris had toured Balochistan and had sanctioned the construction of Spin karez earthen dam and reservoir in the Uruk valley, near Quetta. That reservoir was designed to impound the run-off of about 40 square miles of catchment area including part of the lofty Takatu range of mountains. In fact, the annual run-off in the Spin karez reservoir was practically negligible. The sanctioning of the Spin karez dam and reservoir, without checking the run-off data of its catchments, did not signify a high degree of engineering talent on the part of Mr Harris.

I recall that, during those days, I had begun to get the distinct impression that there was no intention on the part of the British Indian government to develop Waziristan. This lent credence to the allegations of the Indian National Congress that the Government of India was deliberately maintaining a distinct political entity of the tribal areas of the NWFP, as a training ground for the Indian Army.

14 Players of the Great Game

Sheikh Mahbub Ali Khan

In the autumn of 1928, my friend Abdul Latif Khan, civil judge, D.I. Khan, invited me to jointly call on Sheikh Mahbub Ali Khan, a friend and former colleague of his, who had recently been posted as assistant commissioner, D.I. Khan, on his promotion to the IPS. We went to the Circuit House in the evening where Sheikh Sahib was temporarily residing, to pay a courtesy call. A remarkably obese, youngish man of thirty-five or thereabouts, Sheikh Mahbub Ali was immaculately dressed in an English style suit. He had a round face with a fairish complexion and pleasant features, except for small shifty eyes. He sported a moderate-sized moustache, and his black glossy hair was parted in the middle. He was seated all alone at the head of a large dining table, which was covered from end to end with exotic Indian and English dishes exuding a delicious aroma, which filled the large sitting-cum-dining room of the Circuit House. The cook, who had prepared all those delicacies, was being tested for employment. He was respectfully standing on one side, while the big man tasted each dish in turn, and pronounced his judgment. The dishes covering the table had enough food for a feast for twenty people. To my surprise, Sheikh Sahib greeted my friend with a cool and distant air while utterly ignoring me. He did not even invite his friend or me to be seated. After a brief exchange of greetings, and after congratulating Sheikh Sahib on his promotion, my friend returned, apparently gratified with his reception. I remained his silent appendage during the call. Sheikh Mahbub Ali Khan's reputation as one of the architects of the ouster of King Amanullah Khan, had already preceded his arrival at D.I. Khan. Amanullah Khan, the king of Afghanistan, was not in the good books of the British government since he had wrested his country's independence from their control, after the Afghan War of 1919. However, in order to wean him away from Russian influence, the British government adopted a policy of conciliation and gave him and his queen, Soraya Tarzi, an enthusiastic welcome everywhere they went during their official visit to Great Britain in 1929. However, despite the warnings by the British government, when King Amanullah visited Moscow, where he also paid homage before the tomb of Lenin, he became a marked man.

Before Amanullah's return to Kabul, Sir Henry Dobbs, the British envoy in Afghanistan, had speedily gone into action, and prepared the grounds for a rebellion by Habibullah Kalakani—aka Bacha-i-Saqao (the son of a water carrier)—a brigand Afghan chief who operated in the vicinity of Kabul. For the purpose of organising an army of rebellion under Bacha-i-Saqao, considerable British gold in pound sterling was clandestinely distributed among Afghan tribal chiefs, religious leaders, and others, through the agents of the British embassy. It was widely rumoured that Sheikh Mahbub Ali Khan, who was then working as oriental secretary to Sir Henry Dobbs, at Kabul, was one of the masterminds behind the successful rebellion against King Amanullah. On his return to Kabul, Amanullah had to flee the forces of the Bacha-i-Saqao to Kandahar and thence via Karachi to Italy, for good. He declared that he had decided to take that step to prevent bloodshed among the Afghan people. The brigand Bacha-i-Saqao and his gang looted Kabul city and created mayhem, to the extent that the Afghans came to despise him and his followers.

The feeling against Bacha-i-Saqao also ran high among the Indian Muslims, who resented the intrigues of the Western Imperial power against the independence of a Muslim country. A Muslim poetess from the United Provinces (India) published an impassioned poem against Bacha-i-Saqao. Among other sentiments, she declared that 'As soon as [the current] holy month of Ramadan was over, she would go to Kandahar, and personally gouge out the eyes of the brigand from his face!' The British government also soon discovered their blunder in having selected an ignorant and ruthless brigand to replace the enlightened King Amanullah Khan.

We learnt that General Nadir Khan, the hero of the Afghan war of 1919, who, in a lightning move across the hills, had captured the fortified British post of Tall in Kohat district of the NWFP, and was a kinsman of King Amanullah, was on his way to oust the usurper. He was then serving as Afghanistan's ambassador to France. The Muslims of India welcomed the reported move of General Nadir Khan and hailed him as the hero of the Afghan War of 1919 and a faithful servant of King Amanullah, and the saviour of Afghanistan. General Nadir Khan, in due course, arrived in Karachi and thence moved to Peshawar, where he was reported to have borrowed a paltry sum of Rs 80,000 from a local chemist friend and after forming a formidable force of Wazir, Mahsud, and Afridi tribesmen, marched on to Kabul. The usurper Bacha-i-Saqao was defeated, captured, and hung on 1 November 1929, along with his chief henchmen. At Peshawar, an Indian Muslim had met General Nadir Khan and expressed the hope that, after defeating Bacha-i-Saqao, he would restore his suzerain, King

Amanullah Khan, to the throne of Kabul. General Nadir Khan had made the cryptic reply that he was concerned chiefly with the welfare of the people of Afghanistan. So, we learned with no great surprise that General Nadir Khan had ascended the throne of Afghanistan, adopting the title of King Nadir Shah. Where he got the requisite funds from, to equip an invading army, was not a difficult guess. His very landing at Karachi lent credence to the speculation that he had assumed the royal mantle with the blessings of the British government.

For me, however, the consequence of this poignant drama which had been played in Afghanistan was the presence, in D.I. Khan, of Sheikh Mahbub Ali Khan, who was said to be closer to the Indian government in New Delhi than some Britishers themselves, and who displayed royal extravagance by tasting a feast for twenty people merely for checking the quality of the cooking, and who behaved arrogantly with fellow Indian officers. Sheikh Sahib's gratuitous dislike for me soon became apparent to everybody. We, half a dozen senior Indian officers of different government departments, invited him, by turns, to dinner. I came away from each dinner offended by some cutting remarks by him which were made without any provocation on my part. When my turn came for entertaining Sheikh Sahib, he ate chicken *pulao* on the table, with much relish. One of his admirers commented on this fact. 'My uncle, may his soul rest in peace, gave me a valuable piece of advice,' said Sheikh Sahib in reply to the comment. 'Eat well at the table of both friend and foe; the friend will be pleased, while [the] foe will be grieved to see you eat well from his table.' A muffled titter of amusement went around. Feeling incensed by the insult and forgetting, for the moment, the rules of Pathan hospitality, I said, 'Sheikh Sahib, your revered uncle (may his soul rest in peace) had indeed given you a valuable piece of advice but, unfortunately, he omitted to issue an important word of caution.' 'And what is that?' enquired Sheikh Sahib, with defiance, 'To mind your tummy.' I said. An icy silence followed my remark. Obesity was Sheikh Sahib's weak point, and his late uncle, a provincial civil service officer, through whose political influence he was said to have been appointed, was also reputed to be a remarkably obese person. In fact, obesity ran in Sheikh Sahib's family.

With that unguarded remark, I realised that I had made a potential enemy. However, a few days later, I was pleasantly surprised when Sheikh Sahib greeted me with marked cordiality. And, as time went by, he made efforts even to cultivate friendship with me. He recalled that we both had been contemporaries at Aligarh, where he was a school student studying in the Matric class while I was a BA student. And we both belonged to the North West Frontier group at the Aligarh college. Sheikh Sahib was evidently an

Grandfather Fazal Din Khan.

Father Mohammad Din Khan.

Uncle Khairuddin Khan.

Family photograph, Abbottabad, 1902. (Standing at the back, from left to right) elder brother Abdullah Khan and cousin Zafaruddin. (Sitting, from left to right) cousin Ferozuddin, uncle Khairuddin Khan, and father Mohammad Din Khan. Abdur Rahman Khan is seated on the floor.

Cricket First XI, Government College, Lahore. Abdur Rahman Khan standing extreme left, 1909.

Abdur Rahman Khan, as a senior scholar at MAO College, Aligarh, 1911.

Abdur Rahman Khan, apprentice engineer, Sylhet, 1917.

Abdur Rahman Khan, assistant engineer, PWD,
Dera Ismail Khan, 1922–3.

Family photograph, Dera Ismail Khan, 1923. First wife Noor Khanum Begum with her children Zubaida (standing left), Aslam (standing at the back), Salim and Tariq (standing right).

Family photograph, Dera Ismail Khan, 1934. (From left to right) Badruddin Khan (cousin), Nigar, Zubaida, Awais, Abdur Rahman Khan, Tariq, and Salim.

Family photograph, Abbottabad, 1961. (From left to right) Faris, Abdur Rahman Khan, Shahid, Iqbal Begum (Bibi Gul), Farid, and Asad (front).

With second wife Iqbal Begum (Bibi Gul).
Celebrating Pakistan's second independence
anniversary, Quetta, August 1949.

Iqbal Begum (Bibi Gul) conversing with the Governor General of
Pakistan Khawaja Nazimuddin. Pakistan's second independence
anniversary celebrations, Quetta, August 1949.

Kurram Garhi Headworks, Kurram River, Bannu, Khyber Pakhtunkhwa.

The Lansdowne Bridge, Indus River, Sukkur.

Gomal Zam Dam, Gomal River, South Waziristan, Khyber Pakhtunkhwa.

Warsak Dam, Kabul River, Valley of Peshawar, Peshawar, Khyber Pakhtunkhwa.

Mangla Dam, Jhelum River, Mirpur District, Azad Jammu & Kashmir, Pakistan.

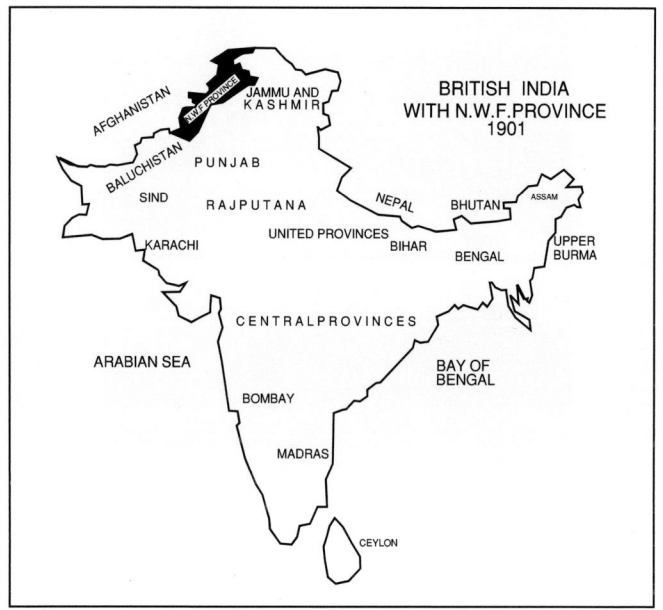

Map of British India, 1901

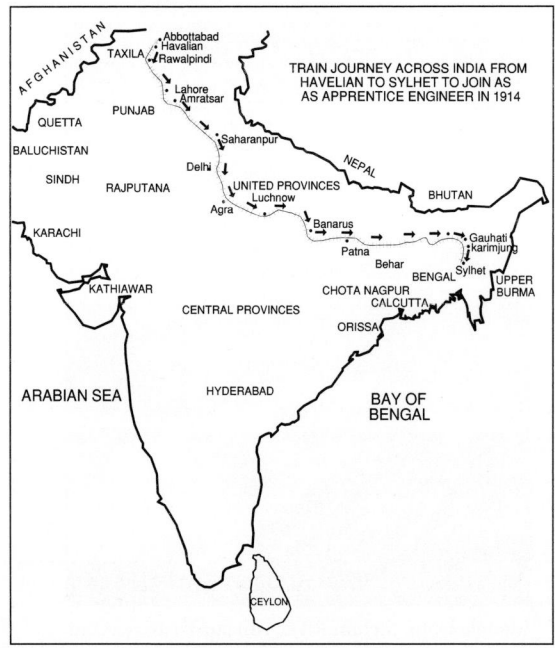

Train Journey, 1914

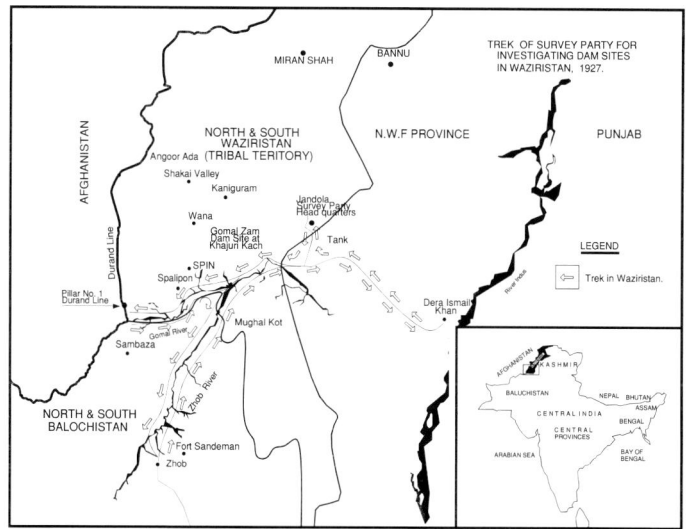

Trek of survey party for investigating dam sites in Waziristan, 1927

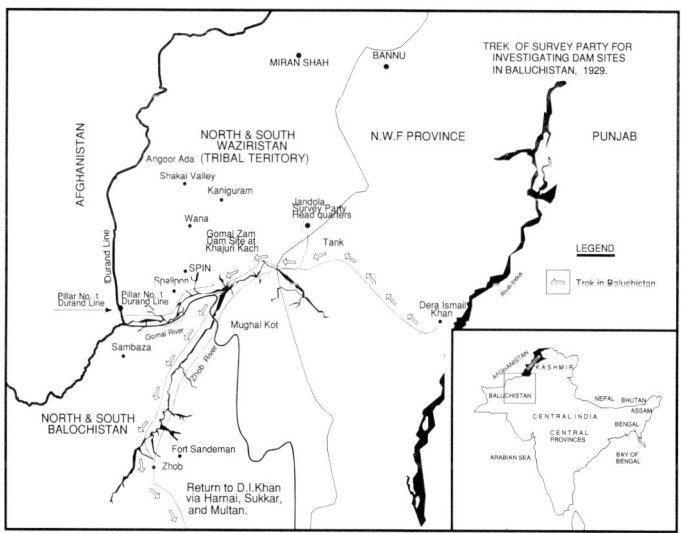

Trek of survey party for investigating dam sites in Balochistan, 1929

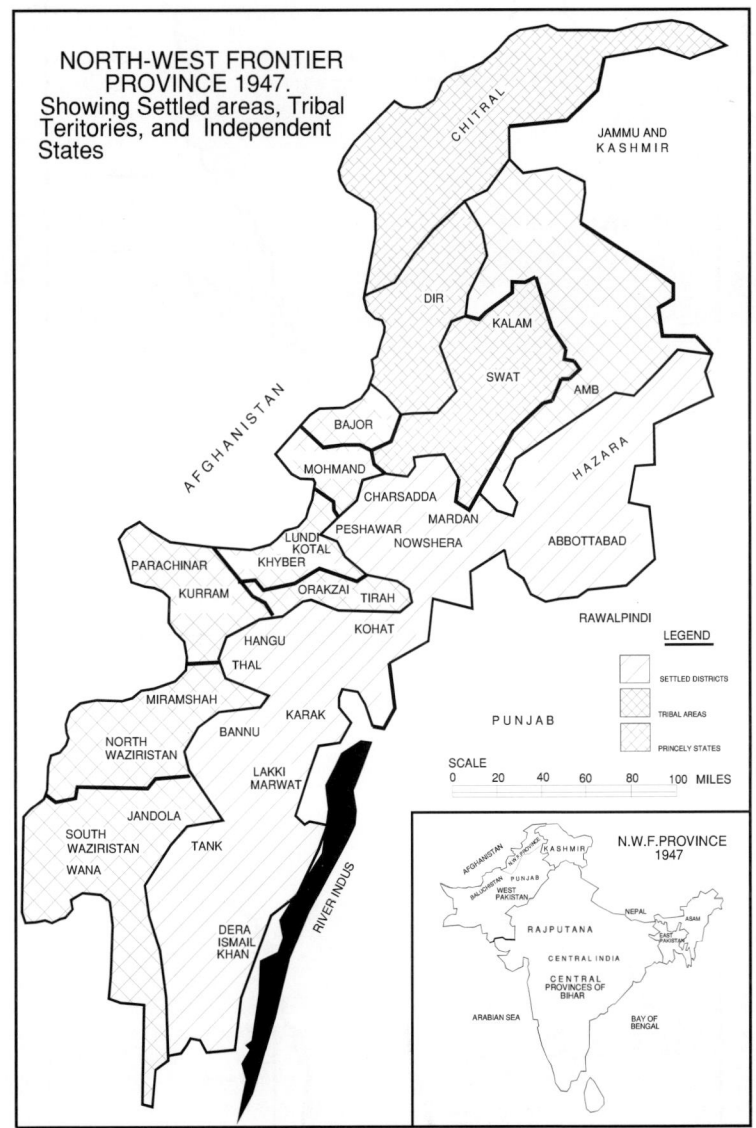

NWFP Tribal States 1947

accomplished diplomat. He probably realised that, in the matter of services, my status was higher than his. Moreover, I too enjoyed considerable prestige and social standing. Sheikh Sahib, perhaps, calculated that cordial relations would prove useful, so he decided to put his feud with me to rest, for the time being.

One day, Sheikh Sahib, along with his young second wife who observed strict purdah and belonged to a family from Kabul, called on me and my ailing wife. Surprisingly, both the wives became fond of each other, and exchanged many visits thereafter. Three years later in 1931, my wife passed away due to heart complications following chronic rheumatism at the early age of thirty-eight, Mrs Sheikh was deeply grieved at her death. I have given a lot of space in this chronicle to my first encounter with Sheikh Mahbub Ali Khan, because I consider him to have been a most remarkable man. I shall have more to say about him later.

In 1921, Edwin Montagu, secretary of state for India, came to India to study the political condition in the country. The result of this visit was the introduction of a step towards self-government by the Indians, known as 'diarchy', under which certain development departments, such as irrigation, roads, public health, education, and forests were made 'provincial subjects' and transferred to Indian ministers. While others, such as the finance and railways, were treated as 'reserved subjects' and were administered directly by the provincial government through European secretaries to government. However, for geo-political reasons, the NWFP was ignored in these new administrative arrangements, and any kind of reform for the province was considered to be out of the question. These developments threw Dr Khan Sahib and Abdul Ghaffar Khan, popularly known as the 'Khan Brothers' and the leaders of the Red Shirt Movement (aka Khudai Khidmatgar), into the lap of the Hindu-dominated Indian National Congress. Contacts were made by the Khan Brothers with the Hindu Congress leaders when they were fellow prisoners in the Indian jail.

The Congress party rejected the diarchy and put in a demand for full 'self-government.' So, in 1928, another high-powered commission, under the leadership of Sir John Simon, an eminent lawyer, came to India to report on the feasibility of granting further concessions to the Indians. The Simon Commission reported against the grant of further reforms, presumably in view of the political rivalry between the Hindus and the Muslims. The Simon Commission toured the whole of India and, in due course, also visited the NWFP. During the course of its visit to D.I. Khan, which was then considered an important station, being the base of operations against Waziristan tribes, the first thing the commissioners saw, as they landed from the steamer on the west bank of the Indus River which was swollen with summer monsoon floods, was a huge placard, inscribed with the words 'Welcome to the land of Majors and Colonels'.

Khan Bahadur Maulvi Ahmad Din

The placard in question had been hoisted a few minutes before the arrival of the commission, after eluding police vigilance, by my young friend, Abdul Rahim, a budding lawyer, (later Dr Abdul Rahim) turned 'revolutionary' against the British Indian government. He was the son of K.B. Maulvi Ahmad Din, who had rendered unique services to the Indian government in the political arena of the tribal belt of the NWFP, and recently retired as political agent of South Waziristan.

In 1893, Maulvi Ahmad Din had worked as Mir Munshi (personal assistant, vernacular affairs) to Sir Mortimer Durand during the latter's demarcation work on the famous Durand Line, which was subsequently accepted as the international boundary between Afghanistan and India. Maulvi Ahmad Din, recounted to me that, during the demarcation of the Durand Line, it was soon evident that Sir Mortimer was proposing boundary pillars at strategic points advantageous to the British without regard to the fact that the proposed line was cutting across the territories of the Wazirs, Shinwaris, Mohmands, and other tribal clans and, thus, dividing the tribes. His Afghan counterpart kept on protesting and objecting at every proposed site for a pillar, reinforcing his protests with obscene abuse hurled at the ever-smiling Sir Mortimer, but ended in attending Sir Mortimer's sumptuous dinners given in his honour each night. The Afghan representatives also sent regular reports of Sir Mortimer's proposal to his suzerain, Amir Abdur Rahman, the British-sponsored king of Afghanistan, who regularly lodged protests with the British viceroy in India, but the protests proved futile. And so, the demarcation of the line was completed up to Chitral.

Maulvi Ahmad Din was also one of the original sponsors of the 'Forward Policy' in Waziristan, which resulted in the construction of the Circular road connecting D.I. Khan with Bannu via Razmak in Waziristan near the Afghan border and at the dividing line between the Wazir and Mahsud tribal territories. A magnificent cantonment protected by a barbed wire fence was constructed and occupied by the regular Indian army. Fortified posts, occupied by the mobile and irregular force of scouts, were constructed at intervals of 10 to 12 miles along the long strips of the Circular road lying respectively in the South and North Waziristan territories. The roads between the scouts' posts were patrolled by Mahsud and Wazir Khasadars recruited from the tribes and were operated from temporary pickets built along the roads.

The political authorities appointed gangs of a dozen Mahsud and Wazir Khasadars under the command of a Malik from among them, to supervise the work of the gangs for road maintenance. The Malik was paid three times the wages paid to an ordinary Khasadar. The Wazir and Mahsud tribes, however, were intensely democratic. They did not comprehend the role of a superior person from among themselves to supervise their work without doing any work himself. So, one day when a petty political official happened to inspect the work of the Khasadars he found the so-called Malik doing three times the duty performed by ordinary Khasadars. Normally, only about half the sanctioned strength of Khasadars turned up for the parole duty. But wages for the full strength were withdrawn monthly from the treasury, and the savings were passed on, under the table, to the highest levels. It was not an easy task, however, to construct the Circular road through Waziristan. A lot of clandestine money was also passed on to the tribes concerned, particularly, to their spiritual leaders, the *mullahs*. Mullah Pawindah[1] was the most notable among the holy men in Waziristan. Many a *patay khabri* (secret talks) were held with him, chiefly by Maulvi Ahmad Din, in which much 'worldly wealth' was transmitted to him. Maulvi Ahmad Din had other interesting anecdotes to tell about his services in Waziristan. He said that whenever the British garrison of the scouts were overwhelmed and routed at a post by the Mahsud or Wazir raiders in Waziristan, the post in question was promptly abandoned and a new one was built in its vicinity. The reason for this was a display of psychological defiance.

Maulvi Ahmad Din sported a beard, in keeping with his avowed title of a 'Maulvi', and dressed in plain clothes, Indian fashion. Even when he was the political agent of South Waziristan, and occupied an elegant residence within the Razmak cantonment, which was protected partly by British troops, he could easily be mistaken for a local primary school teacher. One evening, while going out for a walk, Maulvi Ahmad Din casually strolled out of the barbed wire fence round the Razmak cantonment. When he returned, he found a British Tommy guarding the entrance to the barbed wire gate. He told the misbehaving sentry that he was the political agent but all in vain. He was kept at bay at bayonet point, until someone from inside the barbed wire recognised him and told the flabbergasted Tommy that the person whom he was threatening with the bayonet was no other than the political agent of South Waziristan!

The Enigmatic Lt. Col. Edward Noel

During that period (*c*.1931), I met and became good friends with Col. Noel,[2] who was transferred to D.I. Khan, as the deputy commissioner. He had previously served as the political agent of the Kurram Agency, with headquarters at Parachinar, situated in a high, picturesque plateau flanked on the north by the snowy range of Safaid Koh. He was reported to have introduced electricity in Parachinar town, by instructing the local officers of the Military Engineering Service to construct a small hydel power station on a nearby perennial hill stream. He had also introduced the cultivation of the medicinal herb Artemisia on commercial basis, in the Parachinar plateau which brought immense profit to the local cultivators. In the biography of Mustafa Kemal Ataturk of Turkey, titled the *Grey Wolf* (1932), the British author Harold Armstrong has quoted Ataturk as having protested the subversive activities of Col. T.E. Lawrence in Arabia and Captain Noel in Iran, during the First World War. Noel's activities, however, were apparently of a lesser dubious character. He had remained a vice consul in Ahvaz, Iran, during the First World War and subsequently was consul in Kerman and Persian Balochistan. Captain Noel's task appeared to have been to keep a watchful eye on the Iranian tribes around Shiraz, and to intercept and prevent the entry of pro-German elements into India and Afghanistan. He had been decorated with a CIE and DSO. At the close of the First World War, Captain Noel was taken into the Indian Political Service (IPS) and posted in the NWFP.

Col. Noel was of medium height, broad at the shoulders, with a nicely proportioned figure. He had a thick crop of curly russet hair which usually fell in pleasing disorder over his broad brow. He had a pleasant ever-smiling face, and was calm and deliberate in his manners. He was never flustered or annoyed at provocations and appeared oblivious to any sense of danger. He was extremely hardworking in a dedicated way and, although an administrator, was more at home with development projects of all kinds, particularly those connected with agriculture and engineering. Although it was rumoured that he had suffered imprisonment in Iranian jails, from where he had picked up the habit of using opium, he had a pronounced pro-Muslim attitude.

Shortly after his arrival in D.I. Khan, Col. Noel sent me a note proposing a week's trip with me to the Paharpur Canal area, offering me his hospitality during the tour. I accepted the kind offer. So, one afternoon we left on the first leg of the tour as far as the Paharpur rest house, where we were supposed to spend the night. He drove his own car and I sat beside him in the front seat, his orderly occupying the back seat. My driver followed in my car with the colonel's cook. Our luggage was also loaded in my car. Under a premonition, or call it the sixth sense, I had put some cotton wool, lint, and a small phial containing the tincture of iodine in my pocket in case of an accident before starting.

Col. Noel was so deeply engrossed in the discussion on the problems of the Paharpur Canal that he drove his car inattentively that led to its swaying alternately from one edge of the tarmac to the other, to my alarm. However, since he was restricting his speed to a comfortable 30–35 miles an hour, I did not mind his erratic driving. When we were a few miles out of D.I. Khan, an old farmer tried to cross the road about a hundred yards ahead of us, holding a rope that was tied to the neck of an unwilling goat that he was dragging along behind him. Right in the middle of the road the goat stopped dead and refused to budge despite frantic tugging by the old man. We were now only about 50 yards away from the pair. Col. Noel, anticipating that the goat would not move, swerved the car to the right in order to pass the stationary pair. Unfortunately, as we drew nearer, the goat too moved slowly forward a little bit at a time, so that the colonel went on swerving the car more and more to the right, to save the old man. The old farmer was standing at the extreme right edge of the road, completely blocking our way, while his stubborn goat was still standing fast on the tarmac. Without batting an eye, the colonel swerved the car down the 10-foot high road embankment in order to save the old man. The car completely overturned at the foot of the slope.

While sliding down the slope, in a detached state of consciousness, I thought that my last hour had come. I was amazed, however, to find that while buried under the car, I was not only alive, but seemed not to have sustained the slightest injury. Col. Noel's seat beside mine was empty. The car's windshield was completely shattered; pieces of glass were all over me as well as on the adjoining empty seat. The steering wheel had disappeared. Only the vertical rod, supporting the steering wheel, remained. I crept out of the narrow opening between my seat and the collapsed canvas roof of the car, expecting to see Col. Noel outside. But he was nowhere to be seen and was presumably still inside the overturned car. In answer to my repeated calls, Col. Noel silently emerged from the rear of the car, with the usual smile on his face. There was a cut on one of his forearms and another on his abdomen. Blood was seeping from both the cuts. I hastily took out swabs of cotton wool and the iodine tincture and vigorously rubbed it on the cuts. He allowed me to do so good-humouredly, remarking that the cuts were only slight, and it really did not matter to fuss over them. We were startled to hear the moans of Col. Noel's orderly. He was lying on his back, with his face protruding outside the car. His chest was pinned down under the heavy car; a part of his chest and the rest of his body lying inside the car. We both tried to lift the car, but it was too heavy. Luckily, a bus full of strong Marwat tribesmen arrived from the direction of Bannu. I signalled the bus to stop and told its passengers to help us extricate the poor fellow from beneath the heavy vehicle. The car was soon lifted and again put on its wheels. A deep dent was visible on the orderly's chest below his shirt. We put the poor fellow on the bus; I wrote a chit to the civil surgeon D.I. Khan to admit him in the hospital, and gave it to the bus driver.

We then turned our attention to Col. Noel's damaged car. To our amazement it promptly started again. But since it was badly damaged Col. Noel told my driver to take it back to D.I. Khan for repair. We both got into my car and continued our journey to the Paharpur rest house, where we spent the first night of our tour. The next morning, I got up at 5 a.m. Col. Noel was still asleep. I took my seat at the breakfast table at 5:30 a.m. sharp, as decided the previous night. Col. Noel got up 15 minutes later. I was amazed to see him come to the breakfast table precisely at 5:30 a.m. Within 15 minutes he had shaved, bathed, and dressed! Our breakfast consisted of generous portions of canned fruit salad, accompanied by an equally generous portion of excellent curds. This was the sole breakfast for the colonel during the hot summer season. But in consideration of his guest (i.e. myself) he had also ordered bread, butter, and tea.

That day we arrived at Bilot at the head of the Paharpur Canal. The Indus River and the Bilot Creek were in spate. The canal was in full flow. We climbed up the hill to the Bilot rest house and a further 200 or 300 feet to see Kafir Kot, the ruins of a fortified Hindu habitation dating back to about the eighth century AD, and the seat of the Hindu raja, Bil. The raja used to extract toll from the boat traffic plying on the Indus River. The ruins of a similar Kafir Kot on a somewhat bigger scale existed 25 miles further upstream, on a prominent spur of the Mali Khel hills. We spent the night in the Bilot rest house. Although the Paharpur Canal was in the direct charge of the PWD since 1928, nevertheless Col. Noel took a keen interest in its working. I soon discovered that both Col. Noel and I were kindred spirits. Both of us shared a keen interest in development work and looked for innovative ideas to address the problems of the peasants. We spent many hours discussing a wide range of subjects and found that we had many similar traits of character and interests, such as carelessness about our personal appearance, frugal eating habits, and a passion for development. A sense of companionship sprang up between us.

One day, during the course of an altercation between a Hindu shopkeeper and a Muslim customer in the D.I. Khan bazaar, the former used indecent language against our Holy Prophet [PBUH]. A Muslim passer-by overheard the objectionable language and stopped to protest. In a twinkling, a dozen or so neighbouring Hindu shopkeepers rushed to the aid of the Hindu shopkeeper and beat the protesting Muslim. Some Muslim passers-by came to the rescue of their co-religionist. The news of the incident spread like wildfire in the town. Muslims started dragging the Hindus out of their shops and beating them. The news soon spread to the surrounding villages, from where more infuriated Muslims surged into the main streets and lanes of the town. Soon the main streets were choked with crowds of Muslims attacking Hindus, wherever they found them. The Hindus hastily closed their shops and climbed up to the roofs of their double-storey houses overlooking the main streets and, in the lanes, from where they started pelting the Muslims in the streets below, with brickbats, empty soda-water bottles, etc. Many Muslims were seriously injured. Since the Muslims could not break open the closed and barred houses of the Hindus, they set some of them on fire.

At that stage, Col. Noel, arrived at the scene at the head of a mounted contingent of the Frontier Constabulary and forced his way through the Muslim crowds in the main street of the town. He signalled to the Hindus to cease throwing lethal missiles (stones, glass bottles, etc.) from their houses on crowds below. But his warning was not heeded and, on top of that, a missile thrown by a Hindu from above grazed his forehead, which bled a little. With his usual nonchalance, he wiped the blood off his forehead with the back of his hand and repeated his order to the Hindus to stop pelting the Muslims with stones, etc. As his warning again went unheeded, he shouted the order to the crowd of Muslim with a single word, *looto* (plunder). The Muslims forthwith broke open the shops of the Hindus and started looting merchandise and carrying it away to their homes. Col. Noel turned his horse about and left the city, along with his mounted constabulary guard. After about an hour of looting, the constabulary returned to the city and cleared the streets of the Muslim mob. The Hindus came down from their fortified houses and salvaged their remaining goods. The next day Col. Noel sent out search parties of the police and constabulary to the poorer quarters of the Muslims in the town, and in the surrounding villages and retrieved most of the previous day's loot and had it restored to the Hindu shopkeepers.

The Hindus of D.I. Khan were a very well-to-do and a highly vocal lot. They created a furore against Col. Noel for having allegedly connived with the Muslims allowing them to loot the property of the Hindus. Col. Noel had to appear before the Lucknow High Court to explain his conduct in the affair.

His simple statement was that in his order to the Muslims he had not used the word *looto* (plunder) as alleged by the Hindus, but *lauto* (return to your homes). He got off scot-free! The Hindus of D.I. Khan had, however, failed to realise that, in permitting the Muslims to plunder their shops, Col. Noel had managed to save their lives from the wrath of the injured and infuriated Muslims and, at the same time, had most of the loot retrieved and returned to them.

A few months before the date of my retirement, in 1946, I entered into a three-year contract of service in the Bahawalpur state. I learned later that the prompt acceptance of my appointment had been due to Mr Oram's strong recommendations to Sir Richard Crofton, prime minister of the Bahawalpur state when they had chanced to meet in New Delhi, where Mr Oram was working in the MES after his retirement from the NWFP. Sir Richard had enquired whether Mr Oram knew me, and he had simply answered, 'Grab him!' While I was considering the Bahawalpur state position, I received an unexpected letter from Col. Noel, who had retired as director of agriculture, NWFP, opting for the post of a commissioner in the political service. The letter bore the post mark of Ahvaz in Iran. Col. Noel disclosed the following amazing story.

In AD 242, a great battle was fought between the Sassanian king Shapur I of Persia, and the Roman emperor Valerian, in which the latter was totally defeated and taken prisoner along with the bulk of his army. The defeated Roman emperor and his army spent many years in captivity in Persia, in the town of Ahvaz on the banks of the Karun River, in south-west Persia. At last getting tired of inaction, Emperor Valerian petitioned King Shapur to be allowed to build a bridge across the Karun River, at Ahvaz. The petition was granted. The Roman army under the direction of Emperor Valerian accordingly constructed a masonry bridge across the Karun River.

That ancient Roman bridge was intact, wrote Col. Noel, in 1946, except a gap of a few hundred feet in the middle. The colonel somewhat naively suggested that the ancient bridge, after closing the gap, and with other suitable modifications, would serve as an excellent weir-barrage, for taking off a couple of irrigation canals from the right and left of the weir-barrage for the development of irrigation of the Ahvaz plain. Col. Noel further added that he had formed a British company with a nominal capital of £500,000 to be increased as and when required to undertake the development of irrigation of the Ahvaz plain. He had, moreover, nominated me as the chief engineer of the enterprise and asked me to come to Ahvaz to take charge of design and construction work as soon as practicable. At that time, the southern half of Iran was in the possession of the British, and the northern half was held by the Russians, and none of these powers appeared willing to relinquish their stranglehold on the rich prize. My activities at Ahvaz, I figured would have been tantamount to helping to consolidate Britain's hold in southern Iran. I had, therefore, no intention of accepting Col. Noel's offer. Moreover, I also knew that Kamal Ataturk had bitterly complained about the clandestine activities of Col. Noel in Iran and Col. T.E. Lawrence in Arabia during the First World War.

15 Political Turbulence

In early 1931, I took leave in order to complete the construction of my house in Abbottabad, which I had started building in 1927. My wife, Khanum Begum, and I had been married since 1909, when I was eighteen and she was sixteen. She was my first cousin. In 1923, she had an attack of acute rheumatoid arthritis which had left her a permanent invalid. Two more children had followed in 1925 and 1927, mainly on the recommendations of doctors who were of the opinion that she might improve during her pregnancies. In October 1931, she died of heart complication due to kidney failure; she was thirty-eight.

My world suddenly became very dark. I looked upon my house, and on the surroundings, the encircling hills, the sky, and the clouds, in an utterly detached and bewildered way. So, this was to be the end of everything that signified 'life'! I had read history with considerable interest and satisfaction, without giving any thought to the fact that in due course I, too, was destined to join the numerous people who had crossed the portals of death and entered the realm of the past. True, such a notion had indeed crossed my mind on rare occasions, but I had regarded my own death as a very remote possibility. But here was my life's companion snatched away at the young age of thirty-eight! Some of my relatives asked me to have a last good look at her features, calm in death. But a thought promptly crossed my mind, 'These features will soon be obliterated forever within the grave.' One evening, two days before her death, she had suddenly said: 'Hello, who are all these people who have come to see me. I can see all the seven Heavens opened up before me.' 'Hush,' I had tried to soothe her, but she had continued, 'I will not be here, and you will not be with me.'

My leave was till after the Christmas vacations of December 1931, but I reported back to duty at D.I. Khan only a few days after my wife's death, at the end of October. My home had looked too desolate after her death. There, I found Kirpa Ram who was officiating in my position during my absence, firmly in the saddle. Thus, I was placed 'on special duty', and was attached to Mr Oram's office, while information was transmitted to the chief engineer about my premature return from leave. Mr F.A. Burkitt, the new chief engineer, however, sent strict orders that I should be re-posted to my old post as executive engineer, and Kirpa Ram should revert to his previous post as personal assistant to Mr Oram.

During the early days of my long leave at home in Abbottabad, when my wife was alive and well, I was one day going down the steeply slanting portion of the road near our house in my car, when I saw Dr Ziauddin, our mathematics professor at the Aligarh college, laboriously ascending the climb on foot. I stopped my car, and alighted, and greeted him with respect, as of old. I felt great pleasure in the fact that he instantly recognised me after a lapse of twenty years! He said, 'You were the lone candidate from our college who had successfully passed the competitive examination for entry into the Thomason Civil Engineering College, at Roorkee.' He said, he was residing with Sir Sahibzada Abdul Qayyum Khan as his guest, but his house was too public, and he was constantly bothered by visitors. I invited him to come and stay at my house. Sir Ziauddin (for he had since been knighted) had come to Abbottabad as a member of the Niamatullah Committee (presided over by Mr Justice Niamatullah of the Lucknow High Court) to enquire into the law pertaining to the Frontier Crimes Regulations. (It was, in fact, a lawless law). Sir Ziauddin spent two or three weeks at my house.

Tumultuous Times

During the years, 1930–2, tremendous political upheavals had taken place in India. The rift between the Hindus and Muslims had widened. The Muslim League under Mr Jinnah's leadership was emerging as an effective platform for voicing the demands of the Indian Muslims. In 1930, presiding over a meeting of the Muslim League at Lahore, Dr Iqbal, the philosopher-poet, outlined a vision of a separate homeland for the Indian Muslims in the Muslim majority provinces of India. Two years later, a tangible form was given to Dr Iqbal's vision by Choudhry Rahmat Ali, a student at Oxford University, who was also president of the Oxford Union. He proposed Pakistan—the Land of the Pure—as the name for the proposed Muslim homeland. The proposed name was the result of a rare inspiration of Choudhry Rahmat Ali; it referred to the names of each of the Muslim-majority province of India.

Mr Gandhi, on the other hand, enjoyed the solid support of the Pathans of the Muslim province of the NWFP, as well as other Congress-oriented Muslims of India, under the leadership of Maulana Abul Kalam Azad. In 1930, he launched his famous Civil Disobedience Movement against the British rulers in India. The campaign crystallised into the deliberate violation of one specific law, namely the illegal manufacture of salt from sea water. Since salt was a taxable commodity, it was a penal offence for private individuals to manufacture salt from sea water. The apparent reason put forward by Mr Gandhi for breaking the law in question was that the law was unjust because it hit the poorest citizens of India.

So, one fine day, after an impressive display of silent devotion and contemplation, Mr Gandhi marched from his *ashram* to the seashore near Bombay city. He was carrying a small copper vessel for filling with sea water, and also a bundle of fuel wood for lighting a fire under it in order to distil the sea water for extracting illegal salt in a minute, symbolic quantity. Mr Gandhi was followed by a long line of about 3000 Hindu followers, similarly equipped for breaking the salt law. The procession was accompanied by a large number of policemen. As soon as the long line of fires was lit along the seashore and the vessels containing sea water were put over the fires, the salt law had been technically violated. Mr Gandhi and his followers were taken into custody and marched off to jail. Similar demonstrations were carried out by Hindus at many other places in India, resulting in peaceful arrests.

Mr Gandhi's civil disobedience campaign, however, took a grave turn in the NWFP. A huge crowd took out a protest procession in the famous Qissa Khwani Bazaar of Peshawar city. A British police officer in an armoured car arrogantly forced his way through the crowd and callously crushed a small Pathan boy to death under the wheels of his car. The infuriated mob surrounded the armoured car and set it on fire, burning its occupants along with the car. A contingent of the Garhwal Rifles, an infantry regiment of the Indian Army, composed of Hindu Rajputs was hastily moved in to avenge the death of the police officer and orders were given to fire on the crowd. The Garhwali soldiers refused to fire on an unarmed crowd. They were withdrawn and ordered back to the barracks, where they were disarmed and interned. A contingent from a British regiment from the Indian Army was moved in who started firing indiscriminately on the crowd. Instead of running away from the firing squad, the Peshawar citizens rushed forward and, with bared chests, invited the fire of the British troops. The British Tommies, of course, promptly obliged them. The people of Peshawar were demonstrating their adherence to Gandhi's message of non-violence with an incredible sacrifice of their lives. The carnage was frightful, since the crowd refused to retreat from the firing line. There were, reportedly, 200 to 300 deaths and many more were injured before the British troops' lust for revenge was satiated. To this day, there are a couple of inconspicuous memorials to the martyrs of the 1930 firing, built on the narrow footpath of the Qissa Khwani Bazaar. One of them built by the Khudai Khidmatgars, the original followers of the Indian National Congress of Gandhi, who had suffered in the firing, and by their subsequent successors in power—the Muslim League.

Another astonishing example of non-resistant behaviour under fire was displayed by the Wazir tribesmen of North Waziristan. During the course of a non-violent demonstration at a post, 9 miles from Bannu, near the border of the tribal territory, the Wazirs, with rare courage and spirit, offered no resistance to the firing by the Indian Army, and suffered fifty or sixty fatalities. In the summer of 1932, while inspecting the Ahnai Tangi dam site on the Tank Zam River, I engaged our Mahsud gauge reader in friendly conversation. 'I cannot understand why you Mahsuds and Wazirs have become the followers of Abdul Ghaffar Khan and Gandhi?' I asked. I was referring to the Wazir's allowing themselves to be slaughtered by the Indian Army without offering resistance at the Domel Post, during Gandhi's civil disobedience campaign of 1930. 'Do you wish to be dominated by the Hindus, if India gains independence?' He pondered over the reply for some time, and then said: 'Well, the Hindus are not as strong as the British. We will strike a better bargain with the Hindus!'

Thus, while Gandhi's Hindu followers in the rest of India mostly got away with light sentences of imprisonment, the Muslims of the NWFP paid a heavy toll in blood. This barbaric display of power by the Government of India in the NWFP, however, created great ferment in India. The government appeared to have been thoroughly alarmed. Lord Irwin, viceroy of India, forthwith released Mr Gandhi, and successfully persuaded him, and the Muslims, to attend a round table conference in London, to discuss the question of self-government in the form of dominion status for India. There were two round table conferences, the first in 1930 and the second in 1932. Differences arose amongst the Hindus and the Muslims about the measure of power to be transferred to the two major communities of India. In one of the round table conferences, Dr Iqbal, the philosopher-poet, was also roped in by Muslims as a representative of the Muslim League. He, however, failed to contribute in the field of diplomacy. Speaking on the issue of the levy of land revenue by the Indian government, the British representative remarked that they had inherited the right of ownership of all the land in India from their predecessors, the Mughal emperors, and while the Mughals collected one-third of the gross agricultural product from the *raiyats* the British government in India had adopted the more human and just procedure of claiming only 'half net assets' in agriculture. In response, Dr Iqbal exclaimed, 'Where are the Mughals now who claimed ownership of all the land in India. The land belongs to God, and to the people. The claim to ownership of the land, both by the Mughal and by the British, is ridiculous.'

The outcome of both the round table conferences was inconclusive. The Hindus representing the Indian National Congress wanted to grab all the power, while the Muslims, ably led by Mr Jinnah and backed, among others, by the mercurial genius, Maulana Mohammad Ali, demanded their share of power as an important religious entity in the subcontinent. Unfortunately, Maulana Mohammad Ali, who had developed a heart condition due the stress of the long struggle for independence of Muslims, and the rigours of a prolonged life in prisons, had a fatal heart attack in London. According to his last wish he was buried in the graveyard attached to Masjid Al Aqsa, the third holiest site in Islam.

One thing, however, clearly emerged from the aftermath of Gandhi's Civil Disobedience Movement, i.e. the Montagu-Chelmsford reforms which had been introduced in the rest of India since 1921, and denied to the NWFP on various pretexts, could no longer be withheld. Those reforms were introduced in the NWFP in early 1933. The NWFP was raised to the status of a governor's province. Sahibzada Nawab Sir Abdul Qayyum was asked by the Indian government to form a ministry for the 'transferred subjects', i.e. the departments of agriculture, health, education, forests, and the PWD, and to nominate ministers to these departments. Two ministers were nominated from the Hazara district. One of them, representing the Khanate class, was my former classmate, with an extremely poor understanding of English. The other was a former police officer of the provincial cadre. Both made pathetic figures; they were at a loss as to how to deal with their highly experienced British heads of department.

Rabindranath Tagore, the Bengali Hindu poet, was knighted by the Indian government. Another Hindu, Satyendra Prasanna Sinha,[1] a prominent lawyer and first governor of Bihar and Orissa, had been promoted to British peerage, and sat in the British House of Lords as Lord Sinha, and was made a member of the Privy Council. Both these honours were political measures to appease the Hindu community. On a representation made by some members of the influential Muslim community of Punjab, Dr Iqbal was also knighted, presumably for similar reasons.

The Price I Paid for the Montagu-Chelmsford Reforms

As a result of the implementation of the Montagu-Chelmsford reforms of 1921, the B&R department of the NWFP which, since 1901, had been administered by the Imperial Engineering Service (IES), was transferred to the provincial PWD. Taking advantage of the change, the Government of India 'smuggled' six junior officers of the RE, and one British sergeant—who was promoted to the rank of an assistant engineer—to the NWFP PWD. The IES stood abolished since 1921. The sanction of the secretary of state for India was invoked as a special case to reopen the recruitment to the Indian Service of Engineers (ISE), in order to absorb the RE officers of the Military Engineering Service (MES) in it. My own confirmation in the ISE had been recommended to the Government of India much earlier, and was expected to be approved as a matter of course, but was still pending. However, by this controversial act and underhanded decision, all the RE officers imported from the MES became senior to me. Major Lang Anderson, the senior-most RE officer among the lot, was drawing substantially less salary than me. It had also become obvious that the reason for maintaining my officiating status in the ISE was that, if made permanent, I would become senior to Mr Brown, the son-in-law of Mr Burkitt, who was being prepared to eventually succeed Mr Oram as chief engineer of the NWFP after Mr Oram's expected retirement in 1940.

Sahibzada Sir Abdul Qayyum, the NWFP chief minister, was either powerless or unwilling to intervene on my behalf. As a counterpoint to the import of the six RE officers in the PWD, he recruited six junior Indian assistant engineers in the NWFP electricity department. I was repeatedly asked to give my consent to being confirmed in the new Class-I service, but I refused. The upshot of the constitutional reforms in the NWFP, for me, was that I was, henceforth, designated as executive engineer, D.I. Khan, PWD division, and the buildings and roads of the district, and the ferry steamer and motor launches plying across the Indus River, between D.I. Khan and Darya Khan, were placed in my charge, in addition to the Paharpur Canal. Mr (formerly Sergeant) Fairs, an Englishman, was appointed to serve under me as the sub-divisional officer in charge of the ferry steamer 'Jhelum' and five motor launches. Mr Fairs was, reportedly, unhappy over his appointment, and promotion to the 'officer' rank. 'What will happen to my 5 per cent?' he was reported to have asked.

A lot of corruption was prevailing among the subordinate rank of the MES. The MES sub-divisional officers, generally recruited from the subordinate ranks, reportedly took a 5 per cent cut on the gross value of a contract, from the contractors. A few months later, a Muslim police inspector, an acquaintance of mine, came to see me. During the course of our conversation, he casually remarked, 'The case of corruption against your Mr Fairs is complete. He will be prosecuted shortly for using petrol for the motor launches in his private car.' I was alarmed. If Fairs was prosecuted, it might well go against me also, with the British high-ups in the government. I was well aware that the British in India, from the junior-most to the highest officials in the province, were all members of one clan that ruled India by backing each other, through thick and thin. I asked Fairs to see me and disclosed to him the imminent peril he stood in. Thanking me profusely, he hurried to the owner of a petrol pump and opened a predated petrol purchase account with him. He thus evaded prosecution. I noticed that, henceforth, the attitude of my British superiors towards me became extra cordial.

My step-brother had completed his BSc in agriculture from the Islamia College, Peshawar. He remained unemployed for a year and was living with me during that period. His application for the job of agricultural assistant in the agricultural department of the NWFP, addressed to the Chief Minister Sahibzada Sir Abdul Qayyum, had brought no result. I mentioned the matter to Col. Noel, deputy commissioner of D.I. Khan. He told me to remind him when he next visited Nathiagali, the summer seat of the NWFP government. He said that his friend Mr Hopkinson, ICS, was secretary to the chief minister, who would manage to obtain sanction for my brother's appointment by placing his application among the routine files to which the chief minister usually put his signatures without going through them. However, when Mr Hopkinson placed my brother's file with the sanctioning memo typed on it, Sir Abdul Qayyum, surprisingly, read it slowly and then, looking up with a knowing smile at Mr Hopkinson, signed it!

In 1932, Mr Oram (superintending engineer) was transferred to the Punjab Irrigation Department. Before he left, he became overfriendly and started confiding his personal matters like the injustice done to him in his transfer. He told me that Mr Burkitt, the chief engineer, had deliberately manoeuvred his transfer to bring in Mr Ross, his relative, to NWFP so that he could succeed him in becoming the chief engineer when he retired in 1934. This meant that Mr Oram could only become a chief engineer for two years, instead of six years, without this change, before he retired in 1940. I found it very unusual for a British officer to share his grievances against a fellow British officer with an Indian.

Mr Ross duly arrived and took over from Mr Oram. During the course of our tour to the outlying areas of D.I. Khan, he put me this amazing question: 'Who is your best friend?' Since we had met for the first time, I felt like telling him he wouldn't know who he was; instead, I said, 'He is Minhajuddin, a fellow engineer who was two years my senior at Roorkee.' 'Oh, I know Minhajuddin very well,' he said, looking at me with renewed respect. 'He is the straightest Indian I know!' Minhajuddin had been a brilliant mathematics student at Aligarh college and had secured a permanent position in the IES after graduation. He was a saintly character, thoroughly honest and tolerant with a wry sense of humour. He was appointed the first Muslim chief engineer of the Punjab province, but unfortunately died before his retirement due to a heart ailment that ran in his family.

The year 1932 had ushered in the Great Depression. Perhaps due to over-production, world over, the prices of consumer goods, including gold, came crashing down. Agricultural products suffered the most. In the summer of 1933, I toured the cultivated areas under the Rod Kohi (flood) irrigation from hill torrents in the D.I. Khan district. I rode through a large banked field in which the tall millet crop, fully ripe, was growing. The stalks were tall enough to conceal me while riding through them. 'Why don't they harvest this magnificent crop?' I enquired. 'Because the price in the bazaar will fetch less than the cost of cutting it,' I was informed. The government treasury was said to be empty. There was not enough money to pay the salaries of government servants. It was said that in the USA wheat grain was being used as fuel to fire steam engines.

16 Farewell to D.I. Khan

A Helpful Shikar Trip

Since my wife's death (*c*.1930/1931), I was left with six children on my hands. My eldest son, Aslam, was studying civil engineering at the University of Sheffield in the UK and the son younger to him was studying at the Islamia College in Peshawar. But I still had four minor children to care for, two boys and two girls, of whom the eldest was a girl, Zubaida. She was growing up as a tomboy, playing cricket with me and my friends on an improvised pitch on a side lawn of my bungalow. In the absence of my wife my household fell into a chaotic state and I became depressed with symptoms of neurasthenia.[1] Besides, I had been working at D.I. Khan continuously, for the past eleven years, and had become tired of the place. Luckily, an opportunity for transfer from D.I. Khan came up in November 1933.

During the summer of 1932, the main current of the Indus River invaded the Bilot Creek half-way down from Chashma spur. During the succeeding winter, however, it had receded back to Chashma, but had left the mouth of the Bilot at Chashma high and dry, thus cutting off the winter supplies to the Paharpur Canal. The matter was reported to Mr Burkitt, the chief engineer at Peshawar. Mr Burkitt accordingly announced his programme for the inspection of the Bilot Creek at Chashma. The tour would take three days.

Since Mr Ross (superintending engineer) was living without his family, and was staying at the D.I. Khan club, I suggested to him that Mr Burkitt should be my guest during the tour. On Mr Ross's agreement, I obtained Mr Burkitt's acceptance to my invitation, by phone. I had an excellent cook who was an expert in preparing Western cuisine. With the help and suggestions of the English wife of Ikramullah Khan, the sub-divisional officer at my headquarters, I purchased the necessary quantities of tinned and fresh foodstuff, including beer, wine, and spirits. The party consisted of Mr Burkitt, Mr Ross, Captain Stuart (personal assistant to Mr Burkitt), Mr Ikramullah Khan and myself. Swiss Cottage tents equipped with galvanised iron tubs for taking daily baths, and separate small toilet tents, were provided for each officer. A large separate large tent was set up as a lounge with a dining area. There was also a large number of *chholdaris* for the kitchen and servants. A small camp for the party was also set up at Chashma. During the two nights we spent at Chashma, we met for dinner in formal evening attire. Arrangements were made for the daily supply of fresh bread, butter, and soda water from D.I. Khan by car.

On the first day of the tour we arrived by car at Bilot, the head of the Paharpur Canal, 40 miles away from D.I. Khan. Arrangements had been made for a picnic lunch there. Unaware of the kind of messing arrangements I had made, Mr Burkitt somewhat petulantly asked for lunch. He was astonished to see a generous spread of cold cuts, assorted vegetables, bread and butter, cream crackers with cheddar cheese, and cake pudding. From Bilot we boarded a motor launch and reached Chashma in the afternoon. The afternoon tea was served on the riverbank with raisin cakes. When we assembled for dinner in the mess tent at night, we were shivering with cold. The appearance of steaming plates of tomato soup as the first course came as a pleasant surprise. After a satisfying dinner, my guests lingered over their liqueurs and wine.

The next day was reserved for shooting black partridge in the vast island that lay between two branches of the river at Chashma. We moved in a row, some 50 yards apart from each other, each attended by a couple of beaters. The island was a couple of miles long and was thickly covered with tall grass. As the birds were flushed, each one of us was required to shoot only those birds that arose directly ahead. The morning bag consisted of 130 birds with Ikramullah Khan making the biggest contribution of fifty. According to the established convention the entire bag was taken across the 10-mile wide river to Kundian railway station and dispatched by special messenger to Mrs Burkitt in Peshawar.

The following day Mr Burkitt inspected the Bilot Creek at Chashma. He decided to 'canalise' the 20-mile length of the creek between Bilot and Chashma, and to construct a permanent head regulator from the Paharpur Canal on the bank of the Indus River at the point of off-take of the Bilot Creek. I ventured to express my doubts about the efficacy of a head regulator without a weir across the river. Mr Burkitt, however, replied: 'Rahman, this is a cast-iron scheme. The main river always hugs the Chashma spur where we plan to put the head regulator. And there is enough slope in the Bilot Creek to feed the Paharpur Canal by flow.' After Mr Burkitt's inspection, we returned to D.I. Khan by motor launch. On the way, we shot by turns whatever game came our way, such as ducks, cranes, water snipes or crocodiles. Mr Ikramullah Khan in his turn, was lucky to score a bull's eye with his rifle on a crocodile basking on a small sand bar at the edge of the water about 200 yards away. The victim wriggled with pain on the sand for some time, and then slid into the water. When another crocodile was sighted, Ikramullah Khan got so excited that he took a shot out of turn, but his gun misfired. Mr Ross remarked: 'Ikramullah, your gun has more sense than you.'

The tour was evidently a great success. Mr Burkitt declared that the previous day's *shikar* was one of the best of his life, during which he was completely relaxed from the cares of his office. Clearly, the time was ripe for me to broach the subject of my transfer from D.I. Khan. Addressing Mr Burkitt, I said, 'Sir, I have been serving continuously in D.I. Khan for the past eleven years. During this long period, I have paid as much rent for my residence as I would to buy me the bungalow. Sir, if I am to retire from this station, would you advise that I buy a house in D.I. Khan?' Mr Burkitt smiled. 'I see you do need a change. Remind me next April,' he said and, addressing Captain Stuart, his personal assistant, continued, 'when orders will be issued for the mutual transfer of Anderson, executive engineer, Mardan division, and Abdur Rahman.' So that was that.

In April 1934, orders for my transfer to Mardan duly arrived. During my long period of service in D.I. Khan and, perhaps, in view of my successful construction of the protective embankment on the bank of the Indus River during 1923–6, to protect the town from river erosion, I had become well known amongst the local Muslims, Hindus, and the British officers. So, a civic reception was arranged at the D.I. Khan Club, after obtaining permission of the provincial government, to bid me farewell. In addition, a meeting of the Muslim and Hindu members of the D.I. Khan Municipality, presided over by Col. Bruce (deputy commissioner), was held in which I was publicly thanked for the services rendered to the D.I. Khan city and district. A copy of the resolution was forwarded to the provincial government. The farewell party turned out to be a huge affair. It was attended by hundreds of Hindu and Muslim notables of D.I. Khan, and also by many British officers. One of them also played some tunes on the club's piano. Mrs Zacharpal, the attractive wife of Major Zacharpal, superintendent of jails, brought me a present of 'chocolates and pastries' and the citizens of D.I. Khan presented me with an elaborate silver tea set.

And so, one fine day in April 1934, I left for Mardan. I could hardly believe that I was actually the executive engineer of Mardan division and was to occupy the same bungalow at Mardan in which I had reported my arrival to Mr Burkitt in February 1917—the first day of my appointment in the NWFP Irrigation Department. The walls of my residence, constructed in 1891, were festooned with creepers consisting of purple bougainvillea, white and yellow Jasmine, and tiny-sized white, yellow, and pink roses. The sight and scent of the creeper roses, in particular, fascinated me. The bungalow stood on 4 acres of land. It had a fully developed spacious front lawn and a couple of tennis courts. The lawn also contained a small rose garden, and other flower beds. On the side of the house was a large fruit orchard with a variety of mature fruit trees. On the other side was a large vegetable garden.

The Lower Swat Canal, which was to be my principal charge, was already 35 per cent developed. There were only routine problems of water distribution involving numerous daily applications from cultivators and large landowners for increase in the size of irrigation outlets. I made myself accessible to all applicants, and personally received applications from an average of fifty cultivators daily. The fertile Lower Swat Canal region of the Peshawar district was the stronghold of feudal Khans who owned most of the land. The Khans were generally at loggerheads with one another, aligning themselves to specific *para-jambas* (groups) hostile to each other. They vied with each other to excel in hospitality to strangers in general, and to government officers in particular, especially British officers, in order to tactfully extract favours from them. Notable among the Khans were the Nawab of Hoti, the Khans of Mardan, Khan Mahi, and the numerous Khans of Hashtnagar (eight towns) in the Charsadda tehsil, to name a few.

Locking Horns with the Khans

Shortly after my arrival at Mardan, Mr Ambrose Dundas, deputy commissioner, Peshawar, paid an official visit to Mardan, which, in those days, was part of the Peshawar district. He stayed at the canal rest house, which was next door to my residence. I paid a courtesy call on Mr Dundas, who had been my contemporary in D.I. Khan. However, to my surprise, his behaviour towards me was distinctly cold although, in an official sense, I was not his subordinate. The reason soon became apparent. He told me, frankly, that he had received numerous complaints against me from the landowners and cultivators on the Lower Swat Canal and that he would have called me had I not visited him myself. I felt my temper rising, but kept a firm hold on myself. After all, Mr Dundas, as the head of the district, was ultimately responsible for the well-being of the people. Mr Dundas said he was inspecting the Lower Swat Canal the next day and requested me to accompany him. I could sense that he had succumbed to the pressure tactics commonly used by the feudal landlords to get their way.

The next day we drove along the main line of the Lower Swat Canal. April being a month of slack demand for canal water, because only the mid Kharif crops of cereals and alfalfa were under cultivation, the main canal was running half supplies. Since the canal ran mostly in cutting, the direct outlets on the main line were consequently dry. When we arrived at the first dry direct outlet, our way was barred by a large number of protesting cultivators. Mr Dundas got down from the car and requested for an explanation for the dry outlet and for no water being provided to the cultivators. I explained that all direct outlets on the canal were so designed as to start running when the canal was running three quarters full and the cultivators were perfectly aware of the fact. We could not afford to run the canal and distributaries on full supply, and allow large volumes of water to be wasted at the tail of the canal, merely to oblige the cultivators of small areas of land on the direct outlets.

The next complainant was an individual cultivator whose crop had practically failed, i.e. its yield was under 20 per cent. He complained that although due to failure of his crop he was entitled to full remission of water rate, the canal department was charging him full water rates. Mr Dundas wished to know why this manifest injustice was being done to the cultivator. I informed Mr Dundas that the complaint of the cultivator was quite genuine, and I had full sympathy with him, however, I was powerless to help him as I was bound by government regulations. Explaining further, I said that in order to curb the alarming amount of corruption which prevailed amongst the petty canal revenue staff of the irrigation department, who allowed large amounts of fictitious *kharaba* remissions (remission of water rate for failed crop) to big landowners, the government had decided that the total amount of the remission of water rates for failed crops for each village be divided amongst all the villagers, in proportion to the size of their holdings, irrespective of whether the crop of an individual cultivator had failed or not. 'But the regulation is preposterous,' exclaimed Mr Dundas, 'Why do you follow it?' I replied that the regulation in question had been duly gazetted, so I was bound to follow it. But if Mr Dundas was interested in having it repealed, he should persuade the revenue commissioner to take up the matter with the chief engineer. If they both agreed, they should approach the governor to cancel the gazette notification in question.

Arriving near Tangi village close to the head of the Lower Swat Canal, Mr Dundas was received by Mir Alam Khan, the Khan of Tangi, along with some of his retainers. Mir Alam Khan was an important and influential member of the Khanate class and Mr Dundas met him cordially. Mir Alam Khan of Tangi was over six feet tall and had a commanding personality. He complained to Mr Dundas right in my face, that I was a friend of his rival, Khan Bahadur Saadullah Khan of Umarzai village situated near Tangi, because of which I was prejudiced against him and had, therefore, refused to entertain his request for excavation of drains in his water-logged lands.

Mr Dundas looked at me for an explanation. Addressing Mir Alam Khan, I said, 'Khan Sahib, we have never met before, and are seeing each other for the first time in our lives. You never came to my office to make a request for drains in your lands, nor to my recollection did you send me an application in that connection. It is true that I know Saadullah Khan of Umarzai because we were colleagues at D.I. Khan for a number of years, but I do not allow my personal relations to influence my professional duties. For your information, Saadullah Khan has not made any request for excavation of drains in his lands. If you care to put up your problems to me, you are always welcome.' My spirited speech silenced Mir Alam Khan as well as Mr Dundas.

A few weeks later, I visited Charsadda to attend the farewell party for Captain Leeper, assistant commissioner, Charsadda, on his transfer to another station. It should be recalled that as Lt. Leeper of the South Waziristan Scouts, he had accompanied my survey party to Gul Kachh during the autumn of 1927. He had since been recruited in the IPS. It was a large crowd. Most Khans of Hashtnagar were there. Mr Dundas also attended the party. While I was talking to a small group of guests, Mr Dundas came to meet me, and we greeted each other. He then beckoned Mr Alam Khan of Tangi to join us and, in the presence of everybody, reprimanded him for having made unfounded allegations against me during Mr Dundas's tour of the Lower Swat Canal. Despite his fault, I felt embarrassed by this public chastisement of a local Khan by the British district officer.

Sheikh Mahbub Ali Khan succeeded Captain Leeper as assistant commissioner, Charsadda. So, we had come together again; moreover, his official residence in Charsadda was under my charge. He called me to arrange to have a purdah compound wall constructed at the back of his official residence, because Mrs Sheikh observed strict purdah. The provincial government, however, had issued strict orders against construction of purdah enclosures attached to official residences of officers of all Indian status, who were generally British, and they strongly objected to them. So, I expressed my inability to oblige Sheikh Sahib in that matter. After a few days Sheikh Sahib again called to inform me that he had arranged to have a purdah enclosure, consisting of wooden trellis work, constructed at a cost of Rs 475 which he had paid from his own pocket. He also said that, sometime in the future, I will have to repay the amount with compound interest!

The change from D.I. Khan to Mardan had done me a lot of good. I joined the lawyers club, played tennis regularly, and made many friends among the Indian officers. My next-door neighbour was my old friend Faqir Mohammad Khan, executive engineer, Swabi division, of the Upper Swat Canal. We, together, regularly called on the Nawab of Hoti. The Nawab had numerous visitors and entertained them with tea, cakes, sweets, fruits, and mutton tikkas. On one such occasion the Nawab requested Faqir Mohammad Khan and me to stay back for dinner. He was interested in Muslim history and had a sizeable library of historical works. After dinner, the conversation drifted to the Third Battle of Panipat, fought between the Marathas and Ahmad Shah Durrani in 1761, in which 50,000-strong army of Ahmad Shah routed the 300,000-strong Maratha army. Najib Khan Yousafzai (aka Najib-ud-daula), a commander in Ahmad Shah Durrani's army, had [according to legend] personally cut off the head of Sadashivrao Bhau, the supreme commander of the Maratha army, and retrieved the large pearl necklace worn by the Maratha chief. As a reward for his valour, Ahmad Shah allowed him to retain the necklace. The Nawab of Hoti casually remarked that the necklace in question was in his possession, by inheritance. Faqir Mohammad Khan and I expressed great interest in the necklace and requested the Nawab to allow us to take a look at it. The Nawab called his Hindu *diwan* (caretaker of the Nawab's finances) and ordered him to open 'the treasury' and bring out the necklace for our inspection. The treasury was a small chamber in the basement of the Nawab's drawing room in which we were seated. It had a stout iron door, which was closed with a couple of huge padlocks. The casket containing the necklace was brought out. The necklace was a magnificent piece of jewellery, consisting of a double rope of pearls about 3 feet in length, with a gold clasp. In addition, the casket was chock full of literally hundreds of gold rings and earrings, bracelets, armlets, etc. which the Nawab said, he had collected during his frequent visits to Bombay. He spilled the contents of the casket on the carpet before us, and, perhaps deliberately, turned his back to us and remained in that position for a considerable time.

The Pathans of the Khanate class were known to have a weakness for preventing their *tarburs* (cousins), by hook or crook, from prospering and raising their status. I had noticed this phenomenon with some surprise in the case of the Khan of Shewa in 1921, who had expressed open hostility to one of his poor second cousin who occupied the comparatively humble status of a Zilladar[2] (revenue official) in the irrigation department. The Khan, evidently, was apprehensive that the advancement of his humble relative in the official bureaucracy would pose a threat to his status in future. The same phenomenon surfaced in a more serious form in the case of the Nawab of Hoti at Mardan. The following story about the incident was in circulation in Mardan.

Two sons of the Nawab's deceased brother were living in a house in Hoti town, adjacent to his own castle-like residence which was situated in the most densely populated area of Hoti town. The Nawab was displeased with his nephews and wanted them to leave their old home in Hoti. He placed *chowkidars* (guards) on the lofty roof of the third storey of his house, overlooking the courtyards of the ladies' quarters of his nephews' house. In desperation, the Nawab's nephews started constructing a new house for themselves on a plot of land located on the comparatively quiet road connecting Hoti town with Mardan. Unfortunately, the site they had selected for building their new house was situated on low ground, which lay along the path of the overflow of heavy flood waters in the Kalpani Nullah, which passed through Mardan and Hoti towns. For that reason, they built the plinth of their new house about 5 feet higher than the ground level. It so happened that the cluster of huts belonging to the Nawab's tenants lay on the same low ground, but on the upstream side of his nephews' new house under construction. The Nawab promptly put in a complaint to the deputy commissioner, Mardan, requesting for a stay order against his nephews' partially constructed house on the ground that it would obstruct flood waters and lead to an overflow from the Kalpani Nullah damaging his tenants' huts, which lay upstream.

Major Laughton, the deputy commissioner, Mardan, was apparently reluctant at the prospect of displeasing the influential Nawab of Hoti. He, however, found a way out for dodging the issue and phoned me suggesting that, as the issue was a technical one, I was the most suitable person to handle it. I refused to take up the matter, partly for the same reason, and partly because I was on friendly terms with the Nawab. The astute Major, however, phoned the chief engineer, Mr Burkitt, who promptly sent orders that I should take up the case of the dispute between the Nawab and his nephews and report on its technical aspects. At that stage, an influential jirga of the Khans of Mardan, headed by Sardar Mughal Baz Khan Afridi, political agent of Mohmand Agency and a friend of mine, called on me in this connection. They said that they knew that I was a friend of the Nawab of Hoti, but they were also aware of my reputation as a just and impartial officer. The dispute in question they declared would prove a test case for my sense of justice!

With forebodings of trouble in the offing, I ordered a contour survey of the land in question to be carried out and assess the maximum possible volume of overflow from the Kalpani floods which would spread over the low-lying area. Calculations showed that at the most not more than 3-inch deep water would spread over the wide area surrounding the new house of the Nawab's nephews. The house would, therefore, cause practically no heading up of the thinly spread overspill from the Kalpani floods. I submitted my report with the above conclusions to the government. The Nawab of Hoti was naturally surprised and highly offended. He promptly lodged a civil suit against my award in the court of Mirza Fazal Rahman, civil judge at Mardan. In the face of my 'expert' opinion the Nawab's suit was dismissed by the civil court. He then appealed against the decision, which was dismissed by the high court at Lahore.

In 1946, on my retirement from government service, the Nawab of Hoti came to see me and, after complementing me on my long and distinguished service, mentioned that he had been greatly impressed by my integrity. He then made me a curious offer. He said that he was in possession of surplus cash of a large amount of money and asked if I would join him as an equal partner in setting up an industry in Mardan which would be managed entirely by me. I thanked him for his kind offer but declined, saying that I had little experience in business.

Settling Down Again

The conditions of service at Mardan were near idyllic. However, I still had fits of depression now and then which generally took the form of a feeling of the futility of all existence. My eldest daughter, Zubaida, was another concern. She was growing up fast. Although she did not observe purdah and was my constant companion, I could not offer her the kind of advice and assistance which a mother could. At that time, a young Pakhtun lad, Mohammed Said Khan, belonging to Charsadda, who had recently returned from England after obtaining his degree in civil engineering from the University of London, was placed in my charge as apprentice assistant engineer. In 1935, to my surprise, he requested the hand of my daughter in marriage. I told him that my daughter was just too young for the responsibilities of marriage and advised him to repeat the request after six months through his father, stipulating further that his father must also bring along a jirga of the local Khans of Charsadda, who should make a joint request to me. Punctually, after six months, his father brought along the jirga of Khans, which included among others, Abdul Ghani Khan, the son of Ghaffar Khan, and made his formal request for the marriage of my daughter to his son. So, the *nikah* ceremony of my daughter was held with Mohammed Said Khan according to the Pathan custom, at Mardan. Although she was now Said Khan's legal wife, it was agreed that the marriage would be consummated only after my daughter had come of age. And so, during the Christmas of 1935, Said Khan asserted his right to take home Zubaida, his legal wife. He had also diplomatically sent a message that he was not interested in any dowry and would prefer a car as he did not own one. I ordered a brand-new Plymouth Sedan from a dealer in Peshawar on which the newlyweds drove off.

Thus, with the recent death of my only real brother and the departure of my eldest daughter, a pall of gloom fell on my house. The late Khanum Begum, my first wife, was my first cousin—my father's elder brother's daughter. My uncle's younger son had three daughters of marriageable age, I had thought of marrying the middle one, by name Iqbal Begum. She was many years younger to me, but since she had been known to me, I planned to try my luck and propose for her hand. On the evening of 22 July 1936, I and my three younger children were sitting in gloomy silence in front of the fireplace of the drawing room in Mardan. All of them surprisingly asked me to marry their cousin, Iqbal Apa, and bring her home to restore cheer to the house. The next day was Sunday, on an impulse, I picked up the phone and called my boss, Mr Oram, to take a few days leave. 'It is raining outside,' he said, 'are you sure you want the leave urgently?' 'Yes, my business is somewhat urgent,' I replied. 'Very well,' he said.

The next morning, we packed up and left by car for Abbottabad and arrived there in pouring rain. I saw my cousin and talked to him about my desire to marry his daughter, Iqbal Begum. My cousin was only three years senior to me in age and we had been life-long companions and friends. He agreed and accepted my need for an immediate wedding. My elder cousin, Ferozuddin, however, did not approve and raised a fuss because he wanted Iqbal Begum for one of his sons. He was only pacified when my younger cousin agreed to marry his two remaining daughters to Ferozuddin's sons, and so all three sisters were married, in a quiet ceremony, the same evening.

Iqbal Begum found the mantelpiece of our bedroom like a regular chemist's shop, containing all the existing brands and varieties of laxatives. 'What are these for?' she inquired. I told her that those drugs were absolutely indispensable to me. She ordered all the drugs to be dumped in the store. From that date onwards, I became a stranger to laxatives and my neurasthenia also disappeared. I did not celebrate my second marriage by flute and drums, with lights and bunting. It was a quiet wedding. The news, however, gradually leaked out. A week later, the wife of a friend of mine invited Iqbal Begum to tea and also invited fifty or sixty ladies from among the elite of Mardan society to meet her. Among the invitees was the attractive wife of my British colleague Mr Lawley (executive engineer). Mrs Lawley enquired, 'Who is the guest of honour?' 'It is Mrs Rahman,' someone replied. 'Which Rahman?' asked Mrs Lawley. 'Of course, Mr Rahman, executive engineer, your next-door neighbour.' Mrs Lawley was surprised and spread the news amongst the British circles.

On 1 June 1936, the government conferred on me the title of 'Khan Bahadur'. I called on Nawab Sir Abdul Qayyum, the chief minister, to thank him for his kindness in having nominated me for this honour. He said that the honour conferred on me was rightly deserved for, during my term of service in the Mardan division, wherever he went he had received expressions of praise for me and my work and further that he had not received a single complaint against me either from the petty tenants or the Khan landowners. That, he considered, was a great achievement. He had been receiving numerous complaints against officers of all the government departments serving in that area containing allegations of the officers' partiality for one group of rival Khans or the other.

My wife was expecting soon after our marriage and the baby arrived late in December, a beautiful boy. We advertised for a nanny-cum-governess. A young Sikh lady from a respectable family replied to our advertisement and was promptly engaged. We found out that she was a well-to-do lady with an adventurous spirit and had accepted the job in the spirit of a 'voyage of discovery'. She remained with us for a year-and-a-half as my wife's companion and became a trusted member of the family.

When I joined my wife in Abbottabad during a summer vacation, I seemed to have received a new lease of life. My life became very vital and absorbing. Come sunshine or snow, I made a daily trip, walking up to the top of the 500-foot high Brigade Circular hill and back—a round trip of over 2 miles in half an hour. My health was back to normal. On my return from the walk at around 11 a.m., lunch would be ready which we ate together in a state of peace and bliss.

17 Bannu

In April 1937, I received orders for my posting as executive engineer, Bannu PWD division, with headquarters at Bannu. The work mainly consisted of maintenance of roads and buildings of the district, along with the general supervision of two indigenous canals dating from the Buddhist period, taking off from the Kurram River, which were maintained by the civil department. The Kurram River had a perennial winter discharge of about 600 cubic feet per second, which fed the larger of the two canals, namely the Kachh Kot canal. The river was subject to heavy floods of up to 100,000 cubic feet per second during the summer monsoon period. All the summer floods went to waste in the Indus River, of which the Kurram River was a tributary.

The government was at that time at war with the tribe of Tori Khel Wazirs of North Waziristan, whose territory ran along the boundary of Bannu district. Two brigades of the Indian Army were deployed for the expedition. The only way to access the Tori Khel territory lay along the 16-mile long Tajazai–Tajori dirt road. From Tajori village, the Tori Khel hill-top on which they had erected their bunkers, was situated a few miles away. The Tajazai-Tajori road lay within the administered district of Bannu and consequently, was in my charge. I received orders to spread and consolidate a 6-inch thick blanket of river shingle on the road surface by steam and diesel road rollers, while the road was under heavy military traffic. A Hindu contractor of Bannu obtained the contract. The construction work was placed in direct charge of Kirpa Ram, my experienced Hindu sub-divisional officer. Scores of trucks and hundreds of camels and donkeys were employed for the transport of the shingle obtained from a neighbouring nullah. The actual work of spreading the road metal and consolidation was carried out at night. The work was satisfactorily completed within the stipulated period of a month.

In the meantime, new elections had taken place in NWFP, in which Sir Abdul Qayyum's pro-government party had been badly defeated, and a 'Red Shirt' ministry under the leadership of Dr Khan Sahib had been returned to power with an overwhelming majority. The landslide had occurred due to the immense personal popularity of Dr Khan Sahib who was known to be a dedicated patriot devoted to the welfare of the Pathans. This was in spite of the fact that his ministry was heavily tilted towards the Hindu-dominated

Indian National Congress. In his farewell speech to the outgoing assembly, Sir Abdul Qayyum made a graceful admission that he was passing on the reins of the government to a younger and more vigorous leader. He, however, was deeply disappointed, and survived his defeat by only a few months.

Dr Khan Sahib, the new premier of NWFP, was usually clad in cotton pyjama, a cotton shirt, and a short cotton coat along with a white Gandhi cap. His receptions in the palatial chief minister's residence were marked by extreme simplicity in which tea was served with plain biscuits. He dealt with even-handed justice to both the Pakhtun and non-Pakhtun citizens of NWFP and was honest to a fault. Shortly after his swearing in, he paid a visit to Bannu for inaugurating the Tajazai-Tajori road. He had been informed by secret intelligence that my Hindu SDO and the Hindu contractor had been in collusion. He called me and requested an explanation. 'Was I not aware of the fact that my SDO was a partner in the contract? And if I was aware of the fact, why did I fail to take action against the SDO?' My explanation was a candid admission of my knowledge of the fact that the SDO and the contractor were working as partners. But since the work was of urgent military importance and had to be completed in a specified period of time, and in view of the fact that Kirpa Ram, the SDO, was highly competent in the design as well as construction of road works, I had ignored the situation but made sure that the work was carried out strictly according to the specifications, as regards quality and quantity. So, the government did not lose by this illicit arrangement. My explanation was accepted. In fact, my prestige increased by the successful completion of the work.

Jani Khel Expedition

A clear proof of my rising prestige was soon forthcoming. A few weeks later, I received orders from Sir George Cunningham, governor of NWFP, to make a rapid reconnaissance survey for a new road from the Jani Khel Post along the stony and barren no-man's land to the foothills of the territory of the Tori Khel Wazirs, in order to open a second front against the tribes. Although that work was the legitimate responsibility for the engineers of the MES, yet, the governor had entrusted it to me, a civilian engineer. After due deliberation, the British brigadier of the Bannu Brigade proposed that I should make a surprise dash for reconnaissance for the proposed 11-mile long road, on horseback escorted by cavalry, starting from Jani Khel Post and return before the Tori Khels could catch wind of my presence. Jani Khel Post was situated on a lonely spot along the strategic road that ran along the foothills of the tribal belt of the NWFP, from Jamrud in the Khyber to the border of Dera

Ghazi Khan district of the Punjab, a distance of about 500 miles. The road was meant to serve as a springboard for attacking the tribal territory at any point. It was a dirt road but passed through gravel and shingle-strewn country in the foot hills. The post itself was situated about 30 miles from Bannu. It was arranged that I should reach Jani Khel Post at 8 a.m. sharp on the appointed day where a squad of cavalry of the Indian Army would be waiting for me. I arrived at 8:45 a.m., my car having got stuck in the mud in the way as it had rained the previous night. The British colonel commanding the Indian Army contingent stationed in the Jani Khel Post, eyed me with disdain on my late arrival. He informed me that the cavalry squad would be under my command but advised me to make a hurried reconnaissance.

The stony ground, across which we had to ride, rose gently from the Jani Khel Post up to the line of the bordering hills and one had a clear view. The colonel informed me that our party would be under constant observation through telescopes. If we were ambushed and attacked by the Tori Khels, a contingent from Jani Khel would rush to our help. Furthermore, we would also be similarly under observation from the Girni Post, which was situated near the border hills a few miles to our right. A beautiful thoroughbred bay, saddled and ready, was waiting for me. The Indian non-commissioned officer (NCO) in charge of the squad of cavalry rode by my side presumably to take my orders. A couple of Tori Khel Wazir 'loyal' Maliks also rode by my side. The thirty-six cavalry *sowars* divided themselves into four groups, forming the front, rear, and the flank as moving guards. They rode keeping a distance of a hundred yards on all sides while I rode in the middle. We started at a slow pace. The NCO, the Wazir Maliks, and the moving pickets did the same. It then dawned on me that the whole party was absolutely at my command. I put my horse to a smart trot, then a canter, everyone silently followed suit.

During the onward journey, I had to make frequent stops at drainage crossings to take notes, and also to note where rock cutting was required. After a short while we had covered 10 miles up to the foothills, where there was a couple of miles wide gap in the hills, leading up to a mile-long valley further inland. The hills receded there and at the head of the valley we could clearly see the bunkers of the Tori Khels on the top of the low hills, a mile away. Providentially, a circular outcrop of rock about 100 yards in diameter and about 50 yards high rose up from the middle of the gap in the hills. I dismounted at the foot of the spur, remaining under its cover. Borrowing the binocular of the NCO, I ascended the top of the spur and under the cover of rocks and bushes scanned the remaining one-mile length of the country up to the Tori Khel hills. It was flat, even ground, involving no problems. Completing my notes, I came down, mounted my horse, and ordered my escort to return. My escort had also dismounted behind the cover of the rock outcrop and had noted with audible exclamations of relief my intention to reconnoitre the remaining distance from the top of the outcrop. They gladly re-mounted their horses for the welcome return journey. There were now no temporary stops en route and, moreover, we were now familiar with the way. So, we reached Jani Khel safely at a smart canter, shortly after 1 p.m. I phoned my friend Major Laughton, deputy commissioner, Bannu. 'I am back!' 'Well done,' was his brief reply. I then phoned my wife of my safe return and told her I would be returning home at around 4 p.m., as a local Khan whose village lay along the route had invited me to lunch.

Misfortunes Compounded

My wife was rather weak in health and depended on the nanny to take care of our son. His nanny, unwittingly, did not clean his feeding bottles properly and my son fell sick. The baby got severely dehydrated with diarrhoea and convulsions. Despite our frantic efforts to save him, he could not survive. He was only seven months old. My wife suffered a grievous shock. In order to divert her from her grief, I took her on a fortnight's trip to Kashmir. I had read numerous books on Kashmir. In particular, Sir Francis Younghusband's inimitable work titled *Kashmir* (1909), with illustrations in colour by Major Edward Molyneux, that had fascinated me. It was incredible but true that I had not yet visited Kashmir although it was barely 160 miles away and there was an excellent road from Abbottabad. Now, however, the need to pay a visit to Kashmir became imperative. So, one day in August 1937, packing up our luggage, the whole family piled into our car and we left for Kashmir.

Emerging from the Jhelum River gorge below the town of Baramulla, the valley of Kashmir in all its unearthly beauty lay before us. At long last my dream of visiting Kashmir had turned into reality. Stopping the car and lifting a clod of earth from a nearby ploughed field, I brought it reverently to my lips! We arrived at Srinagar in the evening and hired a houseboat named Queen Victoria. The next morning, we rose with the sun on a clear day. The slanting rays of the sun had turned the broad placid, silver surface of the Jhelum River into burnished gold. There were rows of houseboats moored along the high riverbank overlooked by rows of old houses along both the banks, that looked fresh and cheerful in the morning sun. The services of an efficient cook and a bearer went with the houseboat. We also hired a light boat called shikara along with its boatman. The shikara was at our disposal for shopping and trips farther afield. In due course, we visited the Dal Lake and the various Mughal gardens on its shores in the environs of Srinagar. We purchased excellent lotus honey from a vendor outside the Nishat Gardens on the shore of the Dal Lake. We also went to Manasbal Lake and saw the wonderful vegetable plantations in floating plots of soil in that lake.

My wife responded wonderfully to the change and, in large measure, appeared to have overcome her grief. Brushing aside the objections of the attendants against women visiting the shrine, my wife and I paid homage to Kashmir's patron saint, the Shah-i-Hamadan, whose shrine-cum-mosque lay along the bank of the Jhelum River in Srinagar town. After praying for the soul of the saint at his tomb, we visited the adjoining Shah-i-Hamadan mosque and were shown many rare manuscripts, the most remarkable of which was a copy of the Holy Quran handwritten by the Mughal Emperor Aurangzeb, with his seal and date stamped on it. When we came out of the shrine, the attendants, and others who had assembled outside the mosque, taking us to be some important visitors from outside the Kashmir state, bitterly complained about the inhuman treatment being meted out to them by the Hindu Maharaja Hari Singh's Dogra regime. This was confirmed the very next day at Tangmarg where we hired horses for riding up to the top of the green plateau of Gulmarg. The police minions mercilessly lashed some over-eager horse-owners who stepped forward from the row of horses, to offer us their mounts. We strongly protested, and in reply the policemen said, 'Sir, it is only to save you from unnecessary trouble.'

We returned to Bannu refreshed from the Kashmir trip. Early in 1938, Major Stuart arrived from a long leave and took over the charge of the Bannu division from me. To my amazement, I was thereby reverted to the post of assistant engineer. The Kachh Kot and Landidak canals alone remained in my charge. My unprecedented demotion was undoubtedly a deliberate move. As long as I remained a member of the ISE, even in an officiating capacity, I was a potential threat to the six British interlopers from the MES, who had superseded me through the back door. The threat to them was more real now, as the government was implementing reforms and administrative decisions of senior officials which fell within the purview of the chief minister. Mr Burkitt (chief engineer), had retired and Mr Ross had replaced him. He ordered my stepping down for a brief period of three months to 'teach me a lesson' and to make me accept Class-I service, which I had consistently refused for many years, claiming my right to be confirmed in the Indian Service of Engineers in view of my excellent service record. I took my demotion in stride and with dignity. I left the fine new bungalow which was the official residence of the executive engineer and found a house in Bannu city. In the open space outside the house, a clay pottery vendor had spread his wares over a large area. In the small kitchen attached to the house my young wife cooked our meals, while I ate my meals on a kitchen table overlooking the bustling courtyard outside. The period of three months thus turned out to be one of the happiest periods in my life.

I now turned my undivided attention to the two indigenous canals in my charge and came to the conclusion that a lot of new areas could be brought under irrigation if the summer floods of the Kurram River could be harnessed by means of a dam and a reservoir. I, accordingly, wrote a note on the subject and submitted it to the higher authorities. That report was the basis of the new Kurram Garhi Canal taking off from the Kurram River about 7 miles upstream of Bannu along with the Baran Dam for storage of the Kurram floods, which I planned and designed myself, and partly supervised its construction as irrigation advisor to the Government of NWFP many years later.

During the period of my stay in Bannu, Pandit Jawaharlal Nehru paid a visit to the NWFP, presumably at the invitation of Dr Khan Sahib's Congress government. In due course, Nehru also visited Bannu. It so happened that arrangements for his speech were made in the open space adjoining my house in the Bannu city, close to the pottery vendor's outdoor store. Nehru made his public speech in the evening. Although government officials were prohibited from attending political meetings, I joined the crowd to hear the speech of this eminent personality. The loudspeaker went out of order many times during the speech. I do not remember the details of Nehru's speech, but I do remember that he delivered his speech in chaste and eloquent Urdu. It made me wonder on what grounds the Hindus of the United Provinces (to which Nehru belonged) were clamouring for the adoption of Hindi as the official language of India.

Somewhat disheartened by my demotion to the rank of assistant engineer, I sent in my written consent for being absorbed in the Class-I service. I was promptly promoted to the rank of permanent executive engineer in Class-I service, with retrospective effect and with all perquisites, and transferred to Swabi division of the Upper Swat Canal in Mardan. In doing so, five out of the six British REs from the MES who had been transferred to the PWD with the introduction of reforms in the NWFP in 1933, became senior to me.

Early in 1938, Dr Mohammad Allama Iqbal passed away. He had been ailing for the past few years, and clearly foresaw his approaching end. I received a great personal shock on learning of his untimely death. I had not met him since 1915, when he had given me a letter of introduction to Mian Fazl-i-Husain,[1] KCSI, and a letter of recommendation to the finance minister of Hyderabad state. I was planning to pay my respects to him personally, sometime or the other, to tell him that his poetry had, in a very great measure, moulded my outlook on life. In fact, ever since the first publication of his poems like *Sitara* (The Star) published in the *East and West* magazine, Bombay, I had read most of his subsequent works, including *Goristan-e-Shahi* (The Royal Cemetery); *Siqaliya (Jazeera-e-Sisli)* (The Elegy on the Island of Sicily); and other poems. Also, his famous philosophical works like the *Masnavi, Payami Mashriq, Bal-i-Jibril, Zarb-i-Kalim, Javed Nama, Zabur-i-Ajam* and others, during the 1930s. I had committed to memory many of his verses which caught my fancy. In fact, even after his death, I had made comprehensive studies in philosophy in order to understand his work.

My eldest son, Aslam, returned from the UK in September 1938, after qualifying as a civil engineer from the University of Sheffield which was a matter of great satisfaction to me. Shortly after, Lord Linlithgow, the new viceroy of India, arrived at Mardan to perform the opening ceremony of the recently constructed hydroelectric power station on the Upper Swat Canal at Jabban near Malakand. As a member of Guides Cavalry Club, we were introduced to the viceroy along with our British fellow members. The function at Jabban was a colourful one. The viceroy made the customary complimentary remarks about the engineers of our PWD who had worked on the hydroelectric project. Lord Linlithgow brought a gift of Rs 10 million for the improvement of livestock in India. It was a great deal of money in those days. This gesture was mainly aimed at improving the breed of livestock and was also, transparently, intended to please the Hindus of India who held the cow sacred.

The incident reminded people of a joke about his immediate predecessor, Lord Willingdon. Shortly after he assumed the office of viceroy, Lord and Lady Willingdon went to inspect a Hindu *gowshala* (a large establishment that bred cattle) near Delhi, and also served as a shelter for aging cattle. Lord and Lady Willingdon were taken by different groups of officials to different parts of the establishment. Lady Willingdon was taken to the stall of a prize bull and the official escorting her remarked with pride: 'Your Excellency, we are happy to inform you that his bull has been with 500 cows during the current season.' 'How remarkable!' exclaimed Her Ladyship. 'I think His Excellency should be informed.' An official hurried and informed His Excellency. 'Give my compliments to Her Excellency,' retorted Lord Willingdon and tell her, 'but not with the same cow!'

The Rise and 'Fall' of Faqir Mohammad Khan

Dr Khan Sahib was now firmly in the saddle as chief minister of NWFP. He recalled his friend, Faqir Mohammad Khan (executive engineer), from the Punjab and appointed him as superintending engineer in the Peshawar PWD. Thereby, all the RE officers of the ISE who were junior to him in service became his subordinates. He issued strict and irksome orders to all executive engineers serving under him, including the RE officers, to personally tour the drains in waterlogged areas on horseback. He was my close friend and we went to Malakand together to attend the inauguration function at Jabban. He was in high spirits. Faqir Mohammad Khan's tenure of office as superintending engineer was, however, short lived, and came to a tragic end. The roof of the Peshawar Museum was old with cracks in the supporting jack arches. So, it was decided to replace the dangerous roof by steel battens and brick tiles over the steel girders. One afternoon Faqir Mohammad Khan, while having tea with his executive engineer, Majidullah Khan, suddenly expressed the desire to inspect the dismantling work of the jack arches of the museum. So, both the officers went to the museum, accompanied by a Hindu subordinate. In sheer bravado, Faqir Mohammad Khan started walking on a girder on one side of which was the intact but cracked jack-arch roof, while there was nothing on the other side. He was followed by the executive engineer and the subordinate and while all three were still on the girder, it suddenly caved in onto the floor below. All three, unfortunately, fell with the falling material. Faqir Mohammad Khan was pinned down underneath the fallen girder and was killed on the spot. Majidullah Khan and the Hindu subordinate were both seriously wounded and expired a few days later, in hospital.

Faqir Mohammad Khan's subordinate RE officers heaved sighs of relief. Major Hainsworth was appointed superintending engineer at Peshawar. 'It was divine judgment,' he remarked audibly in my presence, referring to Faqir Mohammad Khan's tragic death. He had evidently in mind Faqir Mohammad Khan's order issued to him for touring the drains on horseback. Some months before Faqir Mohammad Khan's death, his only son Nisar Mohammad Khan had also died under equally tragic circumstances. Young Nisar was a brilliant and promising doctor, having qualified from the United Kingdom. He was posted in the Indian Medical Service. A fair Pathan, he was generally indistinguishable from Europeans. He met his death from the bullet of a Mahsud sniper in South Wazirastan, while going to Sarwakai Post in a convoy of armoured cars along with other European military officers, some of whom were also killed in the ambush. The convoy was ambushed at Chagmalai, the very spot where, in 1928, a Mahsud Malik had brandished his knife-scimitar close to my face, while explaining his irrigation problem to me by drawing imaginary lines in the air. Faqir Mohammad Khan, at the time of his son's death, was posted as executive engineer at D.I. Khan. The

Mahsuds who had shot his son apologised to him, for having mistaken his son to be a European officer. Faqir Mohammad Khan was, however, reported to have partially lost his mental balance since then, and had become reckless.

And now Hainsworth, formerly my junior, was our superintending engineer. It was customary for the executive engineers to submit approximate forecasts of their annual requirements of roads building, and canals maintenance programmes in the beginning of the fiscal year in July, for purpose of budgeting. These were followed in September by detailed estimates of the works proposed. The differences in the forecast and detailed estimates, if any, were adjusted or modified by the superintending engineer, in the budget. My predecessor, Lawley (executive engineer), had submitted a forecast for only Rs 76,000 for the maintenance of roads in the Swabi division. Lawley was by nature a miserly fellow. He smoked Red Lamp cigarettes, the cheapest brand in the market. He had, as usual, sent an over-economical forecast. When I prepared an estimated in September the actual amount came to Rs 88,000, a difference of only Rs 12,000 which was trifling.

On receipt of my detailed estimates Hainsworth made a surprise tour of the roads in my division. During his inspection which he made in his own car, we were seated in the back seat while his driver was at the wheel. Hainsworth abruptly asked me about the 'large' difference of Rs 12,000 in the forecast in the detailed estimates of the proposed roads programme. Somewhat taken aback at the tone in which the query had been made, I explained politely that Mr Lawley was responsible for the forecast figures which were not based on detailed estimates and the difference of Rs 12,000 between the forecast and the estimates was trifling.

Hainsworth replied, in a haughty tone, that I would be allowed only Rs 76,000 for the roads programme. I suggested that in that case I would delete certain items from the budget in order to bring it down to the sanctioned amount. 'You will do nothing of the sort,' snapped Hainsworth. 'You will complete your programme as proposed, but at a cost not exceeding Rs 76,000. Losing my temper, I replied: 'Does it mean that I should make up the difference of Rs 12,000 from my own pocket?' Hainsworth flushed, but noting that I was also angry, he controlled himself; there was complete silence between the two of us for the rest of the tour. He sent me a letter from Peshawar sanctioning the extra amount of Rs 12,000 for the roads programme. Hainsworth was soon reverted to the post of executive engineer and Mr Brown, the permanent incumbent, took over from him. Surprisingly, only a week after taking over, Mr Brown sent me intimation of his four-day tour of my division. Arriving at Mardan, he told me that he had received unfavourable reports about my division; that I was lagging behind the target in the completion of works in my division; and that the standard of maintenance of works in my division was poor, and so forth. It was plain that Hainsworth was taking his revenge through Mr Brown.

Mr Brown was among the most brilliant engineers to be posted in the NWFP. After obtaining a Tripos in mathematics from Cambridge, he took an engineering degree from University of London. He was a brilliant mathematician and a talented engineer, with a capacity for original research in irrigation sciences. He was also the son-in-law of the more famous F.H. Burkitt, our former chief engineer, who had gained international standing with the publication of his classic *Experiments on Broad Crested Weirs*, on which he had read a paper in the Punjab Engineering Congress in 1919. A reference to his work was subsequently made by R.B. Buckley in his famous and monumental work, *The Irrigation Works of India* (1893). Mr Brown had been my superintending engineer when I was posted as executive engineer, Bannu division, in 1937 and we had been on friendly terms. His officious attitude now was undoubtedly due to Hainsworth's attitude. This was, of course, to be expected. Whatever the private differences amongst themselves, the British in India invariably backed a fellow British officer, against an Indian.

Our first destination was the Jagannath Canal rest house on the Maira branch of the Upper Swat Canal, where we arrived late in the evening. I had taken my cook along with me with rations for a four-day stay in various rest houses in my division. Mr Brown's 'bearer' came to me with a message that I was to be Mr Brown's guest during the tour. I sent a curt reply that I had brought my cook along, with adequate rations, and would prefer to have my own meals. Back came the message that Mr Brown expected my agreement to the arrangement. I sent the same reply again. Back came a longish note stating that Mr Brown was not used to receiving refusals from his subordinate officers and that according to the established convention whenever two officers happened to tour together, the junior officer was expected to be the guest of the senior officer. I agreed to accept his hospitality on this basis and wrote back that I would gladly be Mr Brown's guest during the tour.

Mr Brown, I discovered, possessed a lot of knowledge on a vast range of subjects apart from engineering. I myself had reached an age when only serious subjects interested me. In those days, I was deeply interested in the philosophy of science, particularly in the idealistic view of scientific laws. The perusal of Sir Arthur Eddington's *The Nature of the Physical World*, in particular, had made a deep impression on me. I was also interested in the theory of relativity and Eddington's clear exposition of the subject. I had also read Iqbal's famous work, *The Reconstruction of Religious Thought in Islam*, and was struck by the similarity of views—although expressed in different terms—of both these thinkers on the nature of reality.

Mr Brown appeared to be equally interested and conversant with all these subjects. So, despite his somewhat cold and distant attitude towards me, we had long and interesting discussions on these and allied subjects after dinner each night during our four-day tour of my division. In the course of our discussions on relativity, Mr Brown remarked that it was not difficult to visualise a four-dimensional universe. 'Imagine a world of three dimensions,' he said, 'wound like a spiral round a vertical dimension of physical time.' Noticing my sceptical smile, Mr Brown remarked, 'You don't seem to realise that I took a Tripos in mathematics at Cambridge!'

Mr Brown sent a somewhat chilling inspection note on his tour of my division. The note alleged that I was lagging behind in the progress of road repair work. It was not difficult to submit a rebuttal of the charges as the working season had hardly begun and would last throughout the winter months. Moreover, the total amount of road work in my division was the trifling amount of Rs 88,000 which would only require a couple of months to accomplish. One good aspect about the British rule in India was the constitutional protection provided to gazetted officers by the government. I knew that although my superiors could damage my career, which they had already done, by various devious mechanisms to promote their own interests, they could do little else. While taking a stand against my British superiors, I was perfectly aware of this fact.

One Sunday morning, I happened to visit Dr Khan Sahib at his house in Charsadda. I found his younger brother, Khan Abdul Ghaffar Khan sprawled on a bare string charpoy reading an Urdu pamphlet. 'What are you reading,' I asked. 'It is Bolshevik literature,' he replied. During 1938 and 1939, my relations with Dr Khan Sahib, the chief minister of NWFP, had become fairly close. He appeared to have been impressed with my work and integrity and placed me in charge of the construction of the sugar mills at Takht Bhai, although the mill fell within the jurisdiction of an adjoining division. I also designed a two-storey private residence for him, in his village, at his request. He extended an open invitation to me and my wife to visit his home and family at Utmanzai, whenever we pleased. His beautiful daughter, Maryam, was about my wife's age and they became good friends. Dr Khan Sahib's English wife, Mary, a good-natured lady, also befriended my wife. Mrs Mary Khan was somewhat unbalanced in mind, it was said, because of her husband's frequent and long confinements in Indian jail, and also partly due to the untimely death of her young son, a graduate of the

University of Oxford, from typhoid fever. She gave me sensible advice on diet as a prescription for longevity, suggesting moderation in eating habits, by eating only small quantities of meat and that only with vegetables, and that too with less fat. However, on the contrary, Dr Khan Sahib was extremely fond of *chapli kababs*. He called them Pathan cutlets. My wife and I often joined the family during their informal lunches.

During those days, I had many frank and informal discussions with Dr Khan Sahib about the pro-Hindu Congress stand taken by his and Abdul Ghaffar Khan's party of Red Shirts in NWFP. Dr Khan Sahib's view in those days was that since the Hindus formed an overwhelming majority in India, the Muslim minority had, perforce, to come to terms with them. Asked about the Red Shirts' demand for the Gandhi inspired 'Pathanistan' by uniting the Pakhtuns across both sides of the Durand Line, which in practice meant the amalgamation of NWFP with Afghanistan, and in answer to my specific question: 'Would you consent to join the backward country of Afghanistan, which was under an autocratic and primitive rule?' Dr Khan Sahib exclaimed, 'But we shall rule them, rather than be ruled by them.' 'And what about the Punjab and Sind?' I asked. 'I am not concerned with the Punjab and Sind,' was his somewhat disconcerting reply.

In the summer of 1939, Mr Gandhi visited NWFP on the invitation of Dr Khan Sahib. I saw Mr Gandhi for the first and last time, in the Company Garden at Abbottabad. He had just delivered a speech in Hindi, and a crowd of people, mostly Muslims, were milling around, to catch a glimpse of the great man at close quarters. I found Mr Gandhi a pleasant and a likeable person. I recall that a Muslim asked him: 'Sir, why don't you bring about unity between the Hindus and the Muslims?' Mr Gandhi replied in mild smiling tones, 'This rests in the hands of God Almighty.' I admired the frankness of his reply.

18 The Second World War, 1939–1945

Since April 1938, I and my son, Aslam, had remained glued to our old HMV box radio, listening to Hitler's unintelligible speeches which were usually delivered in thundering tones followed by thundering applause, usually from Munich. We both had forebodings of another war in Europe, to the amusement of our friends and relatives who thought we both, father and son, were obsessed with war phobia. Sure enough, events in Europe moved with lightning speed. Soon after Prime Minister Neville Chamberlain's parleys at Munich with Hitler, which he triumphantly brandished for all to see outside his official residence at 10 Downing Street, London, Hitler moved swiftly into Poland, and shortly thereafter into Belgium as a prelude to his invasion of France.[1]

On the evening of 3 September 1939, I and Aslam were startled to hear the voice of our King Emperor George VI from London: 'My beloved subjects,' declared the King Emperor, in what seemed like stuttering tones, 'We are at war with Germany.' So that was that.

On a Sunday in May 1940, I went to Peshawar on a shopping errand. A clerk from the chief engineer's office met me in the street, and said, 'Do you know you have been transferred to Kohat.' I hurried back to Mardan to find the transfer orders waiting for me, in my office. Dr Khan Sahib's government had issued orders for all executive engineers to regularly visit the village *hujras* and to spend ten nights every month there to study and discuss the problems of the farmers. Some of the other executive engineers were reportedly faking their stay in the village *hujras*, and thus successfully evading the government's orders. I frankly refused to obey the order which appeared to me to be impracticable and devoid of sense. I wrote to the superintending engineer explaining that my division was small and compact, and a cultivator or landowner could reach my office at Mardan, and after transacting business with me, return to his village in a single day. I, therefore, expressed my inability to comply with the orders requiring executive engineers to pass ten nights per month in the village *hujras*.

Here was the opportunity to set an example! I received a stinker from Mr Oram (chief engineer), my former boss and friend, censuring me for insubordination, for non-compliance with the government orders! 'Sir, you need not have come to the office at all,' said my head clerk, as I reached my office punctually at the proper hour after having taken charge of the Kohat division. 'I will send you any paper requiring your signatures at your home!' It was now crystal clear; I had been transferred to Kohat as a measure of punishment! The Kohat district had until then comprised a PWD sub-division, attached to the Bannu division. There were no government canals in the district. There was a small annual road repairs programme only which the sub-divisional officer at Kohat used to complete within a couple of months. The SDO Kohat lived in a decent old PWD bungalow. With the upgrading of the Kohat sub-division into a division, another sub-division was constituted at Hangu.

There was no official residence available for the executive engineer at Kohat. I was expected to construct a new residence for myself. The new building construction programme in my division comprised, in addition to the construction of the executive engineer's residence, the construction of the civil surgeon's residence at Kohat and a police training school at Hangu. In addition, there was the usual annual maintenance work of all the government buildings in the small Kohat district. It was altogether a pitifully small volume of work for a division. I rented a century-old, almost dilapidated bungalow, built of massive sundried brick walls cracked with age, on the Mall road in Kohat. Luckily, Sheikh Mahbub Ali Khan was the deputy commissioner at Kohat. I invited him to lunch and showed him the dismal abode I was living in and asked for his help in securing better accommodation for myself. He promised to speak to the British brigadier in charge of the Kohat Brigade to sanction allotment of a bungalow to me in the cantonment. I was consequently allotted a magnificent bungalow, one of the four bungalows that had recently been constructed on a short, but picturesque, road in the cantonment, named the 'Happy Valley'. The allotment of the bungalow was subject to the condition that I would vacate it within forty-eight hours if required by the military.

I brought my family to Kohat in September. We passed some happy times in our fine house. I relaxed and got used to having little official work on my hands. We were grateful to Sheikh Mahbub Ali Khan for having managed to get us our nice house. The Second World War was on, and the Kohat cantonment was overflowing with the arrival and departure of numerous Indian regiments. Sheikh Sahib held an almost royal court in his palatial residence, the famous Cavagnari House[2] entertaining lavishly his Muslim,

Hindu, Sikh, and British guests from the army and the civil departments, and other distinguished persons. I soon got tired of inaction and looked about for some irrigation development schemes in the Kohat district. Unfortunately, there was no large river within the Kohat district. The mighty Indus River flowed in a deep gorge along its eastern boundary; the Kurram River only touched its western corner at Tal village. There were only large and small hill torrents with their waters heavily laden with silt.

The only river carrying sweet water in the district was the Kohat Toi, which had its source in the Samana hills near Hangu and flowed in a west-east direction, along a 70-mile course to join the Indus River near Khushalgarh. Kohat Toi had a considerable perennial flow with which intensive irrigation was carried on en route. It, however, carried large volumes of floods during the summer monsoon season which, if stored, could have additionally irrigated a much larger area. My problem was to find a suitable site for the storage of the summer floods. The only direct storage site on the river was at the Khwaja Khizr gorge, but the lands upstream were under intensive cultivation and considered so valuable that building a dam and reservoir there was out of question. So, the best solution appeared to be to find an indirect storage site. This was located at a narrow gap in the hills running at right angles to the course of the Kohat Toi and parallel to the Kohat–Bannu road, about 5 miles from Kohat town.

A weir was proposed across the Kohat Toi River a few miles upstream of Kohat town for dividing the flood waters of the Toi along a supply channel which was to cut across the low chain of hills running parallel to the right bank of the Toi. These channels could be dammed at the gap in the hills where a sizeable reservoir would be created to store the floodwaters of the Toi. I formulated this scheme and submitted it to authorities of my department.

The scheme remained dormant until the 1950s, when it was taken up by the Pakistan government as part of its irrigation development programme. Said Khan, my son-in-law, who succeeded me in that position at a later date, had located an alternative site for indirect storage at a place called Tanda not far from the site that I had selected. Geological investigation showed Said's choice of the Tanda site to be better than my choice. So, an indirect storage dam and reservoir for the floods of the Kohat Toi was constructed at Tanda during the 1960s. The weir on the Toi and the alignment of the supply channel as I had originally proposed were retained. But I am marching ahead of time.

Another interesting scheme was sent to me for study by Sheikh Mahbub Ali Khan (deputy commissioner, Kohat), under special confidential instructions of Sir George Cunningham, governor of NWFP. An area of 1000 acres of land consisting of fine soil lay in the vicinity of Tal village where the Kurram River emerged from the Parachinar gorge into the Kohat district. The land in question had long lain fallow but could easily be irrigated from a water course from the Kurram River from a point upstream. This valuable land had long been a bone of contention between the Talwals of the Kohat district and the Buland Khel Wazirs of North Waziristan, whose northern boundary touched that area. A fifty-year-old voluminous confidential file on the dispute existed in the office of the deputy commissioner of Kohat. I received confidential instructions to study the file and submit a report on the old dispute. After an intensive study of the case on the file, I submitted my recommendations in favour of the Talwals of the Kohat district. Sheikh Mahbub Ali Khan transmitted my findings to the governor. In due course, the file was returned to me for perusal of the governor's comments on my report. To my surprise the governor had disagreed with my findings. There was a curt remark. 'I do not agree,' in the margin of my report! I learnt about the governor's decision on the case the next year, after I had left Kohat division. My successor was ordered to construct a watercourse from the Kurram River to irrigate the land. After the watercourse had been constructed the land was divided into two equal parts and a line of demarcation was excavated on the ground. A small mosque, consisting of a single room and a veranda, with one minaret, was constructed at one end of the line of demarcation. A force of the Frontier Constabulary was temporarily posted on the land, under whose armed supervision the Talwals and the Wazirs were invited to occupy the half portions of their land, which adjoined their respective territories. Both the parties to the dispute accepted the division, and the constabulary was withdrawn!

Our stay in the Happy Valley bungalow was brief. One fine morning, we found the Kohat barracks crowded with Sikhs, all wearing bright orange turbans. A Sikh regiment had arrived at Kohat. I received intimation from the military authorities to vacate the Happy Valley bungalow within forty-eight hours. Sheikh Mahbub Ali Khan again came to my rescue and very kindly placed the vacant bungalow of the assistant commissioner, Kohat, at my disposal. And so we moved into that more comfortable house, with its mature lawns, ornamental shrubs, flower beds, vegetable garden, and fruit orchard, and enjoyed its amenities till my transfer from Kohat.

There was an avenue of magnificent hardwood shisham trees growing on one side of one of the main roads in Kohat cantonment, opposite the Royal Air Force complex of buildings. Sheikh Mahbub Ali asked me to auction off those trees, as he wanted them for construction of his house in his village. The felling and auction of green trees along roads and canals was forbidden, so I expressed my helplessness to oblige him. Sheikh Sahib said, 'You need not bother. I will manage this affair myself.' A few days later I received orders to auction off a dozen of the green trees. The Royal Air Force authorities had written to the government that the row of trees in question obstructed the take-off and landing of planes at the Kohat airport!

Sheikh Sahib then came out with the unexpected demand reminding me of the 'debt' I owed him for his having spent the sum of Rs 475 from his own pocket in the construction of the purdah enclosure for his official residence at Charsadda. He requested me to add a brand-new suite, equipped with bathroom and modern sanitary fittings to the outer guest room attached to his official residence, the Cavagnari House at Kohat, for the convenience of Sir Arthur Parsons, advisor to the governor, who was coming to stay with him as his house guest. Sir Arthur would be on leave during that period and would be relaxing from his onerous duties of state. Since Mrs Sheikh observed purdah, he added persuasively, the construction of the outside suite and room had become necessary! I was caught in a fix. Sheikh Sahib's request amounted to alterations to a government building, which could not be undertaken without the express approval of the government. In such situations, special funds were required and I had neither the approval nor the funds to meet Sheikh Sahib's demand. Providentially, I received a sizable sum of Rs 30,000 as the share of Kohat division from the funds sanctioned by the Government of India for 'extraordinary special repairs' to buildings in the NWFP. Although the sum could not strictly be spent on 'extensions and improvements', the proposed work was for the convenience of the advisor to the governor. Those were economical times, the total estimate for the work amounted to Rs 15,000 only. My sub-divisional officer, Sultan M. Naeem, later chief, engineer, Tarbela and Mangla dams, made a good job of it and completed the work to the satisfaction of Sheikh Sahib.

Some Good News

Sheikh Sahib invited me to have tea with him and Sir Arthur Parsons, as a gesture of thanks. He eulogised my professional proficiency to Sir Arthur, who said he already knew my work and me. The evening was spent in congenial conversation about the proposed water supply schemes in Kohat district. And now there arrived an unexpected piece of good news. The government appointed me superintending engineer, Public Works Department at Peshawar. Following my friend Faqir Mohammad's brief and tragic tenure, I was the second non-Britisher to be appointed to this senior position in the NWFP.

The war in the East had taken a critical turn. The Japanese had occupied Rangoon and were knocking at the gates of India. As a consequence, all the former REs, inducted into the NWFP PWD, were recalled, and had joined the Burma front. Mr Oram (chief engineer) sent me a flattering demi-official letter. 'The government,' he stated, 'were fully satisfied with my professional efficiency, and administrative ability to hold the senior appointment of superintending engineer.' With a smile, I remembered that, only a few months earlier, I had been censured by Mr Oram for insubordination. My appointment orders created a stir in the civil and military circles of my Muslim and non-Muslim friends in Kohat. Sheikh Sahib regarded me with added respect and threw a splendid farewell party in my honour. Lavish farewell parties were also arranged by other friends. My Hindu and Sikh military friends went to great lengths to outperform each other in arranging lavish farewell dinners. And so 'in a blaze of glory' I took over the Northern Circle, at Peshawar.

I took over the position from Col. Lang Anderson, the senior-most and the last of the former REs of the PWD, to leave for Burma front. 'I am handing over to you the best bungalow in the Racecourse Gardens. Being situated at a height it commands a superb, uninterrupted view of the racecourse, and the hills and snowy ranges in the north. Many European officers will offer to exchange their house with yours, on the plea of better purdah facilities existing in their own houses. But please don't listen to them and hang on to your splendid house.' Thus, the colonel offered me his sage parting advice. The very next day the superintending engineer, electricity and his wife paid me a visit. They offered to exchange our respective houses. Their house, they said, had a sheltered back yard protected by wooden trelliswork. 'Your wife being in purdah will surely appreciate the better purdah facilities of our house,' they remarked. I invited my wife to join us in the drawing room and introduced her to the couple. I said, 'This house has a fine view and I would not exchange it under any circumstances.' They went away disappointed.

The war had taken a dramatic turn. While the Japanese were poised to enter India through the Kohima front in Assam, Gandhi launched his third and final 'quit India' campaign against the British in India. The Congress ministries in India, including Dr Khan Sahib's ministry in the NWFP, had resigned. It was common knowledge that, in 1937 or thereabouts, Mr Mohammad Ali Jinnah, leader of the All-India Muslim League, met Sir Sahibzada Abdul Qayyum, chief minister of NWFP, and sought his help for the Muslim League, but his overtures were politely rejected by the Sahibzada. In November 1939, a coalition ministry was formed under the aegis of the Muslims League, with Sardar Aurangzeb Gandapur, as the chief minister.

Both he and I were former students of the Aligarh college, where he was my junior. I had taught him mathematics for his FSc examination. He freely advertised the fact of having been my 'pupil' at the Aligarh college. A Hindu and a Sikh minister were added in the ministry. The Muslim League under the leadership of Mr Jinnah celebrated the resignation of the Congress ministries as a 'day of deliverance'.

The British government, harassed by successive defeats on the western and eastern fronts, sent a mission[3] to India, headed by Sir Stafford Cripps, offering self-government to Indians under full dominion status on the successful conclusion of the war. The Indian National Congress rejected Cripps's offer of dominion status, boycotted his mission and started their 'Quit India' campaign. In reply to the Congress's demand of quit India, Mr Jinnah responded with 'divide and quit India'. His statement, however, was not backed by any anti-British campaign. He believed in constitutional methods, so all that mattered at that time was the Congress's quit India campaign, which was launched in the form of a 'civil disobedience' movement.

19 Peshawar

My PWD Circle, with headquarters at Peshawar, comprised five divisions, each under an executive engineer. It included all the principal canals in the NWFP and included buildings, roads, and public health works of Peshawar and Mardan districts, and Malakand Agency. It was one of the heaviest PWD circles in India, with a very large technical and revenue establishment.

I took up my place amongst the British heads of department. My next-door neighbour was Mr Price, finance secretary, and a British snob of the old order. He had a wife and a grown-up daughter. On the occasion of his daughter's wedding, a few months later, besides British guests, he only invited Nawab of Junagarh's nephew, Nawabzada Mir Saeed Alam, who was sessions judge at Peshawar at that time, and his charming wife and daughter, from among the Indian officials at Peshawar. He confided to me later, saying, 'Mrs Saeed Alam is a princess in her own right.'

However, not all of my British colleagues were snobs. We had very cordial relations with Mr Dundas, revenue commissioner, and his charming wife; and Col. Leeper, secretary industries, and Mrs Leeper. The former became governor of NWFP in 1948. The latter, as Lt. Leeper, had accompanied my survey party to Gul Kachh in 1928 and had since joined the Indian Political Service. And, of course, there was my immediate boss, Mr Brown, chief engineer and secretary to the government. His relations with me were good and sometimes even cordial but always formal. A good-looking person himself, his spouse, Mr Burkitt's daughter, was plain looking but an immensely proud lady. Mr Brown enjoyed tremendous prestige both among the British and Indian official circles due to his reputation as a brilliant engineer. He was a heavy smoker and kept an expensive cigarette case stocked with an expensive brand of cigarettes. When he was advised to give up smoking, he took to drinking whisky, sometimes immoderately.

Captain Johnson, who had taken my party to Gul Kachh in 1928 as head of the escorting South Waziristan Scouts force, had similarly been taken into the Indian Political Service. He was now a full colonel and posted to the senior position of resident in Waziristan. Meeting me at Peshawar, he had remarked: 'Khan Bahadur, if you ever chance to visit Waziristan again and go to Gul Kachh I shall arrange the whole length of the road from Jandola to Gul Kachh to be picketed by South Waziristan Scouts in your honour!' He was evidently referring to the road link connecting NWFP with Balochistan, which was a product of my inspiration.

One sunny Sunday morning in winter, I was strolling in the front lawn of my house enjoying the fine view of the green racecourse in the foreground below me and the range of purple hills in the background. An Englishman was taking a brisk walk in the racecourse. Presently, he opened the iron gate of the perimeter fence around the Racecourse Gardens and approached my house along the narrow depressed footpath between the toe of embankments of my lawn and the barbed wire fence. He stopped, and with 'Good morning Abdur Rahman,' clambered up the 10-foot high slope leading up to my lawn where I stood. It was Sir George Cunningham, governor of NWFP.

'You have a fine lawn,' remarked Sir George, looking around at the flower beds, particularly the bushes of sweet peas planted in rows. 'But the slope of your lawn facing the racecourse looks bare, so does Price's (finance secretary). You should cover your slopes with evergreen bushes.' 'Sir George,' I replied, 'This work would involve engaging extra gardeners for planting and daily watering the bushes which, I regret to say, I cannot afford.' Sir George smiled. 'Very well,' he said, 'I will send a team of gardeners from the Government House, to plant rows of mesquite bushes along your and Price's slopes. They will also water the bushes for six months. 'Tell Price.' The next day a team of gardeners duly arrived and planted four rows of bushes along the 100-yard length of my slopes, and six rows along Mr Price's slope which was 5 feet higher. The Government House gardeners regularly watered the bushes for the next six months, till our slopes were carpeted with green mesquite.

In 1941, the fateful Pearl Harbour event occurred, heralding the America's entry in the Second World War. Japan's swift victories in the Far East, culminating in its conquest of Rangoon, and knocking at the gates of India on the outskirts of the Indian province of Assam, left the Indians stupefied. Mr Gandhi openly talked of negotiating terms with the victorious Japan. In NWFP, the Quit India Movement took the form of public exhortations to government servants to quit government service, for children to stop attending government schools, and hoisting the Indian National Congress flag on government buildings.

One morning, while I was in my office doing official business, Dr Khan Sahib, our chief minister, walked in unannounced. I stood up and greeted him. Ignoring my greetings and looking straight at the wall opposite, he brought out a paper from his pocket and started reading the Indian National Congress manifesto exhorting government servants to quit service. Having finished reading, he walked out without looking at me for a single instant.

I found out that he had similarly entered the office rooms of my staff and had also made a round of numerous offices of other government departments and had read out this message.

We received a government directive that whenever Dr Khan Sahib entered a government office, the staff should rise and listen to him in silence and take their seats only when he had left. However, the activities of Abdul Ghaffar Khan and the younger brother of Dr Khan Sahib, were far more effective. He established himself in a serai in Peshawar city, from where he organised and sent out processions of his followers into the streets, exhorting the British to quit India and planting the flags of the Indian National Congress on government buildings. The government resorted to *lathi* (baton) charge to disperse the processions, but the Pathans bore the cruel blows on their heads and bodies without offering any resistance, with maddening self-composure. The police thereupon resorted to firing on the crowd. One morning, I saw a procession of the Red Shirts, headed by four men, bearing on their shoulders the bier of a Pathan killed during firing, moving along in dignified silence in front of my office. From time to time, they shouted the single slogan: 'Death to the tyrant government!' By that time, the NWFP government was at its wits' end. They dared not repeat firing on unresisting, but non-yielding, crowds of Red Shirts for fear of alienating the whole Pathan nation of whom thousands were fighting with the British Army in Burma and other fronts.

Iskander Mirza, my old friend and cricketing teammate during my earlier years at Mardan, was then holding the important post of deputy commissioner at Peshawar. He was also, like Sheikh Mahbub Ali Khan, a blue-eyed boy and enjoyed immense prestige with the British due to his rare political acumen, and his claim to an illustrious ancestry. He claimed descent from Sayyid Mir Mohammad Jafar Ali Khan Bahadur, the commander-in-chief of the forces of Nawab Siraj-ud-Daula of Bengal.[1] Iskander Mirza offered to quash the Red Shirt activities provided he was given a free hand to take whatever measures he decided. His offer was accepted. He issued instructions to the police that whenever the Red Shirts should attempt to hoist the Congress flag on a government building, the police instead of resorting to *lathi* charge should help them to do so by providing ladders, etc. The Red Shirts were at first mystified at the unexpected behaviour of the police. But when the police acted in a similar manner everywhere the Red Shirts felt highly embarrassed and appeared ridiculous before the general public. The flag hoisting activities of the Red Shirts came to an abrupt end.

Settling the Kurram River Dispute with Afghanistan

The Government of India nominated me to hold negotiations with the representatives of the Afghan government regarding an old dispute over the Kurram River waters between the Afghans and Turi tribe of Kurram Agency living on opposite sides of the Durand Line. The dispute had taken a serious turn due to alleged aggressive intentions of the Afghans. It had alarmed the Government of India into making a formal request to the Afghan government to settle the dispute through negotiation. I was the guest of Mr Emerson, the political agent of Kurram Agency, at Parachinar, during the four days of negotiations. He was also to act as my background assistant in the parleys.

The water dispute was evidently the outcome of a longstanding religious animosity between the Sunni Afghans and the Shia Turis of the Kurram Agency. Taking advantage of their commanding position upstream on the Kurram River, the Afghans regularly diverted the winter channel of the river away from the watercourse of the important Kharlachi village of the Turis which lay a short distance downstream of the Durand Line. This occurred at the beginning of each winter. It was noteworthy that the lands of the Afghan village lay at a higher level and were not irrigated by the Kurram River. Their spur building activities were solely directed towards preventing the Turi lands downstream from benefiting from the Kurram River water. The immediate cause of alarm was the alleged intention of the Afghans to build permanent spurs in cement and divert the river flow.

The first view of Parachinar vale, as one emerges from the gorge of the Kurram River, is breath-taking. One is overwhelmed with the grandeur and beauty of the Safed Koh range[2] stretching away for miles in the near distance, with its snow-clad peaks ranging in elevation between 13,000 feet and 16,000 feet, and its enormous slopes clad in delicate, shimmering, silky green. One is equally bewildered to notice the bare, pebble-strewn light-brown plateau of Parachinar sharply sloping southwards from the foot of the Safed Koh range, towards the pretty bright green and broad valley of the Kurram River carpeted with rice cultivation hugging the low range of bare dark-brown hills in the south. The Parachinar plateau, about 15 miles long by 10 miles in width, is screened on the west by low hills. Their watersheds mark the Afghan boundary with the higher, thickly wooded Afghan hills peeping from the background. However, the gaze continuously turns towards the incredible grandeur of Safed Koh range, dominating the entire valley.

Two small posts, built by the British and the Afghans, stood on the opposite banks of the river facing each other across the Durand Line, which was, at that point, an imaginary line crossing the Kurram River in the plain of Parachinar. The political agent of Kurram, who was also to host the refreshments for the parties during the parleys, erected tents for the discussions on the British Indian bank of the river. The Afghan representative arrived on horseback. He was a tall, young man clad in immaculate English-style riding breeches, a sports jacket, and a white solar helmet. He was as fair as a Western European and was inclined to obesity. After cordially greeting me and Mr Emerson, he informed me in Urdu that he had ridden 50 miles that day and had met many 'beautiful ladies' from the villages en route along the Kurram valley. His name was Mohammad Kabir Lodin, and he was a young engineering graduate from Michigan University, USA. He gave me this information, with a wink!

Sardar-i-Ala, Juma Khan, the governor of Khost, accompanied Mr Lodin. For two days we wrangled fruitlessly. I repeatedly pointed out to Mr Lodin the futility of spending so much time and labour annually on the construction of spurs, which did no good to the Afghans but only harmed the Turis. Mr Lodin, however, was adamant. He persisted in his assertion that the Afghans had sovereign rights to do what they pleased within their own territory. We reported each day's proceeding by phone to Peshawar from where they were transmitted to New Delhi. On the third day, we received peremptory instructions from the Government of India to come to an understanding with the Afghans at all costs, as it was not considered desirable to alienate the Afghan government, while the war was on. So, I proposed to enter into an agreement with Mr Lodin on the following terms: the Afghans could construct temporary stone and brushwood spurs, but they would not replace them by permanent spurs in stone and cement. I pointed out to Mr Emerson that the temporary spurs of the Afghans were sure to be washed away during the spring and summer flood seasons. Mr Emerson agreed with me. I also made Mr Lodin to agree to my suggestion. A formal agreement to the effect was accordingly drawn up in duplicate and signed by Mr Lodin and myself. Both Indian and Afghan governments were satisfied with the successful conclusion of the negotiations. After the agreement was signed, Mr Lodin asked my advice about how to design canal channels, which I provided in the short time that we had available. He then asked me to procure a copy of Kennedy[3] Diagrams for him. I passed on his request to the Government of India. Mr Lodin was eventually supplied a copy through the British Legation at Kabul. He had probably not heard about Lacey[4] Diagrams on the design of channels, which were an improvement on Kennedy Diagrams.

Mr Emerson was a generous and affable host. However, he appeared to be at a loss as to what dessert he should provide along with our lunches and dinners. Luckily, strawberries were in season and to his question whether I would like strawberries for dessert, I said, 'I would love them'. So, we had delicious strawberries and cream after every meal during my four-day stay with him. Mr Emerson requested me to visit Shalozan village to explore the possibilities of development of irrigation from the Shalozan stream. I walked 1,000 feet up the valley, where I saw large chunks of ice, as big as small houses that had fallen from the snowy slopes above. Most of the precipitation (including snowfall) on the Parachinar plateau and on the Safed Koh range occurs in winter. The summer monsoons are weak and short-lived. Moreover, the summer is also brief. There are no storage sites available in the miniature Safed Koh valleys. For this reason, there are poor prospects for Kharif (summer) cultivation. There is the possibility of tube-well irrigation in the lower portions of the arid Parachinar plateau, adjoining the lush Kurram valley, stretched like a broad green rug along the southern edge of the plateau. I also visited the *barani* fields of the medicinal herb *Artemisia* in the vicinity of Parachinar that was introduced by the inimitable Col. Noel, while he was the political agent at Parachinar. There were distinct possibilities of extending the cultivation of this medicinal plant.

Visit to Swat with Lt. Col. Mohammad Ayub Khan

In 1934, while I was working as executive engineer at Mardan, I received a letter from my elder brother, an agricultural engineer then stationed in Ferozepur, in which among other things he had mentioned that he had recently met a charming young man, a Captain Ayub, hailing from the district of Hazara, in NWFP, and who was being transferred to a station in the Frontier. He had suggested to him to look me up whenever he should visit Mardan. Sure enough, a few weeks later Captain Ayub presented himself giving my brother's reference. Captain Ayub was a handsome young man with engaging manners and a warm disposition. I took an instant liking to him and we struck up a warm acquaintanceship. We only met off and on but in 1944, when Lt. Col. Ayub returned from the Burma campaign of the Second World War and was posted to command a regiment in Landi Kotal, in the Khyber Pass, and I was working as superintending engineer at Peshawar, we came to know each other well. During these days, he was a frequent visitor at my house, particularly during the weekends. While posted at Landi Kotal, Col. Ayub was a stranger to these parts and was anxious to visit the beautiful Swat state, which he had not visited thus far. He knew that Malakand and the Swat valley were in my jurisdiction and I was personally acquainted with the ruling family of Swat state. He, therefore, requested me to arrange a visit for him to Swat, which I did. The visiting party consisted of Lt. Col. Ayub, Mohammad Said (my son-in-law), Said's younger brother Mohammad Ayub (later Lt. Gen. and the head of the Pakistan Army's medical corps), and myself. We were to be the guests of the *wali-ahad* (crown prince) of the Swat state. I had planned that, on arrival at Malakand, the party would stay in the beautiful canal rest house at Amandara at the head of the Upper Swat Canal and have dinner with Abdus Samad Khan (executive engineer), before leaving for Swat the next morning.

Col. Packman, political agent of Malakand, and an old friend and colleague of mine, on learning about my proposed visit to Malakand, called to protest that I had ignored him, while accepting Abdus Samad's hospitality. As a compromise, therefore, it was agreed that the party on arrival at Malakand would drive straight to Col. Packman's house for drinks and then proceed to Amandara, 6 miles further away, for our dinner and stay. The 7-mile stretch of road between Dargai in the plain to the Malakand Pass winds up along a spur of the southern face of the Malakand ridge. As one looks at the Malakand Fort sprawling along the rising flank of the Pass, one has no inkling of the beautiful valley which lies hidden on the northern side behind the Pass. Our party uttered exclamations of delight as we reached the top of the Pass and paused opposite the main gate of the Malakand Fort to gaze on the beautiful light-grey northern slopes of the Malakand ridge clothed in clumps of pine and wild olives. The evening shades were deepening into dusk as we started down the other side of the Pass towards the political agent's bungalow, which was perched on top of a low, isolated hill. The house was ablaze with lights as our cars wormed their way to the top.

Col. and Mrs Packman accorded us a warm welcome. I introduced my friend Col. Ayub, whom they met with traditional British hospitality, and made us feel very welcome. Despite the conditions of austerity prevailing due to the war, Col. Packman produced some bottles of genuine Scotch whisky. He and Col. Ayub and Said sat down to drink, while the rest of us, including Mrs Packman, kept company with soft drinks. The small talk gravitated to the inevitable topic of the war. General Bernard Montgomery had just launched the final offensive towards El-Alamein. With exclamations of delight, Mrs Packman repeated the day's news: 'The grip is tightening around Berdia.' The party became merrier as the evening wore on. So many tit bits were served that Abdus Samad (who had also joined us) began to grumble about his guests being left with no appetite for dinner. Then the parlour games started. Col. Ayub and Col. Packman were both about equally tall and broad shouldered. They both decided to pit their strengths against each other in a bout of arm wrestling. The two stalwarts lay down on their stomachs, in the middle of the cleared drawing room floor, tightly gripping each other's hand, in a furious struggle to pull down the other's hand. I watched nervously as the contest vacillated between the two and their faces turned redder and redder. After a strenuous struggle, hands shivering under the strain, Col. Ayub floored his opponent's hand. Mrs Packman appeared embarrassed at the outcome of the contest. Col. Ayub patted the lady on her back and remarked. 'Well Madam, I have beaten your husband!' The lady was startled at this display of familiarity. However, as it was soon apparent that no discourtesy was implied, and the embarrassing moment passed without much notice.

The next morning, after breakfast we started for the Swat state. The overpowering beauty of the Swat valley burst forth into view the moment we left Amandara rest house. It was a broad valley 2 miles wide carpeted with brilliant green and yellowing rice fields. In the middle of the valley, the Swat River tumbled down in a broad meandering channel with a bed of blue gravel and boulders. The flanking hills were clothed in bright green, studded with darker groves of pine, wild olives and dwarf oak. The spectacular road bridge spanning the Swat River at Chakdara, 2 miles upstream, along with the Chakdara Fort picturesquely perched on a hillock, were the conspicuous features of the eastern foreground. From the Chakdara bridge downwards, the main channel of the Swat River is guided through a series of training works to hug the Amandara rock face, where it is caged by a pretty canal headworks, which lay at our feet like a structure in Toyland. The Upper Swat Canal issued forth from the headworks in a majestic sweep westward; downstream, its broad channel mirroring the reflections of the flanking hills.

We entered the Swat state after crossing Thana, the last village of the Malakand Agency. Throughout, the road ran along the south bank of the river. The valley grew prettier as we drove along upstream. We passed several prosperous villages surrounded by orchards and rows of poplars, amidst green rice fields. Extensive Buddhist ruins could be seen at several places and we also passed a number of well-preserved stupas situated in the foothills. At one particular point the river hugged the south bank against sheer vertical rocks, shaped in profile like an elephant's trunk. Hiuen Tsang, the Chinese monk, scholar, and traveller who visited India during AD 629–645, has mentioned this peculiar rock. We were evidently traversing a centuries-old route in the valley. Prince Miangul Jahanzeb, the *wali-ahad* and our host, was waiting for us as our cars drove into the driveway of his palatial guesthouse situated in the middle of Saidu Sharif town. He had kept himself informed about our progress through the state's telephone posts located at short intervals along the way. We walked up the ramp to greet our host, who provided us a courteous welcome. I introduced Col. Ayub to the *wali-ahad*.

After refreshments we were conducted to our suites. The host remarked to me 'Khan Bahadur Sahib, I hope you like your room. It is the one usually reserved for viceroys and governors. Your guest is in the second-best room.' Col. Ayub appeared satisfied with these arrangements. A sumptuous lunch with the host was followed by a brief siesta. In the afternoon, after an elaborate tea, our host drove us to Miana, to present us to his father the Wali of Swat Miangul Abdul Wadud. Tucked away in a miniature valley, sheltered under the northern slopes of the 9,700-foot high Ilam peak, 12 miles from Saidu Sharif, was the summer retreat of the Wali, where he had built a miniature palace of white marble. Miangul Abdul Wadud, who was in the ripe middle age at that time, was a remarkable personality. Barely two decades earlier, he had carved out a virtually independent kingdom in a beautiful yet turbulent country and had established a regime of peace and order, albeit by strong-arm methods, which was the envy of his British friends across the border. The British government in India held him in high esteem. Moreover, the Wali was a romantic figure, being the grandson of the famous Akhund of Swat who had valiantly fought the British in the Malakand campaign of 1895, in which Winston Churchill had reportedly taken part as a war correspondent.

The Wali had been informed of our visit. As we reached, we saw him standing on the terrace on top of the stairs, waiting to receive us. He was wearing a long, plainly cut coat in *khaki* drill. We respectfully lined up below the stairs and ascended the stairs, one by one, to be introduced to him. Col. Ayub was sent up first. He was dressed in military uniform in *khaki*. While greeting the Wali, he affably remarked, 'Well Wali Sahib, I am glad to see that you are also wearing a military uniform. This shows that you like our military profession.' On hearing the above remark, the *wali-ahad*, our host, looked accusingly at Said.

I ascended the stairs next and greeted the Wali. We embraced each other, and he enquired about my health. We all walked up to the centre of the lawn and sat down on the marble seats with the Wali making me sit next to him. I recalled that I was visiting the state after ten years. 'It is nine years,' the Wali corrected me. He was right; what a marvellous memory he had. 'You have a beautiful retreat here,' I said. 'A nice place for rest and quiet.' In reply, he pointed to a steep 1,000-foot high spur towering overhead and casually remarked that he climbed to the top of the spur every morning for exercise! We were conversing in Pashto. Suddenly, the Wali addressed Col. Ayub in Urdu: 'What part of the country do you come from?' 'I come from Hazara sir, the district to which Khan Bahadur also belongs.' Col. Ayub replied. 'Are you a Pathan?' The Wali enquired. 'Yes sir.' 'What type of a Pathan are you when you do not know Pashto,' the Wali retorted. So, the Wali had not been slow to return the rebuff.

Two years later, early in 1946, Lt. Col. Ayub, who was still posted at Landi Kotal, met me at Peshawar, and enquired about my future plans after my impending retirement from service, to take place later that year, at age fifty-five. I informed him that I was considering various options for re-employment. He said, 'Khan Bahadur Sahib, don't go in for service after retirement. I myself am retiring shortly, as there are no prospects left in military service. I suggest that we both start a joint business venture, starting with the construction of a *mandi* (a big market) in Haripur town, for handling wholesale trade in food grain.' Nothing definite was settled, however. A year later, in 1947, the Partition of India took place. Lt. Col. Ayub was promoted to the rank of a brigadier and was posted to Jalandhar, in East Punjab, to command a mixed brigade of Hindu and Muslim forces of the Indian Army. There goes our joint enterprise, I thought to myself.

Late in 1942, I had an acute attack of appendicitis, and was advised to undergo an operation. On the night before my departure for the hospital I had a disturbing dream. I dreamt that my first (deceased) wife was in a square room surmounted by a dome, on a high ground by a roadside. The single door to the room was locked from inside. I swarmed up the slope and pounded at the slammed door and called my wife to come out. She, however, refused to open the door and said from inside the room: 'You cannot enter, go back and arrange jewellery for Nigar!' In despair, I climbed down the slope to the roadside. I awoke depressed in the morning, fearing the worst. I told the dream to my wife. She said 'Congratulations; it is a lucky dream for your wife refused to allow you to enter her room. The operation is sure to prove successful.' 'But what about her request about my daughter Nigar's jewellery?' I asked. 'Oh, that message is for me,' she replied, 'My aunt would like us to arrange Nigar's marriage.'

Nehru's Visit to the NWFP

In 1946, an interim government was set up as a prelude to the Partition of India in which Jawaharlal Nehru was appointed to the governor general's executive council with responsibilities for external affairs and commonwealth relations. His responsibilities included the portfolio of tribal affairs and tribal areas. During those days, the NWFP government was under the Congress ministry with Dr Khan Sahib as the premier. Having recently ousted the

Muslim League ministry by an impressive margin and taking advantage of this victory and support received by the Indian National Congress from many segments of the Pathan society, Jawaharlal Nehru, against the advice of official quarters, including that of Sir Olaf Caroe, the new governor of NWFP, decided to visit the NWFP and the tribal areas. His objective apparently was to win over the sympathies of the Pathans to the cause of a united India and reject the Muslim League's claim to Pakistan. So, visits were planned for Nehru to Wana in Waziristan, and the Khyber and Malakand agencies, where he would meet with tribal Maliks and address tribal *jirgas*.

But Nehru, as well as the provincial Congress party leadership under the Khan Brothers had grossly underestimated tribal sentiments. He started his visit in Peshawar where he faced a large and hostile crowd and had a narrow escape, departing from the gathering with Dr Khan Sahib through the 'back door.' His visits to Wana and Razmak in Waziristan were equally disastrous where, in the company of the Khan Brothers, he was mobbed by tribesmen with shouts of 'We don't want Hindu raj'. But the culmination of this highly unsuccessful mission took place in Malakand Agency. Sheikh Mahbub Ali Khan was the political agent and he, reportedly, advised Nehru that a large tribal *lakshar* (armed force) had assembled to disrupt the visit, and diplomatically advised that he cancel his plans. However, Nehru went ahead and was confronted by a large crowd who blocked the path and pelted his convoy with stones, one of which hit and injured him. The retreat was equally perilous, and it was reported that Nehru escaped with his life. In the inquiry held afterwards, the blame for the debacle was placed on some political officers bearing sympathy for Pakistan, with Sheikh Mahbub Ali being singled out for his complicity in the events and charged by the government with negligence of duty, a charge from which he was acquitted by the Madras High Court.

Two high ranking officers of the US government, General Wilson and Donald G. MacDonald, visited the NWFP. The PWD was to host their visit to the Khyber Pass and adjoining tribal areas. Consequently, Col. Lang Anderson (chief engineer) and I decided to escort them ourselves. While in the middle of the Pass, Col. Anderson pointed out with pride to the Americans, the different types of obstacles that had been placed to block the road, and to the presence of strategically located pickets on hilltops to halt any invader. At Landi Kotal (presumably thinking about the three Afghan wars that the British had fought in the area) the General stopped the car to jokingly enquire from a fifteen-year-old Afridi lad, whether he had won the rifle after killing his man! On our way back, General Wilson enquired from the colonel, whether I was really retiring from government service after a few months. 'Yes, of course,' replied Col. Anderson. 'He is retiring after attaining the age of fifty-five, prescribed by the government.' 'Nonsense!' retorted the General. 'I am seventy-two and Mr MacDonald here is seventy and the US government shall retire us only when we ourselves decide to retire. You are retiring him in the prime of life when he is capable of making the most useful contribution to his profession.' To this, the colonel had no answer. In fact, I looked and felt young for my age; there were very few grey hairs on my head at the time.

20 Retirement and the Partition of India

At about the time of these events, my son, Mohammad Aslam Khan, was posted as assistant engineer, Malakand PWD. His was the sole official residence of a civilian officer within the Malakand Fort premises, which at that time was occupied by a regiment of Hindu and Sikh forces. Daily news of communal riots and massacres of Hindus, Sikhs, and Muslims had raised tensions to a peak, and Aslam and his servants had received serious threats from some Hindu and Sikh soldiers about their intensions to avenge themselves on him and his family. Feeling particularly vulnerable, and not knowing what to do, he visited Sheikh Sahib and reported the evil designs of the drunken soldiers. Sheikh Sahib smilingly 'pooh-poohed' his fears and requested Aslam to stay on for dinner. During dinner Sheikh Sahib excused himself for a visit to the toilet, where he remained for over an hour. Returning, he profusely apologised and leisurely bade Aslam farewell. Full of foreboding, Aslam returned to the fort in his car and saw a curious spectacle. On one side of the road, a large number of military lorries and jeeps were standing in a row. Sepoys were busy transporting kits from inside the fort and dumping them in the lorries. The officers of the regiment were standing in groups and talking in excited tones. Mystified, Aslam approached a group of officers and enquired from a Sikh officer he knew, the reason for this unusual nocturnal activity. With a string of profane oaths, the officer informed him that orders had been received from the General Headquarters (GHQ), Rawalpindi, for the regiment to evacuate before sunrise the following morning. By midnight, the officers had left and before sunrise the next morning the entire regiment had vacated the Malakand Fort.

Aslam hastened to the political agent's house and informed Sheikh Sahib about the unexpected developments overnight. With a twinkle in his eye Sheikh Sahib remarked that he had at once realised the gravity of the situation and the peril which Aslam and his family faced. So, during the course of the dinner, he had devised a scheme to take measures for their rescue. As a desperate move he had resorted to the move that followed.

He had contacted the GHQ on phone, conveying the urgent intelligence that a tribal *lashkar*, many thousands strong, was poised to converge on Malakand with the avowed resolve to wipe out the Hindu-Sikh regiment of the Indian Army stationed there. The tribesmen were reported to be well-armed and were thirsting for revenge having heard harrowing tales of massacre of Muslims in East Punjab. Unless reinforcements arrived immediately, the whole garrison was most likely to be overwhelmed. Since it was too short a time for reinforcements to be sent from any quarter, the GHQ ordered the immediate withdrawal from the garrison to Dargai, situated in Peshawar valley.

Soon after Partition, the Frontier tribes felt a sense of relief at the withdrawal of the restraints of the British rule. Consequently, there appeared signs of unrest in certain tribal areas. There were indications of unrest and revolt amongst the Shia Turi tribe of Kurram Agency, which borders Afghanistan. To cope with the situation, and by happenstance, the Pakistan government ordered Nawab Mahbub Ali Khan, (he had been conferred the personal title of 'Nawab' by the British government for his extensive services) to take over as political agent of the Kurram Agency. Such was the awe attached to Sheikh Sahib's personality that the revolt subsided like a bubble on the mere receipt of the news of his posting to the Kurram Agency. It is said that at that stage of Sheikh Sahib's career, his qualities as a diplomat of exceptional ability were brought to the attention of Quaid-i-Azam Mohammad Ali Jinnah and he was being considered for a high diplomatic appointment, when the hand of death suddenly cut short Sheikh Sahib's exceptional career.

In 1945, Mr Brown having completed his five-year tenure of service as chief engineer, retired from government service. I was the senior-most engineer in the PWD. Now was the opportunity for the Muslim League government to demonstrate their authority and appoint me, an Indian, to succeed Mr Brown as the first Indian chief engineer of the NWFP. So, I met our PWD minister, Sardar Ajit Singh, to solicit his support. He not only promised all support, but also declared that I was the fittest person for the job and said that he was prepared to resign on the issue. But, he said, the chief minister had the sole authority to approve my appointment, and strongly advised me to approach him. Sardar Aurangzeb, the then chief minister, received me with his usual courtesy, and said that he would be happy to appoint me to the post of chief engineer, but there was only the trifling matter of my contributing the sum of Rs 30,000 to the Muslim League Fund. I was aghast and confessed that I did not have that much money. 'Oh no, you need not pay the money out of your own pocket, but you can ask one of your big contractors to pay,' Sardar Aurangzeb said reassuringly. 'But I have never had any underhand dealings with any of my contractors,' I declared with some heat. The chief minister advised me to think the matter over and dismissed the subject. The upshot was that Mr Arthur Oram, who had retired from the NWFP PWD five years ago in 1940, was recalled from his temporary post in the Military Engineering Service and was re-installed as chief engineer.

In a way, this proved good, because Mr Oram had been a kind boss to me with whom I had spent ten happy, carefree years. Mr Brown, the outgoing chief engineer, sensing the injustice being meted out to me, offered to recommend me for the post of chief engineer of Kashmir state which had recently fallen vacant and about which a demand had been received by the NWFP government. Mr Brown held out before me the alluring prospects of the various amenities of the post; pointing out among other things that, as chief engineer of Kashmir state, I would be absolute master of my own department. I considered the offer but decided against it, preferring to retire as a superintending engineer in NWFP.

Momentous events were, in the meantime, taking place outside the NWFP. Unable to shake the rock-like resolution of Mr Jinnah, Britain had, at last, conceded Pakistan, but with serious reservations. Ignoring the fact that Mr Gandhi and the Congress were responsible for their leaving the subcontinent, the British, with their eye to their extensive existing and potential trade with India, secretly made common cause with the Hindus against the Indian Muslims. It was a sad spectacle to see the tangible symptoms of the moral rot setting in amongst our British rulers at the imminent prospects of their quitting India. Lord Louis Mountbatten, the Apollo-like[1] viceroy of India was reported to have been heavily bribed by the saintly Mahatma. Even the beautiful Lady Edwina Mountbatten established over-affectionate relations with Jawaharlal Nehru. The Britishers, generally, started making hay during the short time left for them to rule India. Sir George Cunningham, governor of NWFP, called his brother, a rubber planter in Malaya, to enter into partnership with Khan Bahadur Quli Khan Khattak[2] of Bannu district, a retired political agent widely known for his political acumen, for the contract of supply of dry fruit to the forces, during the Second World War. Quli Khan was reported to have made his pile in the contract.

Early in 1946, Mr Jinnah visited Peshawar. He was carried in a procession through Peshawar Saddar Bazaar and city, followed by a record crowd. A circular directive was promptly issued by Dr Khan Sahib's government, prohibiting government employees from taking part in politics. I promptly ignored the official instructions not to attend Mr Jinnah's public meeting. Towards the end of Mr Jinnah's visit, one afternoon, I went to the chief minister. The first question Dr Khan Sahib asked me was, 'Have you seen Mr Jinnah?' I replied in the negative. 'No, you must see him; he is a great man and is doing a lot for the Indian Muslims.' Thus encouraged, I went that evening to the Cunningham Park situated at the northern post of the high ramparts of the Peshawar Fort to hear Mr Jinnah's last speech.

I was due to retire on 11 June 1946. But, thanks to Mr Brown's recommendations and Dr Khan Sahib's willing sanction, I had been declared a 'required officer' and was allowed to serve till the last date of my superannuation age of fifty-five, and afterwards was treated as on leave on full pay for another three months. My official date of retirement was extended to 11 September 1946. I was also granted extra administrative pension, giving me the benefits of a member of the Indian Service of Engineers—too little, too late!

21 Bahawalpur

Bahawalpur, in 1946, was a city of petty shops, dust, intolerable heat, and multitudes of flies, plastered thick over exposed mutton and beef, fruits, vegetables, *gur* (jaggery), fresh small-sized dates, and anything edible in the shops. The city roads were ostensibly *pucca* but due to lack of bitumen covering, released clouds of dust. Rainy season in Bahawalpur was practically non-existent. One vainly looked for clouds in the sky. The searing summer heat very gradually merged into the tolerable heat of January. I was allotted the spacious vacant old bungalow earmarked for the inspector general of police. Refusing to realise the change in my circumstances due to my retirement, I had brought with me a full railway wagon load of household articles and furniture. I was astonished to note that the residence of none of the other 'officers' of the various departments in the state was cluttered with much furniture, or personal household belongings.

My first official duty was to present myself to the nawab of Bahawalpur, with a large basket of fresh fruit, according to the custom. He met me cordially, declaring candidly that although he was a demigod to his Bahawalpur subjects, he was only an ordinary human being to us 'Frontier' people. He somewhat routinely offered to help me personally, in case of any problem. We, officials, personally met the him on three occasions in the year—twice during the two 'Eid' festivals and once on the occasion of his birthday. On these occasions, all of us officials purchased a small pseudo gold token from the Bahawalpur treasury for Rs 25, which we, one at a time, handed over to the ruler as he sat on a sort of a throne, during a durbar. There were about 600 officials in the state. The accumulated sum was apparently meant for tips for the ruler's personal servants. The real ruler of the state, however, was Sir Richard Crofton,[1] the prime minister. Noticing my evident distress due to the excessive heat of Bahawalpur—I had arrived there on 21 June 1946—Sir Richard declared it was a pity that I had been called early during the height of summer, instead of in the following mild winter.

The Second World War had only recently ended (2 September 1945); so, neither were railway wagons available for transporting road metal, and *bajri* for road consolidation, nor were steel and cement easily obtainable for the construction of buildings. Although the sum of Rs 30 million was placed at my disposal in my annual budget for new buildings and for road consolidation, I was in a fix as to how to utilise this amount. The Bahawalpur state government was powerless to help me, against the all-powerful Government of India in whose hands lay the release of railway wagons, through which most building materials had to be shipped.

As a lucky diversion, I received instructions, in July, to proceed to Karachi, to negotiate the purchase of a 'Bahawalpur House' there. Karachi was a beautiful town of only 300,000 inhabitants in the pre-Partition days of 1946. It had clean wide roads, stately streets with shops crammed with exotic merchandise, many storeyed apartment buildings as well as magnificent mansions like the Frere Hall, the Customs Building, the Seth Goverdhandas Mohatta Market (AD 1895) on Bunder road, and others. Its seashore at Clifton, Keamari, and Hawk's Bay were a pleasure to behold. I selected a spacious, vacant bungalow, with a sizeable compound on Lawrence road belonging to a Parsi gentleman who was migrating to Bombay, that cost Rs 100,000 only, for the proposed Bahawalpur House and on my return reported my find to the Bahawalpur government. However, no action was taken on my proposal.

Tara Singh's Offensive

I had come to Bahawalpur in the heat of summer, while my wife had been vacationing in the cool and beautiful vale of Kashmir, with her father. Our two young boys were enrolled in a boarding school in Srinagar, so, on 3 March 1947, I left for Lahore in my car to fetch my wife, who had come to Lahore, half way by train, and was staying in the house of a friend in the railway department on Empress road. On the way, a small boy, who my driver was giving a lift to, complained of thirst. I stopped the car, and told my driver, who was sporting the typical flower-pot sized, red Bahawalpur fez, to fetch water for the youngster from a road-side Persian wheel being tended by a tall lanky Sikh lad of about sixteen. The Sikh lad was somewhat surprised at the request for a 'tin' of water. On handing over the water, he asked my driver whence we were coming and where were we going. The driver told him our destination, whereupon the Sikh lad exclaimed. 'Then you have not heard the news.' 'What news?' asked the driver. The Sikh lad broke into a cryptic smile and said, 'You will soon hear the news.'

A few miles from Lahore, while the sun was setting, we were astonished to see thousands of black turbaned Akali Sikhs converging on Lahore in sullen silent groups. Each Sikh carried a *kirpan* (a small sword)—a traditional weapon carried by the Sikhs. The Sikhs, to our relief, opened a narrow lane on the road, for our car to pass through. In late evening, we arrived at a barricaded gate on Multan road, Lahore, where our car was stopped by a Sikh deputy superintendent of police and was thoroughly searched by a Sikh police constable, for concealed fire-arms, before being allowed to enter Lahore. All the Lahore roads—the Multan road, the Mall, and the Empress road along our route—were without lights and were strangely silent and deserted. There was great relief at our safe arrival in our friend's bungalow inside the compound of the North Western Railway headquarters office, on Empress road, where my wife was staying.

After some time, I heard rifle shots, and enquired where they came from. I was told: 'Don't you know, Tara Singh[2] (the Sikh leader) has declared war against the Muslims who want Pakistan; the Sikhs and the Muslims are at the moment fighting in the city.' 'Where is Tara Singh?' I enquired with some apprehension. 'His headquarters are on this very Empress road, in a bungalow, not far from here,' I was informed. The situation soon became crystal clear when loud cries of *Sat Siri Akal* resounded from Tara Singh's headquarters, not far away on the Empress road, and the answering Muslim cry of *Allah-o-Akbar* issued from the Qila Gujjar Singh lane, which opened right in front of the main gate of the North Western Railway (NWR) headquarters office, from an opposite road. Thus, I realised with dismay that any confrontation between the Sikhs and the Muslims would necessarily take place on the road just opposite the bungalow where we were staying. I asked a passing Muslim police contingent for the protection of our bungalow. Their leader declined to help with the remark, 'Tonight our hands are full, and so you have to protect yourself.'

The Sikh crowds repeatedly issued forth their war cries from Tara Singh's place, but were deterred each time from confronting the Muslim crowds coming from the Empress road with their cries of *Allah-o-Akbar* from the Qila Gujjar side lanes. As the night advanced, the encounters occurred at longer intervals. In the morning, the Frontier Mail train which left Lahore for Peshawar at 7 a.m. was already chock full of passengers; and the next train likely to leave for Peshawar was the slow 3 p.m. passenger train, on which the railway staff would try to find seats for us. There was no question now of proceeding to Bahawalpur as all the roads radiating out from Lahore were declared unsafe and exposed to roving bands of armed Sikhs. The 3 p.m. passenger train offered our only chance of leaving the embattled city of Lahore. And owing to the mounting chaos created by the riots, that was the last train likely to leave the city for some time.

When we arrived at the Lahore railway station to board the 3 p.m. train, the mile-long covered platform, where the train was standing, was packed with hundreds of black- and yellow-turbaned sullen looking Sikhs carrying *kirpans*. The mountain of luggage brought by my wife had been dumped in a narrow first-class coupe compartment. Two railway officials in uniform were guarding its two doors from both ends. With grave misgivings, but with brave faces, my wife and our small children in tow, pushed our way through the densely packed Sikh crowds, and took our seats in the compartment with a sense of relief. At last the train moved. The railway official guarding our compartment waved us goodbye and good luck. At Kamoke railway station, 26 miles from Lahore, a young Muslim lad trailing four women was moving on the open railway platform from compartment to compartment of the train seeking in vain to find room in the packed train. At last he came and stood before our compartment, and related a piteous tale of how he had managed to escape from the Sikhs with his widowed mother, his two teenage sisters, and a young neighbour's girl carrying a baby and requested us, in the name of Allah, to allow his female relatives into our compartment. He, himself, would travel if necessary outside the train, by grabbing the door handles of any compartment, if he failed to find a seat inside. He said they were bound for Wazirabad, barely two or three hours run from Kamoke.

We willingly allowed his women to clamber into our compartment. At Gujranwala railway station a smart Sikh demanded admittance in our compartment. I pointed out to him that the compartment was already full with six passengers and with mounds of luggage. In any case, the door of the compartment was blocked by shoving a heavy piece of luggage against it, and I stood across the door with no intention of letting him in. After looking at the full compartment, he left. The female fugitives from Kamoke had evidently proved a blessing in disguise. They alighted at Wazirabad. The slow-moving train reached Rawalpindi at 1 a.m. The long, covered railway platform was strangely empty. It usually presented a lively scene even at that late hour with crowds of passengers, and hawkers selling fresh fruit, sweets, tea, and cheap dinners. I beckoned a solitary Muslim railway coolie standing nearby and enquired the reason for the platform being deserted. He said that bloody fighting was going on, at the moment, in the city between the Sikhs and the Muslims and there was great danger of it spreading to the railway station as well. He advised us to keep the windows closed and shuttered and bolt the doors of the compartment. The train stopped there for a full one hour. At last it moved slowly out of the station to our immense relief. We reached Taxila at 4 a.m. The branch train to Havelian (and then 10 miles by road, to Abbottabad) was standing at the other end of the narrow platform. But it was to leave two hours later. The coolies deposited our luggage in the waiting room. The waiting room 'bearer' advised us to keep our door bolted from inside because Taxila was highly unsafe, owing to the nearby Sikh stronghold of Panja Sahib.[3] We started for Havelian in broad daylight. At last, we felt perfectly safe.

Tara Singh's offensive spread all over West Punjab—at Lahore, Amritsar, Rawalpindi, Multan, and elsewhere. The Sikhs had been badly beaten at all those places. Tara Singh had vowed that, since the East Punjab was in his pocket, thanks to the extensive preparations of the Sikhs, through the connivance of the Sikh maharaja of Patiala,[4] he would also 'conquer' the West Punjab by force (the old Sikh dream of conquering the whole of the Punjab). Lord Mountbatten's government, apparently, was letting the 'dogs fight it out among themselves', an indiscreet disclosure by Mr St. John, ICS, deputy commissioner at Abbottabad, during the course of a lunch in 1946. It will be recalled that the Sikhs were with the Muslim League government during 1939–45. In those days, Tara Singh had asked Mr Jinnah to guarantee an exclusive homeland for the Sikhs in the East Punjab. Mr Jinnah had reportedly replied, 'Show me any district or region in the West or East Punjab, where the Sikhs are in a majority, so that I should guarantee them a homeland.' The Sikhs were in fact in a minority even in their own Sikh state in East Punjab. Mr Jinnah probably did not envisage en masse migration of ethnic populations, which actually happened, with a lot of bloodshed and suffering during the Partition. Tara Singh and the Sikhs were subsequently won over by the Hindu leadership.

Pakistan had been conceded both by the British Indian government and by Mr Gandhi's Indian National Congress. A few days after my arrival in Abbottabad, a referendum was held in the NWFP, on the issue as to whether the people of NWFP elected to opt for Pakistan, or for India. Dr Khan Sahib's Indian National Congress government, then in power in the province, adopted a strictly neutral attitude, permitting a free and unrigged referendum. Mr Jinnah had, however, won the day, and the majority of the Muslims of the province were in favour of Pakistan. I, too, cast my vote for the Muslim League and Pakistan at Abbottabad. As soon as I had slid my vote in the slot of the green Muslim League ballot box inside a tent (to ensure secret balloting) I was startled to hear an exclamation of *shabash* (well done) from somewhere inside the same voting tent. On looking back, I saw a policeman in uniform boldly standing behind a semi-transparent curtain within the tent itself, whom I had not seen earlier, so intent was I to cast my vote. It was he who had uttered the muffled exclamation. Since the NWFP government was avowedly pro-Hindu Congress, I wondered who had planted the police constables so blatantly inside the voting tent.

A Risky Trip to Delhi

After a few days, when Tara Singh's offensive against Muslims of the West Punjab had miscarried and peace again returned to the province, I left alone for Lahore, and from there in my own car went to Bahawalpur. In May 1947, I was ordered to go to New Delhi, to try to retrieve Rs 300,000 worth of road consolidation machinery from the Government of India for which the Bahawalpur state government had already paid. It was a dangerous assignment, in view of the reported disturbed conditions in the East Punjab, and in New Delhi. I left Bahawalpur by train accompanied by my personal assistant, Barkat Ali, via Bahawalnagar, Ferozepur, and Bhatinda to New Delhi. A couple of Hindu fellow passengers in the train maintained their traditional polite attitude towards me during the long journey through the Bahawalpur state, but I sensed a change in their attitude as soon as we crossed the Hindumalkot railway station into Ferozepur district of East Punjab. I became cautious, although there was evidently no danger from them. At the Bhatinda junction, we got down and had to wait for three hours to catch the Calcutta Mail arriving from Lahore and bound for Delhi and onwards to Calcutta. It was noon time. Thousands upon thousands of Sikhs were sitting sullenly and silently on the open platform after leaving their homes in the West Punjab, in compliance with the mandate of Jawaharlal Nehru, probably, because of the debacle of Tara Singh's invasion of the West Punjab. The spacious dining room of the railway restaurant was completely full of well-to-do Sikhs.

I took refuge in the kitchen. The cooks, all Muslims, were aghast at my arrival in the Sikh stronghold. They gave me lunch in the kitchen, away from the dining hall. It was stressful to pass three hours among the crowd of scowling Sikhs, but I put up a bold front, loitering fearlessly among the crowd. At last, the Calcutta Mail arrived, and I was lucky to find an empty first-class compartment. In late afternoon we arrived in the neighbourhood of Delhi, slowly passing many small stations on its outskirts. Hindus and Sikhs were standing in continuous rows for miles, all along the line, anxious to hear news of Lahore from passengers in the slow-moving train. A Sikh standing outside shouted at me for news; I shouted back that there was *khair khairiyat* at Lahore, meaning all was well. From my reply he was stupefied to recognise me as a Muslim.

We got down at the main Victoria station of old Delhi. A nearby platform was being hosed down of what looked like blood. We learnt that a short while ago, there had been massacre of Muslims alighting from a train on that platform. Sikhs ran all the taxis outside the station. We hired one for New

Delhi, 7 miles away. My personal assistant, Barkat Ali, sat by me in the back seat of the taxi. He looked apprehensively at the Sikh driver; whispering to me what would happen if the taxi driver drove us away to some other place. I reassured him by tapping on the loaded revolver in my pocket. I would shoot the driver, I said, if we discovered that he had kidnapped us. The Sikh driver, however, took us safely to the Connaught Circus of New Delhi, and to the Hotel Marina, our destination. We had been advised to stay at that hotel. There were only a few Muslim fellow guests in the hotel who expressed deep concern at my arrival, saying that they themselves were leaving shortly.

I was expecting help from Govardhan Lal, an honest and capable engineer belonging to D.I. Khan district of NWFP, who had served under me for five years as assistant and executive engineer, and who had been appointed the previous year to the post of deputy consulting engineer (roads), to the Government of India, at Delhi, on my recommendation. The next morning, I walked to the nearby ramshackle offices of the Government of India, along the small Shah Jehan road, and met Govardhan Lal, sitting in a dingy office room, cluttered with files lying all around, who was greatly surprised to see me. He felt doubtful about being able to help me, remarking that the member in charge of roads in the Government of India was a 'bigoted' Madrasi[5] gentleman, but he would do his best. Bidding me to sit in his office, he went to tackle the Madrasi member. He took a long time to return and informed me that at first the member had flatly refused to consider the request, saying that the road machinery was more urgently required for bigger Indian provinces like Bombay and Madras, but after earnest protests by Govardhan Lal, he had relented and had sanctioned the release of half the machinery requested by the Bahawalpur state. I thanked Govardhan Lal for his special efforts and the help during these difficult times. Even this partial success was an achievement under the circumstances.

I returned to Bahawalpur, boarding the 10:30 p.m. train that was going to Karachi via Bahawalpur the same evening. I was the lone occupant of a first-class coupe compartment. A few minutes before the train's departure, a well-to-do young Hindu, wearing European dress, came and looked inside my compartment. Taking me to be a Hindu, he remarked, 'Hello, you are all alone by yourself in the compartment. Wait, I am coming to keep you company.' Before he had gone many steps, I shouted back that I was going to Bahawalpur. He stood rooted in his steps and, looking at me with evident confusion on his face, went away and never returned. So, I had the whole compartment to myself, and arrived at Bahawalpur comfortably the next morning. Everybody was amazed at my success. To the credit of the member from Madras, the machinery did arrive in Bahawalpur, in due course.

Soon after, I took short leave and returned to Abbottabad. From there, my wife and I drove to Kashmir to see our two minor sons who were studying in Srinagar. We were confident that Kashmir, with its overwhelming majority of Muslims was, in any case, coming to Pakistan. We were, however, bewildered and dismayed to see thousands of Sikhs and Hindus from the Punjab crowding in Srinagar. The beautiful lawns of Chashm-e-Shahi, a Mughal garden built around a copious spring, were reeking with the smell of *pakoras* sizzling in mustard oil in rows of large iron pans. The lawns were packed with Hindus and Sikhs. We were both greatly disillusioned. A few days later, I returned alone to Bahawalpur.

Owing to the unavailability of railway wagons for transport of road material, I had not been able to make much headway in road consolidation work but had managed to construct quite a number of buildings, notably three spacious bungalows on the Noor Mahal road, one of which I occupied. It had a large 5-acre compound in which, with the help of the director of agriculture of the state, a retired director of agriculture of Punjab, and friend of my late elder brother, I planted 150 fruit trees of all varieties of fruit available from the nurseries in the Noor Mahal. Some friends were amused by what they considered to be futile efforts. My reply was, I was doing it for posterity. In later years, after having left the state, I learnt that most of the fruit trees that I had planted had matured, and there was such a competition among officials for the allotment of my bungalow, that the government had converted it into a military mess.

The Pakistan Award

Our rendezvous in the afternoons was the Bahawalpur club. We usually played bridge. Our unofficial president of the club was Sheikh Din Mohammad, chief justice of the Bahawalpur state, formerly, retired chief justice of the Sindh High Court. He was an intolerant bridge player, not very competent and hated to lose, generally venting his wrath on his unfortunate partners. The newspaper of 13 August 1947 carried unofficial news of the award of Pakistan. It was made clear, however, that the award as announced, was not final, and some changes were to be expected. In the 'unofficial' award, issued by a government bulletin, the districts of Ferozepur, Gurdaspur, and Jalandhar in the East Punjab, in which the Muslims were in a large majority, were shown as parts of Pakistan. The final award on the 14 August 1947, however, had cleverly omitted these three districts from the territory of Pakistan. The district of Gurdaspur in particular, gave blatantly a bridgehead to India into Kashmir state. All of us were stunned at the award.

Sheikh Din Mohammad, who probably belonged to Ferozepur or Jalandhar in East Punjab, flew into a rage and started hurling obscene abuse at Mr Jinnah. My Sikh personal assistant, a club-member, kept asking whether the city of Lahore went to India or to Pakistan, and was disappointed to learn that it was to be a part of Pakistan.

The strategy of the British on quitting India was now crystal clear: they had conceded a truncated Pakistan to Mr Jinnah, a Pakistan shorn of East Punjab in the west and Assam in the east, to spite the Muslims; but a bridgehead to the valuable prize of Kashmir state to the Hindus. Moreover, both the British and the Hindu fellow-conspirators believed that Pakistan would not survive for more than six months. Thus, the British were leaving a legacy of permanent hostility between the Hindus and Muslims. Only a few days later, many dead and seriously wounded Muslims started arriving by the passenger train, on the Delhi–Bahawalpur–Karachi route, who were attacked during its passage through East Punjab. Mr Jinnah was profoundly shocked. He expressed sincere regret to the Muslims at the unexpected catastrophic happenings in East Punjab. It was widely believed that Mr Jinnah had again been hoodwinked by the wily Hindus and British diplomacy. It was also rumoured that, after the debacle of Tara Singh's West Punjab offensive on the 3 March 1947, Nehru had sent word to the Sikhs and Hindus of West Punjab to migrate en masse to East Punjab and Delhi.

The bewildered Muslims had allowed the migration to take place more or less peacefully. And now the Muslims of the East Punjab had been trapped in their own homes. A million and a half Muslims were said to have been massacred by Sikhs, having been lured out of their homes through various devices. Another million and a half succeeded in reaching Pakistan in a pitiable state. The post-Partition days in September in Bahawalpur were days of unrelenting horror and anxiety. The post and telegraph communications were disrupted. There was no news of my wife at home in Abbottabad. My two minor sons were stranded in Srinagar. We were also very close to the Indian border at Bahawalpur. There were frequent alarms of 'The Indians are coming.'

Sir Richard Crofton, the prime minister, had retired before the Partition, and Nawab Mushtaq Ahmad Gurmani had succeeded him. He had brought along with him Mr Penderel Moon, a member of the Indian Civil Service, from Multan, whom he appointed as revenue minister. Nawab Gurmani had served as Labour member in the pre-Partition Government of India at New Delhi and appeared to know the Hindu Indian leaders very well. Mr Moon was reputed to be a pro-Hindu, and many wondered why Nawab Gurmani had picked him as a colleague. The dearth of railway wagons for the Bahawalpur state became more acute after the Partition so, one day, I told Nawab Gurmani that according to my information road metal of tolerably good quality was available in the low hills of India's Bikaner state plateau, bordering the low planes of Bahawalpur state, from where it could be transported by camel. Nawab Gurmani was greatly excited at the piece of intelligence and asked me to make further enquiries, and then he would speak to Nehru about the matter. I was astonished at Nawab Gurmani's friendly relations with Nehru, and purposely did not pursue the quest. Then one day Nawab Gurmani called me and asked me to carefully look at the maps as to whether there was, somewhere, a common land boundary between Bahawalpur and Balochistan states. To my astonishment, there was only the Indus River dividing the two states in the vicinity of Guddu, Sindh, and I apprised Nawab Gurmani of the fact. He artfully parried my query about whether there was any significance about the geographical contiguity of these two states, leaving me confused and a little suspicious.

One late evening my servant informed me that thousands of Muslims from neighbouring villages, armed with sticks, axes, scythes, etc., were silently converging on Bahawalpur city. Stepping out of my bungalow on the Noor Mahal road, I realised with a sense of misgiving that this was really the case. We were advised to make our own arrangements for our protection. It was later widely rumoured that there was a proper plan behind the sudden assault of the Muslim villagers on the city of Bahawalpur, but I failed to learn who was behind the scheme. The next day, curfew was imposed in the Bahawalpur city. A regiment of Muslim militia, recruited mostly from the Frontier districts and the tribal territory, surrounded the city. The curfew was relaxed for three hours in the morning to enable people to purchase items of daily food. Much of the fruit, vegetables, and other objects of household use had, however, already been looted. Meat, milk, and cigarettes were unprocurable. The forces soon crushed the feeble resistance offered by the Hindus from their barricaded lanes. There were, however, very few casualties among the Hindus, as a result of that action.

I received orders to build a temporary barbed wire gate for the 20-foot high *pucca* brick skeleton compound wall, within twenty-four hours. I had recently built this compound for the new central jail for Bahawalpur; at a 5-mile distance from the town. The place was thus hastily converted into a refugee camp in which the Hindus of Bahawalpur town were transported by the military. The 5-mile dirt road between the city and the refugee camp had its surface deeply rutted through constant pounding by the military traffic. It was lined on both sides, throughout its length, by hostile Muslims on the lookout for Sikh refugees. The military forces of India had not yet been divided. A contingent from a Hindu Dogra regiment, and another from the Muslim Baloch regiment, were separately guarding the two sides of the main gate, outside the refugee camp. The Hindu refugees were not allowed to carry their heavy steel safes containing valuables, which were carried to designated places, reportedly under instructions from the prime minister's office, and there broken open by civil officials.

Precisely ten days before the Pakistan Award was announced, I had allowed ten days' leave to my Sikh personal assistant for taking his family home to Jalandhar in East Punjab. I was surprised, however, to learn that he was still in Bahawalpur in hiding. The next day I received a phone call from Mr Penderel Moon, requesting me to take my Sikh personal assistant and his family to the refugee camp. 'Khan Bahadur,' he said, 'I understand you are a Pathan from the NWFP, and the Pathan code of chivalry requires you to help those who have served you.' 'Oh, you needn't refer to the Pathan code of chivalry,' I replied. 'I will take my personal assistant to the refugee camp; where is he hiding?' He is hiding in the outhouses of the superintendent of jail,' said Mr Moon. 'And thank you so much, Khan Bahadur.'

I was amazed by Mr Moon's disclosure. The superintendent of jail was a Muslim from Jalandhar in East Punjab. His daughters had reportedly committed suicide recently, by throwing themselves in a well in order to escape abduction by the Sikhs. And he was offering sanctuary to a Sikh fellow citizen. I arrived at the outhouses of the superintendent of jail, in my spacious car, driven by my young 'devil-may-care' driver Muhammad, and knocked on the closed door, calling my personal assistant loudly to come out. 'Oh. Khan Bahadur Sahib has come, we are saved,' said my personal assistant, opening the door which was locked from inside. The personal assistant, his young wife and their three-year-old child emerged from their hideout. I told them to occupy the back seat of the car and tucked in their young Hindu servant below my legs in the front seat. 'See that nobody has been left behind,' I casually remarked; whereupon both the personal assistant and his wife exclaimed that their seven-year-old daughter was still inside asleep. The child was hastily dragged out. I told them to crouch down and had a white sheet spread over them to hide them from view. Pulling my sola hat well down over my face, we started along the dusty and heavily rutted road, at a snail's pace. The heavy crowd of Muslims, lining the road, could not detect what we were carrying. In due course, we reached safely in front of the barbed wire gate of the refugee camp, between the protecting ranks of Hindu and Muslim troops. As soon as we stopped my personal assistant opened the car door and in a few swift bounds reached inside the gate of the camp and safety, leaving his wife and children behind.

I returned by the same route. My driver made a second trip to the refugee camp to deliver the personal assistant's personal baggage. When the personal assistant came to the gate to receive it, he was scarcely recognisable, having shaved his beard and cut his hair short, looking like a typical Hindu. It seemed that Hindu barbers in the refugee camp were doing a roaring business transforming Sikhs into Hindus, because the Muslims were only seeking revenge against the Sikhs. Now I began to realise why Mr Moon was so anxious about my taking my Sikh personal assistant to the refugee camp. He did not trust the Muslim army. All the refugees from Bahawalpur were safely transported by train to East Punjab. Reportedly, Mr Moon also departed with the last batch. The curfew remained in effect for several days after the incident of my Sikh personal assistant's arrival in the refugee camp. A German neighbour of mine, a mechanical engineer, stopped by each morning, to exchange the day's news with me before going to his office on a pushbike. After discussing the usual topics of dearth of meat, vegetables, fruit, and cigarettes in the looted bazaars of Bahawalpur, he mounted his bike with the usual parting remark, 'And no potatoes.'

I felt thoroughly sick, bewildered, and disillusioned, especially at the unexpected turn of events in East Punjab and did not have any feelings of elation over the achievement of Pakistan. I suspect Mr Jinnah's tuberculosis took a fatal turn for the same reason. Mr Jinnah had himself taken over the post of governor general of Pakistan to the chagrin and disappointment of Lord Mountbatten and had proudly declared, 'the pen is mightier than the sword.'[6] He had, however, been checkmated by Hindu–British collusion and the triggering of unheard-of atrocities, by the highly excitable Sikhs of East Punjab.

Owing to the disruption of postal and telegraph facilities I was anxious about the safety of my wife and family, especially my two minor sons, stranded in Srinagar. So, I obtained a month's leave and left for home by train. On the way, I saw harrowing scenes of the hapless Muslim refugees from East Punjab. The spacious covered platform of Khanewal railway station was packed with refugees; sitting close to each other, packed like sardines with vacant, despairing, downward looks. A Muslim refugee woman, hemmed in on all sides by male refugees, urinated. No one took the slightest notice. The Frontier Mail was packed with refugees. The roofs of the train were bulging under the load of refugees crowding over them. The train was moving at the speed of 20 miles an hour. After all, the hapless refugees were our brothers, and citizens of Pakistan.

I found my wife safe and sound in Abbottabad. She had, moreover, also arranged to bring home our two minor sons from Srinagar through a stratagem; she had sent a taxi to Srinagar with an application for eight days' leave for the boys, on the plea of serious illness of their mother. The nuns in charge of the Presentation Convent Higher Secondary School, Srinagar, had allowed leave to the youngsters, sending them home, however, with only a few items of clothing and their pyjama suit, with strict instructions to return to school after eight days, without fail. Since they could not return, I enrolled them in the Burn Hall School at Abbottabad, which had moved to Pakistan from Kashmir after its illegal accession to India.

Tribesmen's Offensive in Kashmir

A few days later, someone knocked at the outer gate of the compound of our house in Abbottabad. I went out and found a Mahsud tribesman with his rifle and bandolier full of cartridges, demanding hospitality. We gave him lunch, which he ate as if by right. On subsequent days we received more such visitors. In fact, it became a familiar sight to see tribesmen visiting houses in Abbottabad, demanding hospitality. They were in fact invading Kashmir via Abbottabad. Both the Kashmir government of the Hindu maharaja and the Indian government were taken by surprise. The Kashmir troops put up some resistance near Muzaffarabad, but were easily wiped out by the tribesmen, who rapidly advanced to Baramulla, within the Kashmir valley. A day or two later, they were reported to have reached the outskirts of Srinagar.

We were jubilant at the news and expected the fall of Srinagar at any moment. On the following morning, I told my wife the dream I had the previous night, that I was closely examining a king-size blue poppy (a rare flower found only in the Kashmir Himalayas) growing in a shady nook of the Takht-i-Sulaiman hill in the vicinity of Srinagar. That meant that Kashmir was surely coming to us. She replied that, on the contrary, she had had a depressing dream. In her dream she had met Mrs Mehr Chand Khanna, wife of a prominent businessman of Peshawar at Srinagar, who had said, 'Mrs Rahman what are you doing here: go back to Peshawar, Kashmir has come to us.' The very next day, we received the news of the landing of Indian troops, equipped with heavy weaponry, in Srinagar. The tribesmen, fighting with rifles only, were no match for the regular Indian troops, particularly in warfare on plain terrain. They were thrown back from Srinagar and were gradually driven out of Kashmir state. Another depressing fact that we heard was that the tribesmen had mercilessly looted their own fellow-Muslims at Baramulla. Nehru was jubilant. 'This settles the Kashmir issue,' he declared. Thus, the half-hearted attempt to rescue the state of Kashmir by Pakistan, through the tribesmen, had dismally failed.

The project of recovering Kashmir through tribesmen was said to have been engineered by Abdul Qayyum Khan, the new chief minister of the NWFP of Pakistan, a Kashmiri hailing from the vicinity of Baramulla. He had publicly declared his enmity for the Red Shirts and was said to have remarked that he would personally put a pistol bullet through Khan Abdul Ghaffar Khan, if the latter dared confront him with his Red Shirt followers. He had, in effect, carried out his threat by ordering the shooting at a Red Shirt meeting being held near Charsadda, in which many people were killed. He had also managed to convince Mr Jinnah of the need to order the shooting in question, in view of the disloyalty of the Red Shirts to Pakistan. At Peshawar, an Afghan Ghilzai tribesman, with whom I happened to share a casual conversation, declared, to my surprise, that the Afghans were not happy at the formation of Pakistan, because their way to India for carrying out their trade had been blocked. On the expiry of my one month's leave, we returned to Bahawalpur, by rail as far as Lahore and thence by car, for which I had telegraphed my driver Muhammad. On Multan road, beyond Lahore, large groups of Hindu and Sikh emigrants were still peacefully moving along the road on their way to the nearest point on the Indian border, as we threaded our car through the silent sullen crowds.

Mr James L. Roy, chief engineer, Bahawalpur state, on deputation from the Punjab Irrigation Department, was to retire in the beginning of 1948 and had plans to go home. He had been on friendly terms with me and we exchanged many presents before his retirement. I was the senior-most engineer in the state; my serious competitor, Hasnain Ahmad, a Muslim emigrant from East Punjab, was an irrigation engineer of average talent, and had very recently been promoted as superintending engineer. So, Mr Roy recommended me to succeed him as chief engineer. To my surprise and disappointment, Hasnain was appointed chief engineer overnight with the rumoured connivance of Makhdoom Miran Shah, revenue minister of the state, who enjoyed the second highest prestige in the stage, and was perhaps the richest in wealth, after the Nawab himself. Mr Roy himself broke the unwelcome news to me on telephone, expressing his deep regret, and enquired whether I would like to stay in Bahawalpur state service under the circumstances. I was stunned to hear the news and replied I would let him know. After carefully considering my situation, I informed Mr Roy that I had decided to stay on, for the time being, and he said it was a wise decision. Before their departure, Mr and Mrs Roy bade us goodbye and Mr Roy, with a cryptic smile, asked me to keep him posted about conditions in Pakistan, especially six months hence.

Winters in Bahawalpur were delightfully mild. The lands irrigated by the Bahawal Canal, a few miles from Bahawalpur, were thoroughly developed by hardworking and polite tenants from the Punjab, growing the usual winter crops of wheat, mustard, and sugarcane. The countryside was interspersed with small bushy, sandy mounds affording cover to partridge and sandgrouse, and provided good shooting. Cholistan is really not a sandy desert at all but has firm fine-soil particles blown over from the neighbouring Rajputana desert by a constant south-westerly breeze. On Sundays, we had picnics on the banks of a perennial distributary canal, cooking our own food for lunch, and shooting partridges and sandgrouse. The polite tenants offered green mustard and sheaves of sugarcane, refusing to accept money in return. Sometimes we went to the canal-cultivated lands on moonlit nights and hid ourselves in the adjoining bushy mounds waiting to shoot deer that came to browse on the young wheat saplings. These deer were being exterminated by the military officers who chased and shot them in jeeps in the trackless but flat Cholistan plain.

Nawab Mushtaq Ahmad Gurmani, prime minister of Bahawalpur state, had taken an inexplicable liking to me. He once invited me to join him on a visit to Karachi without my having any official business in that city. There we met Chaudhry Zafarullah Khan (first foreign minister of Pakistan) at the Mohatta Palace, near Clifton. It was rumoured that Chaudhry Zafarullah Khan was also acting as standing counsel to Bahawalpur state at a monthly honorarium of Rs 3000. He was a staunch Ahmadi of the Qadiani school and had also served on the committee appointed by Lord Mountbatten under Cyril Radcliffe for the division of India, as Pakistan's representative. Chaudhry Zafarullah met me warmly and we reminisced about our college days at Government College, Lahore. Mr Gurmani was evidently surprised by our old association and remarked that he was happy for having brought old friends together. To this Chaudhry Zafarullah recounted how we had exchanged letters for many years until the 1930s but did not have the occasion to meet earlier. I was quite amazed by his memory.

Gurmani stayed at the Palace Hotel in Karachi, where I was invited by him one evening for afternoon tea. He was seated on a massive easy chair before a round table, of roughly 4 feet in diameter, literally groaning under the load of dishes of all varieties of Pakistani sweets, pastries, cakes, and biscuits, and savoury items, such as *pakoras samosas, papars, shish*, and *shami kababs*, etc. I wondered whether all this was Gurmani's daily menu for tea. He was an exceptionally stout person, and preferred to ride in a car, rather than walk, even a few paces. I was astonished to note that I had not performed any professional duty during this 'official' tour.

The nawab of Bahawalpur, who had been in London during the Partition of India, now, at last, arrived in Bahawalpur state. Hindus from different places, other than Bahawalpur city, had not yet left the state. They presented themselves before the nawab in the form of a delegation, bitterly complaining about the looting by the Muslim villagers of the state. They also made serious allegations against prime minister Gurmani, who, according to them, had owed a debt of Rs 2.1 million, before taking over the administration of the state, but had now amassed a large bank balance, of over Rs 30 million. The nawab did not take any action against his prime minister but promised to get back the wealth plundered by the Muslims. For this purpose, he appointed his own father-in-law, with wide powers to search the Muslims of the state, for the money and other articles they had allegedly looted from the Hindus. The father-in-law reportedly extorted Rs 70 million for himself from the state's Muslims.

Makhdoom Miran Shah, Bahawalpur's revenue minister, was a multi-millionaire, reportedly owning 50,000 acres of canal-irrigated lands. He was the sole owner of the township of Jamal Din Wali, which had access to a part of Pakistan's main highway, Bahawalpur–Karachi, through a *pucca* branch road. He was of middle age and had a young son, Hassan Mahmud, by his previous wife who once showed me a shotgun he had purchased for the large sum of Rs 60,000. The Makhdoom and his wife cultivated friendly relations with us and one day he voluntarily expressed regret at having played a role with the Nawab in getting 'that good-for-nothing Hasnain Ahmad' promoted at my expense. Despite Hasnain's continuous provocations, I had managed to construct numerous new buildings in Bahawalpur, notably the bank, the treasury, and many more bungalows for officials. There was a sum of Rs 100,000 in my annual budget for repairs to the beautiful Noor Mahal, vacant for a long time, since the Nawab lived in the fortified Sadiq Garh Palace, in Dera Nawab Sahib, 20 miles away from Bahawalpur. I utilised the funds to carry important renovations to re-establish the grandeur of this historic building. I had also laid out the site plans for the proposed Bahawalpur Satellite Town, and prepared its water supply and sewerage schemes, the latter with the help of Mr Howell, chief sanitary engineer, Lahore. I also vetted the important Abbasia Canal project which had been prepared by Mr Ahmad Hasan (executive engineer, later vice chancellor, University of Engineering and Technology, Lahore) working under Mr Duncan (superintending engineer) on deputation from the Punjab Irrigation Department. Since I was considered a more experienced irrigation engineer, Nawab Gurmani had given me this special assignment. The canal took off from the Panjnad Headworks of the state to irrigate an additional area of 300,000 acres. Mr Ahmad Hasan, a bright engineer, had provided irrigation water for the proposed canal from the existing resources, by cleverly manipulating the existing water allowances of other canals and by proposing the lining of the main Abbasia Canal, to save the excessive absorption losses.

A British army general came to Bahawalpur to present the Nawab of Bahawalpur with two guns as a gift from the British government, as a token of the Nawab's services in the Second World War. A ceremonial durbar was held in the spacious parade ground of Bahawalpur, which was attended by all the officials of the state under a vast enclosure. In addition, thousands of Bahawalpur citizens also witnessed the event. A seventeen-gun salute was given by state gunners in honour of the Nawab, who was entitled to that number by the code of precedence fixed by the British for the rulers of various states. Four state guns stood in a row at a distance of about 1000 yards, in full view of the *shamianas* (marquees). Only one expert gunner was apparently qualified to fire the guns. He fired each gun turn by turn, running from one gun to the other. Unfortunately, one of the guns stubbornly refused to fire. Before the show threatened to become a farce, the sluggish gun suddenly fired, to everybody's great relief.

One day, a prominent Muslim industrialist of Lahore came to see Nawab Gurmani with a proposal to set up a textile mill at Bahawalpur. The mill was eventually set up at Rahim Yar Khan. Gurmani ordered me to select a suitable 100-acres plot outside the Bahawalpur city for the proposed mill. I selected a site and Gurmani personally came to inspect it. In the course of the inspection of the plot, he suddenly addressed me: 'Khan Bahadur Sahib, I have decided to resign from the post of prime minister.' I was flabbergasted. 'Why, sir?' said I. He replied, 'There are allegations against me that I have sent a full goods train load of wheat to East Punjab (India). You know Khan Bahadur Sahib, the allegation is false and frivolous, but I am fed up with this job.' I could only say that they would certainly give him a suitable post at the 'centre', if he so desired. Sure enough, Nawab Gurmani was soon transferred, and assumed an important position of minister in the central government, at Karachi. In the interim period, until a suitable permanent incumbent was nominated, Khan Bahadur Nabi Bakhsh Mohammad Husain, CIE, took over temporarily as prime minister of Bahawalpur state. He had also previously worked as prime minister of Bahawalpur from 1929 to 1943, when the Sutlej valley project was under construction, and was alleged to have taken bribes of millions of rupees from the maharaja of Bikaner and shared the money with the nawab of Bahawalpur, for allowing the maharaja to construct the Ganga Canal from the Sutlej River for lands of Bikaner state, although they possessed no right to the waters of the Sutlej River. Of course, the British Indian government must also have been behind the deal, for they wanted to compensate the maharaja of Bikaner, for his services in the First World War, by providing them with the famous Bikaner Camel Corps.

One afternoon, the personal assistant of the new prime minister called at my house with the message that the prime minister desired to see me. I hurried to the prime minister's house. Khan Bahadur Nabi Bakhsh Mohammad Husain met me cordially in his drawing room. To my amazement, he informed me that he had gone through my case and found that gross injustice had been done to me in having been superseded by a thoroughly undeserving person. He assured me that justice would soon be done in my case, and I would be appointed as chief engineer of Bahawalpur. The next afternoon the personal assistant called again and said the prime minister had agreed to sanction me a three-year contract as chief engineer of Bahawalpur state if I would pay him a sum of Rs 15,000 immediately. I was aghast at this demand. I told the personal assistant that the prime minister's offer was not acceptable to me. Khan Bahadur Nabi Bakhsh Mohammad Husain remained prime minister of Bahawalpur for a short period of only twenty-eight days in which he was reported to have made Rs 30 million, chiefly by the allotment of the cotton ginning mills, abandoned by the Hindus. A few days later, Colonel A.S.B. Shah, secretary, ministry of states and frontier regions, happened to pay an official visit to Bahawalpur. We knew each other very well. He had arrived at a crucial time when my spirits were at their lowest. Himself a thoroughly honest officer, he strongly convinced me to not waste my talents on a comfortable job in Bahawalpur, but to go to Balochistan as an irrigation advisor, where there was a need for competent irrigation engineers, and do some really useful work for the nation.

I was strongly influenced by his advice, and forthwith wrote a demi-official letter to Sir Ambrose Dundas, governor of the NWFP, to recommend me to Mr Savage with the request for the position in Balochistan. Before my retirement from the NWFP PWD at Peshawar, Mr Dundas was revenue commissioner there, and we had also been on friendly terms because I had handed the protective embankment on the Indus River at D.I. Khan to him when he was assistant commissioner. Sir Ambrose telegraphed Mr Savage, strongly recommending me to General Millis Jefferis,[7] engineer-in-chief, Pakistan Army at Rawalpindi. His recommendation was passed on to the chief engineer, Military Engineering Services (MES), in Balochistan.

In, 1948, I went to Quetta to meet Mr Muirhead, chief engineer, Balochistan, and stayed at the Chilton Hotel. It was bitterly cold in Quetta even in November. Muirhead declared he was perfectly satisfied, and advised me to see General Jefferis at Rawalpindi, who would seek central government's sanction for my appointment. So, I went to Rawalpindi to meet General Jefferis. Before meeting him, I thought it expedient to first see my old friend Nawabzada Agha Mohammad Raza, then adjutant general, Pakistan Army. General Raza was busy in a meeting and sent me word to wait for him. 'I have dismissed the meeting with the generals in order to meet you,' he said. General Raza promptly dialled General Jefferis, and started praising my professional qualifications in superlative terms, in his well-known simulated English accent. General Jefferis replied that he had already heard of my credentials and expected to see me the next day. The next morning, I met General Jefferis. He was somewhat past middle age, had a dignified bearing, and both in his figure and facial expressions, resembled Winston Churchill, and we became quite friendly. He promised to write early to the central government to approve my appointment. I returned from Rawalpindi fully satisfied. One last hurdle however, remained, 'What if Hasnain Ahmad should insist that I should complete my three years' contract with the Bahawalpur state PWD?' I had put in only a few months over two years. In fact, I had heard he intended to utilize my services as an irrigation engineer. Here, divine providence came to my rescue in an unexpected manner. Lt. Col. John Dring[8] arrived as Bahawalpur's regular prime minister. He had recently retired from the NWFP as revenue commissioner, having taken over from Mr Dundas. We knew each other very well. He was astonished to find me in Bahawalpur, exclaiming, 'You here!' At my request, Col. Dring issued prompt orders, releasing me from the remaining period of my contract with the Bahawalpur government. Thus, I was appointed commander, MES, with headquarters at Quetta, controlling the buildings, roads, water supply, and electric installations etc., of Zhob, Loralai, and Sibi districts as well as being the irrigation advisor for the whole of Balochistan.

22 Balochistan

When our train arrived at the Quetta railway station, my future office establishment, headed by the garrison engineers, was standing in a line on the platform, to welcome me. An official Chevrolet car and a military one-ton truck for carrying the luggage were standing outside. Evidently the MES enjoyed official perks and a lifestyle that we were not used to in the PWD. My new residence, a spacious bungalow, was situated on Lytton road (now Zarghoon road) with its stately row of massive Chinar trees. The bungalow was furnished with MES furniture, for which the overseer took a receipt from me. It had a large compound of a few acres, with an extensive orchard containing a large number of mature fruit trees. There were apple, apricot, walnut, peach, and almond trees and a long overhead trellis of grapes; a large vegetable garden; a cemented full-size tennis court; lawns and many out-houses. I was the first Pakistani to occupy the house. We could not believe our good fortune.

The clean city of Quetta presented a surprisingly pleasant contrast to the dusty Bahawalpur city. Although it was an extremely cold and cloudy January morning, we were so excited that, after breakfast, suitably clad in warm clothing, we all sallied forth on foot, to explore our beautiful surroundings. We were surprised to note that all the roads of Quetta were neat and dry. The city had an underground drainage system. The elegant Bruce road (now Jinnah road), the main shopping centre of Quetta, was only a stone's throw away from our bungalow. The busy Qandhari Bazar branching off Bruce road was also near to our house. Also located on the Lytton road were the residences of the inspector general of police and the revenue commissioner; and further down to the left, the political agent of Quetta and Pishin. All those elegant bungalows were of the so-called 'hut' type, built in that style in view of the incidence of frequent earthquakes in the Quetta valley. Just opposite our house, across the road, stood the two-storeyed 'residency', the Government House, and residence of the agent to the governor general of Pakistan (AGG).

Col. Muirhead (chief engineer) was a polite officer. From the very beginning he treated me with marked consideration and courtesy. Our first AGG was Sahibzada Khurshid, a nephew of the late Nawab Sir Sahibzada Abdul Qayyum Khan, former chief minister of NWFP. Although the

Government House at Quetta was under the charge of Col. Plaret, Sahibzada Khurshid requested me to have a small, square, 6-inch high marble platform constructed in the bathroom of the guest room of the Government House, for the convenience of Governor General Khawaja Nazimuddin, for his first official visit to Quetta. Khawaja Sahib, a devout Muslim, needed that facility for making his daily ablutions before prayers. Sahibzada Khurshid was soon transferred and Mian Aminuddin, an old ICS officer of pre-Partition days, took over as the AGG. He met me formally in his office, when I called on him for the first time, but we developed very friendly relations, in due course. Mr and Mrs Mian Aminuddin were highly respected among the official circles in the social environment of the newly independent Pakistan.

On 14 August 1949, the second Pakistan Independence Day was celebrated with due pomp and ceremony at Quetta. Governor General Khawaja Nazimuddin was in Quetta on his first official visit (ostensibly on tour, but actually for enjoying the dry cool season). In the ceremony at Quetta's parade ground, a great display of Pakistan's military hardware, in the shape of guns, tanks, and other armament was put up. Khawaja Nazimuddin, slowly riding past in a jeep, took the salute.

The flat portions of the 330 square miles of the Quetta valley, containing Quetta town, were irrigated from numerous karez located at the foot of the Murdar hill of an elevation of 10,000 feet, formed the valley's eastern boundary. When originally constructed, the karez were only high enough for the karez diggers to crawl through in a sitting position. There were also lines of open wells at short intervals along each karez aligned to the rapidly descending slope, until a small perennial discharge emerged at the end of the karez in the flat plain below. The rooves and sides of the new karez were, however, just plain compacted earth. Most of the old karez had their earthen roofs and sides caved in, which required constant and risky maintenance to keep them going. So, I designed an alternative infiltration gallery, with *pucca* but seeping roof and sides to replace the old abandoned karez, and also for new constructions. My strategy for rehabilitating old karez was studied with great interest, especially by neighbouring Iran.

Nushki is a god-forsaken place, situated in a drab, forbidding plain with small, unattractive hills on two sides. The forlorn-looking bungalow of the political agent of Chagai was situated on a small mound overlooking Nushki town. My father must have contracted typhoid fever at Nushki, before he came home after resigning from his service in the railway department, while working on the construction of the Quetta–Nushki Railway in 1910. I had to go through a massive, fifty-year-old file, containing all kinds of proposals for irrigation work for utilising the flood waters of Pishin–Lahore torrent in the Nushki plain, before actually inspecting the site. Only one proposal written in the manuscript, in small handwriting, by an evidently young assistant engineer, appeared worthy of consideration. I was startled to see the signature of the young engineer at the end of the report—R. Cannel. It was the same ageing secretary for irrigation, NWFP, who had boosted my career in the irrigation department of the NWFP by appointing me as in charge of the construction of the protective embankment on the Indus River, at D.I. Khan.

Irrigation Development in Balochistan

The Pishin–Lahore torrent is one of the larger hill torrents of Balochistan. It has its source in the hills on the west side of the Kan Mehtarzai Pass. Flowing through Pishin district in a south-westerly direction, it cuts across a small triangular chunk of Afghanistan territory, to re-appear again in Pakistan's Nushki plain, where it loses itself in the internal drainage depression of Hamun-i-Nushki.[1] The Pishin–Lahore carries a small perennial discharge, irrigating hundreds of acres of orchards of the finest oval-shaped white seedless grapes, through lift, by a diesel engine. In the Nushki plain lower down, its flood waters are utilised for flood irrigation by the Nushki farmers, from a point only 5 miles from the Afghan border. The armed guards accompanying me on the journey to the Kachi flood distribution site on the Pishin–Lahore by jeep, kept looking apprehensively towards the imaginary Afghan border only 5 miles away in the dead flat featureless Nushki plain. The difficulty of designing a distribution work in the flat Nushki plain became all too apparent on inspection of the site. The only feasible proposal appeared to be a low submersible diversion weir, with off-taking canals on either side, but at a disproportionately higher cost compared to the meagre benefits of the scheme. So, the proposal was dropped for the time being.

The elevation of the Quetta and Mastung valleys ranged between 5000 feet and 6000 feet above sea level. Most of the underground water resources of the Quetta valley had apparently been exploited through karez, tube wells, and shallow wells. But water in large quantities was still seeping out of the high zones of Quetta and Mastung, due to the copious Aab-e-Gum (or Abegum) springs, situated at a lower level in the east, and the perennial Khizar Nullah in the west, on the way to Nushki. There was also the evidence of mud volcanoes sprouting near the coast of Makran, as subterranean flow or seepage. However, no means were available for testing my hunch through boring and underground investigation. Khushdil Khan Bund (dam) in Pishin district was constructed by R.G. Kennedy, an executive engineer of the 'silt theory' fame, fifty years earlier. It consisted of a 65-foot high earthen embankment, about a mile in length and built on a gently sloping plain. It was shaped like a crescent with its concave side forming the base of the reservoir. The reservoir was fed indirectly from two hill torrents. Floods were heavily charged with coarse red-coloured silt and detritus, through two supply channels (finally converging into one) which picked up only the finer silt from the upper layers of the floods, allowing the heavier silt and detritus to pass down the torrent beds through sluice gates. Following a hunch, I had the reservoir's capacity surveyed, and was amazed to find that it had reduced by only 19 per cent, during half a century's operation. I have often wondered whether any further progress was made to investigate and tap these potential irrigation sources, in this water-starved part of Pakistan, using unconventional methods and approaches.

The Bolan River is an important river of Balochistan. Rising from the hills near Kalat, it follows a long, easterly course down the Sibi plain into its outfall in the mighty Indus River. On emerging into the Sibi plain, it cuts its way across a gorge, and, thereby, brings to the surface its substantial perennial discharge of about 40 cusecs. I was curious to find out its actual course along the vast Sibi plain down to the Indus. So, on one occasion, I was surprised to locate it, emerging out of a 2000-foot wide gorge among low hills, right within the Sibi plain, at a place called Allahyar Shah about 40 miles west of Sibi town. I had driven to Allahyar Shah along a trackless 'no-man's land'. Here was an ideal site for a low earthen dam, with a large potential reservoir in a flat plain to catch the perennial as well as the flood discharge of the Bolan River.

I reported my finding to General Millis Jefferis, engineer-in-chief of the Pakistan Army. He promptly came to Quetta to inspect the site. It had rained heavily during the previous night when we left Sibi town for Allahyar Shah in a jeep, with General Jefferis at the wheel. The other occupants of the jeep were Col. Muirhead and myself. General Jefferis took a more or less straight course towards the low range of hills containing the proposed dam site, driving at a fast speed with the jeep's wheels almost sinking in the fluid mud. He had contempt for bushes and dry streams with shingle beds and vertical banks en route and pushed through all the obstacles, with our heads constantly bumping against the jeep roof. When we arrived at the proposed dam site, General Jefferis' practised eyes at once realised its possibilities. He straightaway approved my proposal and asked me to prepare a preliminary dam scheme, with approximate estimate of cost, which he personally took to Karachi and delivered to Mohsin Ali, chief engineer of the central government. The scheme was kept in cold storage, till the time both General Jefferis and I had left Balochistan, when it was handed over to a foreign firm, for design and construction. During the late 1970s, however, owing to heavy rains, the Bolan Dam was reported to have been heavily damaged. There was no official response to my anxious query about the size of this disaster to my brainchild.

Zargi Tangi, near Ziarat, was a limestone gorge, about 100 feet across the downstream end of a small elevated plain. A small non-perennial side stream flowed through it during the summer monsoon season, meeting the broad Ziarat valley a few miles downstream. The design of a 95-foot high rock-fill dam across the gorge already existed with a 4-foot thick reinforced concrete curb wall in the middle, plugged into the rocky bed, together with the flanking limestone walls of the gorge. I had to construct the dam. I thought I had made a good job of the construction work, which was carried out under my supervision. The dam was completed in good time, and the reservoir began to fill in rapidly during the August rains. Water in the reservoir rose to a depth of 75 feet, only 15 feet short of spillway level. And there it stood still. Then the water level began to fall mysteriously, and the reservoir fell to dead storage level a few weeks later. Obviously, the leakage had taken place through the bed and flanks of the natural limestone rock. Simultaneously, the discharges of the karez lower down in the Ziarat valley had increased. The benefit from this dam was indirect, in recharging the aquifers and increasing the karez discharge. The same phenomena were observed in the case of the Hanna Lake reservoir, 5 miles from Quetta town, where a stone masonry dam was built long ago in the limestone gorge; the Hanna Lake reservoir, which is bigger in capacity than that of the Zargi Tangi Dam, also gradually sinks down to dead storage level, presumably by leakage through its natural bed and flanks.

Very early after my arrival in Balochistan, I had designed and started construction of a sub-surface weir on the Narechi Rud torrent in Loralai district which irrigated a sizeable area of its surrounding lands. But my principal irrigation task, apart from the control of roads, buildings, and electric installations, was to design and construct the proposed irrigation works on the large Anambar torrent, in Loralai district. The rather innovative design of the dam comprised a 1600-foot long concrete weir in the torrent bed, founded on rock foundation, with an ancillary rock-cut 7-miles long canal, and a 1600-foot long reinforced concrete aqueduct resting on box foundation on the Anambar torrent, 7 miles below the weir. The scheme was a large one for the highland of Balochistan. It was designed to irrigate an area of 15,000 acres. I founded the main weir on watertight shale strata visible throughout the width of the shingle bed of the torrent, with a view to force to the surface the subsurface water flow beneath the shingle bed of the torrent. The work remained under construction for more than a year. The opening ceremony of the project was performed by Governor General Khawaja Nazimuddin. I also drafted his inaugural speech. In his speech, Khawaja Nazimuddin expressed the hope that with the construction of the irrigation works on the Anambar torrent, the Loralai district would rival the Qandahar province of Afghanistan, in producing high quality almonds, grapes, apples, apricots, pomegranates, and other fruits which were imported from Afghanistan. A few years later, after I had left Balochistan, I was disappointed to learn that wheat and corn were being raised at Anambar.

A few days later, I received an official invitation to lunch with Khawaja Nazimuddin, at Ziarat. It was a Friday, and I started from Quetta, after making my ritual ablutions for Friday prayers. At the Government House at Ziarat, I was met by the Khawaja Sahib's daughter, who offered me a seat in the drawing room. After a few moments, Khawaja Nazimuddin walked in, and proposed that we should say our Friday prayers in the Jamia mosque, before having lunch. I sat in the back seat with Khawaja Sahib, in his Cadillac, on our way for prayers at the mosque, via the Ziarat Bazar. We said our prayers seated in the back rows of the mosque among ordinary citizens, under the leadership of a semi-literate Pathan Imam.

At the lunch table, I was given the place of the guest of honour by Khawaja Sahib. The other guests included senior officials and a few members of the governor general's entourage. As a formal gesture, I praised a certain dish placed before me, and had to pay the penalty by having to accept two more helpings of the same dish, pressed on me by my distinguished host! Khawaja Sahib himself was a voracious eater, accepting more than one helping from each dish. After lunch, the political agent of Loralai and his wife left, and Khawaja Sahib invited me out to the lawns for a chat. 'After all, what is the hurry,' said he, after we had seated ourselves on two easy chairs in the beautiful lawns of the Government House. 'We can have a cosy chat here and you can go back to Quetta, after our afternoon tea together.' We exchanged reminiscences of the time we both had been contemporaries at Aligarh college—he was a school student living in the exclusive English House hostel, and I was a BA student living in the Sir Syed Court hostel. We also reminisced about our membership of the Aligarh riding club, but I avoided mentioning the episode of his exasperation when he was given the master mare to ride.

The Ziarat valley presented an interesting contrast in greenery and vegetation. The lower portion of the valley was bare and flat, about 2 miles across, lying between bare flanking limestone hills. The upper part was densely covered with Juniper trees and considered one of the largest juniper forests in the world. The Juniper forest in question appeared dead and somewhat depressing. But it must have been reproducing itself, as evidenced by its very existence as a thick jungle. The forest seemed to sprout out of the bare ribs of the limestone slopes, there being no undergrowth. The Ziarat Pass, leading down to Loralai, is a mile or so further on, at a higher level. The hills flanking the Ziarat Pass attain an elevation of 11,000 feet and the narrow drainage line between the wooded hill slopes opposite Ziarat station remained permanently dry. There was no perennial flow in the channel in summer, due to the lack of undergrowth, and the shallow winter snows falling on the Ziarat hills were entirely sublimated in the atmosphere. The Government House at Ziarat, where Mr Jinnah passed the last days of his illness, was a small country house built in the English pattern. It had beautiful lawns covered with well-trimmed vividly green soft grass and lovely flower beds. The Ziarat club had a small square swimming pool, fed from a tiny spring. Ziarat town had a small bazaar, with a Jamia mosque. In May, the wide and lower barren Ziarat valley was ablaze with orange-red wildflowers of abnormally large size, completely filling the valley between the flanking hills. It presented a fantastically beautiful and almost unearthly scene.

Sibi town is situated in a low plain 80 miles from the Quetta highland. Every year, during February, a Sibi Durbar is held at Sibi, according to an old British tradition, which is attended by the Baloch Sardars and high Pakistani officials. In February 1950, Nawabzada Liaquat Ali Khan, prime minister of Pakistan, also attended the Sibi Durbar. He was himself an 'Alig'—Aligarh college alumni. One evening, he invited all the 'Aligs' among the Balochistan officials—about a dozen in all—for a get-together. He enquired from each one of us in turn the year of our joining Aligarh and appeared somewhat amazed to learn that I had joined Aligarh college three years before him. He was now clean-shaven without a moustache and looked much younger than when I had seen him last during the early 1940s.

While at Quetta, I received an invitation to the annual dinner of the renowned military institution, Command and Staff College. The annual dinner is a notable function held on the occasion of graduation of a select group of Pakistani and foreign army officers on completion of the advanced training course in military strategy and tactics. Local military and civilian officers and their wives and guests from outstations, are invited. The lawns and pathways of the college were tastefully decorated with bunting and illuminated with flood lights. Major General Ayub Khan was attending the dinner. On spying me, he walked to my side of the table and greeted me with his usual effusiveness. We sat down for a long while, chatting on various topics. 'What are my prospects for being selected as the commander-in chief of the Pakistan Army?' he asked me. The post was falling vacant on the imminent retirement of the second commander-in-chief of the Pakistan Army, General Sir Douglas Gracey. After some deliberation, I replied: 'Well, as you are my friend and fellow countryman, my sympathies are strongly with you. But, apart from that, I honestly consider you the best choice.' 'Pray for me.' He said on parting. A fortnight later, General Ayub's appointment as commander-in-chief was notified.

Sometime later, Mirza Bashiruddin Mahmud Ahmad, the sixty-five-year-old son and successor of the late Mirza Ghulam Ahmad of Qadian, the founder of the Ahmadiyya Movement, paid his regular annual visit to Quetta. His followers in Quetta rented a very spacious bungalow for him, with room enough to accommodate his favourite fourth wife, reputed to be a young girl, whom he had recently married, as well as about half a dozen of his numerous adult and adolescent children by his three other wives. In addition, his twelve secretaries had also to be accommodated in the out-houses. A large *shamiana* (marquee) was erected in the front lawn of the rented bungalow, for daily congregational prayers under his leadership. On learning that I was a former college-mate and friend of Chaudhry Zafarullah Khan, Mirza Mahmud Ahmad, invited me to tea.

I met him in his bungalow and he greeted me cordially. We talked for about an hour mainly about the development of Balochistan. He was optimistic about the possibilities of developing the vast province, but I tried to dampen his enthusiasm by hinting at the meagre annual rainfall in the province. He, however, appeared unconvinced by my professional pessimism. He scrupulously avoided any discussion on religion during the conversation. At the end, he requested my permission to stay at my official residence at Ziarat for a week, which I gladly allowed. I kept wondering whether Mirza Mahmud Ahmad, in his detailed queries about the development potential of different areas of Balochistan, was exploring the feasibility of establishing an abode for his followers in that large and scantily populated province.

All the garrison engineers in my circle sent their monthly accounts directly to the Accountant General Pakistan Revenues, at Karachi. I was only supplied details against my sanctioned budget at the end of the financial year. While scrutinising the expenditure against my budget at the end of the year, I was baffled to see an expenditure of Rs 42,000 in the account of garrison engineer Loralai, for which no item existed in my budget. My written approval was required to incur expenditure on an unforeseen and urgent nature. I sent my personal assistant, an officer of the rank of an executive engineer, for site inspection and report. The personal assistant reported that the item in question related to the alleged repair work to the shingle surface roads of the Loralai district, which had allegedly been executed by the garrison engineer, Loralai, under direct instructions from the political agent, Loralai. The latter had received orders from the AGG to 'keep the roads in Loralai district in good repair,' on the occasion of the expected visit of the governor general of Pakistan to that district. The personal assistant further reported that there was no evidence of any repair work having been done.

I, thereupon, made a report against the garrison engineer, Loralai, an officer of the rank of major in the army. He was forthwith suspended. A court-martial was held against him, which found him guilty, and awarded him three years' rigorous imprisonment. He promptly appealed against the sentence to higher authorities at Rawalpindi and returned in a few days, honourably exonerated. I was startled to see him enter my office, one day. After a smart military salute, he requested for his re-posting orders from me. I phoned Col. Muirhead, that I would not have the officer in question working under me. He was, therefore, quietly posted elsewhere!

One afternoon, while I was on a tour of Ziarat, I found the garrison engineer, Sibi, and his wife camping in the MES rest house at Ziarat. They invited me to tea. Sometime later, I received reports that the garrison engineer, Sibi, was incurring heavy expenditure on road repairs, through daily labour muster rolls, without obtaining my prior permission. On investigation, it was disclosed that he had overspent his road repairs budget by Rs 65,000 without any evidence of any repair work having been done! So, I made a report against the officer, and he was promptly suspended. But since he was a civilian officer, he could not be court-martialled and a case of embezzlement of public funds was, presumably, instituted against him. Finally, I learnt that he had been honourably acquitted, and was transferred to some other, possibly more attractive, station.

Early in 1950, General Millis Jefferis retired, and was succeeded by General Veech. A few weeks later, Col. Muirhead also proceeded 'home' (UK) on leave, and I took over as chief engineer, MES in Balochistan in addition to my own duties as commander, MES. It was a barren honour, for it carried no extra allowance. There was a small salient about 3000 square miles in area, inhabited by the Afghan tribe of Sulaimankhel, which Sir Mortimer Durand had included in Balochistan, roughly opposite Fort Sandeman, presumably 'merely for the heck of it'. If he had been intelligent enough to foresee future implications, he would have taken a chunk out of Afghanistan, further north, providing an easier route to the Chitral valley. The Sulaimankhel area in question was reported to be capable of development through tube well irrigation, the sub-soil water level being at an economical depth. I sent a proposal to the Pakistan government, advocating the development of the Sulaimankhel area in question, but no notice was taken of my communication.

By this time, I had decided that I had had enough and was planning to resign from my position, when Col. Hainsworth, chief engineer, NWFP, called me and said, 'I hear you are you leaving Balochistan. If so, would you like to come back home as irrigation advisor, and take over the task of the design and construction of Kurram Garhi Scheme in the Bannu district?' I could not believe my ears! I asked, 'Have you obtained the chief minister's sanction to my appointment?' 'Yes, the chief minister wants you back home for the job,' was his reply. 'I accept the offer with thanks.' Luckily, my friend, AGG Mian Aminuddin, was at Sibi for the three severe winter months, so I would leave Quetta in his absence, as I feared he might have objected to my departure.

I happen to contract pneumonia during my last month at Quetta, probably due to exposure during one of my usual afternoon walks with insufficient clothing. In the meantime, the construction of the Anambar irrigation work, in the Loralai district, which had been completed with a substantial saving of Rs 116,000, required payment to the contractor. The contractor came to my house during my illness, requesting a few minutes' talk with me and offering me a hefty sum in bribe. He was a respected person, being one of the biggest contractors in Balochistan. Declining his offer, I sent him back with the assurance that he would be paid in full for the excellent job he had done as soon as I was fit to attend office, in two- or three-days' time. Before I could attend office, the garrison engineer, Loralai, came to see me at my residence and silently presented a large envelope containing a sum of Rs 35,000 as my share of the savings. I returned the envelope telling him that I would scrutinise his items of expenditure.

On my return to office, I sent my personal assistant to Anambar, to check the extra items put in the bill by the garrison engineer, Loralai. He returned with the report that all the items were fictitious. So, I gave the garrison engineer, Loralai, a sound verbal scolding, and hauled him up on the charge of serious embezzlement, to my boss, Col. Muirhead, with the request that the investigation into the case should be carried out by Lt. Col. Plaret, as I was due to leave Quetta in a few days. On receiving my report against the garrison engineer, Loralai, Col. Muirhead remarked on the phone 'Oh, Khan Bahadur, the moral fibre of the nation has snapped after Independence!' I felt ashamed for my nation, on hearing those remarks from a Britisher! A few months later, while walking one morning along the Peshawar Mall, I unexpectedly accosted the garrison engineer, Loralai, the same officer whom I had hauled up for embezzlement a few days before my departure from Quetta. I felt embarrassed at the meeting and asked him what he was doing in Peshawar. He replied, to my astonishment, that he was working as personal assistant to the commander, MES, Peshawar. And then, he blurted out, 'Oh Khan Bahadur Sahib, you surely proved a simpleton to refuse to accept that money. I offered the same sum to Col. Muirhead, who accepted it, and hushed up the case.' So that was that.

23 Irrigation Advisor, NWFP

My son-in-law, Muhammad Said, superintending engineer, Bannu, had already located a suitable indirect storage site for the Kurram River floods, at the Baran Nullah, situated about 5 miles from the proposed weir site on the Kurram River, at Kurram Garhi. The Baran Nullah had an ideal site for a reservoir of 100,000-acre feet storage capacity, with a 2000-foot long, and 100-foot high earthen dam, abutting against two hills. Said had also already started construction of the link channel connecting the old Kachkot Canal, which utilised the entire dry weather discharge of the Kurram River (600 cusecs) irrigating the 'green oasis' of 50,000 acres, in the Bannu district. Said's alignment of the link channel was faulty, running throughout its length of about 10 miles in high filling, but I could do nothing about it, as construction work on the link channels was far advanced.

My headquarters, as irrigation advisor to the Government of NWFP, was at Peshawar, where I was allotted a spacious bungalow in the Racecourse Gardens. My initial months, however, were chiefly spent in Bannu in Said's posh double-storey official residence designed by an English architect to house the superintending engineer, southern circle, Bannu, who, in pre-Partition days, was always a Britisher. I set up a temporary office and designed the Kurram Garhi Scheme, with the assistance of only a head draftsman who was a clever and hardworking young man who subsequently became adept in the design of irrigation structures and was promoted to the rank of an 'officer'. It was an exhilarating feeling to realise that I was free to plan and design a large irrigation project, in whatever shape I deemed proper, without criticism from anybody. My assistant prepared fair construction drawings from my rough drawings based on my own plans and calculations.

Designing the Kurram Garhi Dam Project

The Kurram River runs in a gorge at the Kurram Garhi site. Here, I located and designed an ogee-shaped (double continuous S-shape) diversion weir in solid concrete on rock foundation. The river's current was known to permanently hug the right bank (looking downstream), so I located the under sluices and the head regulator of the main line canal, also along the right bank. Since the floods in the Kurram River carried large bed-loads of shingle and boulders, I placed the floor level of head regulator 5 feet above the level of the river bed, with proportionate increase in the height of the under sluices, gates, and the top of the weir, thereby facilitating the diversion of the river current into the head regulator. Taking off from the weir, at the head regulator, the main canal, with a capacity of 3000 cusecs, passed through a short tunnel and then bifurcated with a link channel of 800 cusecs capacity, leading to a storage reservoir in the Baran Nullah. Further downstream there was a trifurcation in the main canal with the straight branch carrying a discharge of 600 cusecs, linking with the existing old Kachkot Canal; the right branch carrying a 1200 cusecs flood discharge was the new Kurram Garhi Canal running in a southerly direction along the Waziristan foothills and irrigating the extensive Sara Darga plain in its lower reaches. The third link in the trifurcation took off to the left and, after crossing the bed of the Kurram River through a reinforced concrete siphon, irrigated the area on the opposite bank of the river. The total area under the new Kurram Garhi Canal alone was 170,000 acres.

During the early stages of construction, Ghulam Mohammad, governor general of Pakistan, came to inspect the project. He had had a stroke and could only speak in whispers. As I was the 'father' of the Kurram Garhi Scheme, Khan Abdul Qayyum Khan, the chief minister of NWFP, arranged my meeting with the governor general at the work site. I met him in the picturesque rest house situated on the mound overlooking the imposing Kurram Garhi weir, which was under construction. Ghulam Mohammad whispered in my ear that, as funds were limited, he would consider financing only such development schemes in the NWFP which are profitable.

Dr Khan Sahib was under house arrest and was living in Abbottabad, with his son Saadullah Khan (executive engineer, PWD). We chanced to meet one day on the road near his son's bungalow, while I was home on short leave. He was delighted to learn that I was back in NWFP and was working on the Kurram Garhi scheme. He said, 'You must go back and work under Qayyum (chief minister). He is doing a lot for the NWFP.' I wondered whether this noble man had forgiven Qayyum Khan for ordering firing on the peaceful Red Shirts meeting in Charsadda. On another occasion, Dr Khan Sahib met my wife on the road while she was taking a walk. He remarked, 'Mrs Rahman, you look very pale and weak. Is there any health problem?' Since she gave an evasive answer, Dr Khan Sahib promptly sent his daughter-in-law to enquire about my wife's sickness, and also sent her a prescription.

The Canal Water Dispute with India

In 1948, only a year after Partition, India had suddenly closed canal water supplies to the lands of Lahore district, which were irrigated from the Upper Bari Doab Canal whose headworks was on the Ravi River at Madhopur located in India. The irrigation waters had been deliberately stopped in the crucial Kharif (summer) season, when the Kharif crops of corn, sugar cane, and cotton needed to be watered thrice every week so that they would not perish. The stoppage of irrigation water at the source at Madhopur also automatically affected the canal supplies in the Lower Bari Doab Canal which took off from Ravi River, lower down, at Balloki, in Pakistan. The thought of imminent peril to the Kharif crops on both the canals was frightening. So, in order to prevent disaster to her crops, Pakistan agreed to pay the cost of the irrigation water to India up to 1952. Then, realising that India's claim of possessing sovereign rights over the waters of an international river were false, Pakistan ceased to pay, what was tantamount to a tribute, to India. This created a serious crisis in Pakistan–India relations, threatening to develop into a full-scale war between the two countries.

Since the allied victory in the Second World War, the United States had assumed the role of a policeman to the world, largely because of the collapse of the British Empire due to bankruptcy of her material resources. The US government sent David Lilienthal,[1] an irrigation expert and a reputed trouble-shooter, to India and Pakistan to persuade the two countries to settle their dispute through peaceful negotiations. Mr Lilienthal's proposal was to treat the six rivers of the Indus Basin as a whole from which India and Pakistan would draw their due shares of water according to their respective needs. Since India's share was only 12.5 per cent of the Sutlej River waters, according to the Indus Commission 1941–2 (Rau[2] Commission),[3] which was not put into effect, and since it had everything to gain by gaining access over the entire Indus Basin River system, it readily agreed to the negotiations; Pakistan, being politically and militarily in a weak position, also had to agree. The venue for the negotiations was to be Washington DC, under the auspices of the World Bank, whose representatives would also regularly sit on the conference table but as silent observers.

Since Pakistan's acceptance of Lilienthal's proposals, a number of meetings of the chief engineers of Sindh, Bahawalpur, Punjab, and NWFP were held under the chairmanship of the minister for industries and finance to thrash out the tactics to be adopted at the negotiations. The NWFP chief minister, Khan Abdul Qayyum Khan, called me to his office and said, 'Since you are our irrigation advisor and our most senior engineer, you will have to go to Karachi to represent NWFP in the internal discussions that were taking place on the canal water dispute with India.' So, I had to go on the long and tedious journey to Karachi, by rail, much against my inclination.

The meeting was held in March 1952, in the high court building at Karachi, in one of the vacant court rooms. There were ten chief engineers present and the meeting was jointly presided over by Sardar Abdur Rab Nishtar (central minister of industries), and Chaudhry Mohammad Ali (central minister of finance). Sardar Nishtar rose, and in a grave tone described before the meeting Mr Lilienthal's proposals, which Pakistan had accepted. After his speech, he asked, 'Any questions gentlemen?' I looked around and saw that no one was getting up. So, I slowly got up, and said, 'Sir, I am attending this meeting for the first time. I know very little of the problem but, after having heard the honourable chairman's speech today, may I venture to ask one question: Since the dispute with India relates only to the supplies in the rivers Ravi, Beas, and Sutlej, why are we agreeing to include the purely Pakistani rivers, Indus, Jhelum, and Chenab in the common pool?' Both the presiding ministers looked at each other in obvious embarrassment. Chaudhry Mohammad Ali was probably asking Sardar Nishtar who I was. In reply to my query, Sardar Nishtar got up and said, 'The point raised by Khan Bahadur Abdur Rahman is well put. But we have agreed to these somewhat extraordinary conditions relying on the good faith of the Americans.' The answer did not succeed in dispelling my misgivings.

At the end of the meeting Sardar Nishtar invited all those present at the meeting to his residence in the evening where he would announce the names of the five Pakistani members of the Canal Water delegation who would go to the USA for negotiations with the Indians. Everybody believed that Mohsin Ali, chief engineer of the central government, would lead the delegation. The gentleman himself looked smug and confident. To everybody's surprise, the following five engineers were nominated to go to Washington DC as part of the Pakistan delegation: Khan Bahadur Sheikh Abdul Hamid (chief engineer, Punjab Irrigation Department) as leader of the delegation; Pir Mohammad Ibrahim (chief engineer, Punjab Irrigation) as deputy leader; Hasnain Ahmad (chief engineer, Bahawalpur; he was the same gentleman who had outfoxed me in Bahawalpur); Dr S.M. Qureshi from Sindh; and Khan Bahadur Abdur Rahman Khan (irrigation advisor, NWFP) to represent the Frontier province.

I was pleasantly surprised to hear that I was to be the part of the Pakistani delegation. The next morning, I had my passport-sized photographs taken, by a private firm of Karachi, and by the afternoon of the same day our passports, 'valid for all the countries of the world', complete with visa for the USA, were handed over to us, by the Foreign Office. No visas were required for visits to the UK, as Pakistan was a member of the British Commonwealth. We had to make our own arrangements for procuring the necessary health and vaccination certificates. Since I had come to Karachi utterly unprepared for the new assignment, I requested permission to see my wife and family at Peshawar before proceeding to the USA. It was accordingly arranged that I should leave in a military Dakota bound for Chaklala airport leaving from Mauripur airport early the next morning, and take a car from Chaklala to Peshawar, reaching there before evening. Then I should return the next day by Khyber Mail, leaving Peshawar at 10:30 p.m.

24 Canal Water Dispute Negotiations

Visit to the USA

My wife, who was in poor health, got upset at the thought of my absence from home. I tried to soothe her anxieties by assuring her that it was an important mission and a rare opportunity; and besides, I would be away for only two months. She reconciled herself to the inevitable, and she and the family came to the Peshawar railway station in the evening to bid me farewell. The next afternoon, Pir Mohammad Ibrahim boarded the Khyber Mail from Multan, and chanced to enter my compartment. I was somewhat astonished to note his over-friendly attitude towards me, and his transparent efforts to win my good will!

I had not been able to persuade my friend, Dr Abdul Hakim Khan (civil surgeon, Peshawar) to give me a vaccination certificate. Actually, two certificates were necessary: one predated by a month, and the other predated by a week, certifying that I had been administered vaccinations for typhoid and cholera. Typically, Dr Abdul Hakim, refused to issue a fraudulent certificate, although I told him that I had received orders for overseas travel only two days ago. So, we arrived at the Karachi airport for boarding a BOAC plane, via London to USA, while I had no health certificate with me. My other colleagues had wangled their health certificates somehow. Hasnain Ahmad declared with glee, that I would be hauled down at the very next temporary stop at Bahrain and clamped in quarantine for nineteen days! However, to my good luck the matter was resolved by a helpful airport doctor.

The giant, four-engine Argonaut airliner of BOAC, had a 'cruising speed of 250 miles an hour'. This was my first experience of boarding a large plane, so everything was new and exciting. Shortly after the plane had settled in level flight, the air hostesses started tagging light wooden lunch-trays to our comfortable seats. Lunch consisted of large cups of soup, chicken, mashed potatoes, cheese, and biscuits followed by pudding, a large red apple, and coffee. Sheikh Abdul Hamid, the leader of our delegation, forbade us to touch the chicken, as it was not *halal*. But the remaining dishes were enough to make for a regular feast. The seat belts had been unfastened after we had reached the ceiling height and we were permitted to smoke. After about two hours we were flying over the eastern tip of Arabia bordering Oman territory. It looked like a deep-red coloured dry, sandy desert, devoid of any vegetation. 'So, this is Arabia, the land of the Holy Prophet [PBUH],' I said to myself, with a surge of emotion. The strip of land below us soon passed, and we continued our flight over the sea (the Persian Gulf).

As soon as we struck land at the end of the Persian Gulf, our plane was hit by a sandstorm, the dreaded simoom, an inferno of whirling sand of dark colour, creating an opaque twilight. Fearing that my last hour had come, I decided to meet my end during sleep. Luckily, I did fall into a nap and when I awoke, we were peacefully cruising under a clear blue sky! Looking down the side window, the dark-red inferno was churning a short distance below! In front of us was a bank of white cumulus clouds, many miles ahead fringing the western horizon. From our height, I judged (wrongly) that our plane would pass the cloud bank. The sun was nearing the horizon; we had left the simoom behind and were flying over what looked like a desert plain thinly covered with bushes. This was undoubtedly the north-Arabian desert, mentioned by a German horse-dealer, Carl R. Raswan, in his book, *The Black Tents of Arabia: My Life Among the Bedouins* (1947), concerning a young Arab lad, Faris—the hero of the story and after whom I have named my son, Faris!

The cloud bank on the western horizon had turned pink by the setting sun. Our plane plunged right through it. The cloud bank extended for many miles. Its interior was a fairyland of deep-pink and coloured grottos, caves, and galleries. The plane shivered through air pockets every few seconds. The stewardesses tripped about nonchalantly, in their high-heeled shoes. There was an unearthly splendour in the pink-orange world we were passing through. It was evening when we emerged from the cloudbank which had since turned a dull grey. The exhaust from the plane engines gave the impression as though the engines were on fire. Promptly, a bulletin from the cockpit explained that the exhaust flames were only normal and had been invisible in broad daylight. The sky was overcast; it began to drizzle. Soon we were in complete darkness. I again began to have misgivings about the plane's ability to land safely at night in rain. The lights of Damascus shone along a dark slope. After some time, a line of purple lights became visible along a runway, and we alighted safely at the Beirut airport in pouring rain.

There was a short walk in rain along the tarmac. Dinner was served at the Beirut airport on numerous small tables seating two passengers each, to the passengers and crew of five or six flights. Vain looking male waiters, resplendent in immaculate dark tailcoats and trousers, with white starched shirts and black bowties, were bustling about, serving hundreds of passengers. Forty or fifty passengers of our flight, along with pilots and air crew, occupied vacant seats on the small tables. Bottles of wine with glasses were already placed on each table. I seated myself on a vacant table and ordered a soda to forestall having to drink wine with my meal. Hasnain Ahmad came and occupied the vacant seat on my table. Western cuisine was served, in style, beginning with soup, followed by the main dish of chicken and vegetables, then pudding and coffee. Hasnain wanted a glass of water to drink during the meal and asked me to order a waiter to fetch water for him. 'I suggest you request him yourself,' said I, remembering all his sneaky tricks in Bahawalpur. 'Oh, no friend, I feel shy in giving the order,' said he in a pleading voice. I looked at him critically for the first time. He was a tall, lanky individual, sporting a beard, and with a startling rosy under-lip, due to leucoderma (a disease that results in white patches on the skin). He wore a flowerpot-sized red fez cap of Bahawalpur. His suit was of a drab colour, and primitive cut. He, altogether, cut a comical and somewhat pathetic figure. 'So, this is the man who had caused me so much trouble in Bahawalpur,' thought I, and suddenly felt a surge of compassion for him. I beckoned to a waiter and said, 'This gentleman wants a glass of water to drink.' 'Water?' the stupefied waiter exclaimed, raising both his arms upwards and staring at Hasnain, as if he were some rare species. He rushed to another waiter standing nearby and pointed towards Hasnain Ahmad. At long last, a glass of drinking water was produced for my companion!

As Pakistan was a member of the British Commonwealth, we were received with polite courtesy by the airport officials at the Heathrow airport. Our suitcases were carried to a waiting BOAC bus, and we were whisked to the BOAC offices in London. I was feeling extremely sleepy, having spent an almost sleepless night in the plane, and because it was still night-time in Pakistan. Hasnain Ahmad lay down and fell asleep on a sofa in the comfortable ground-floor rest house in the multi-storey BOAC building. Hamid went off with his son-in-law who was, perhaps, a practicing doctor in London. Pir Ibrahim and Qureshi had also disappeared. I was left alone in London.

Since we had to proceed to the USA the same evening, it seemed a pity not to see some parts of London in the available day time. I stepped out and took a taxi to the Piccadilly Circus. The taxi was a 25-year-old vintage; Britain was exporting all its new manufactures as an austerity measure. Piccadilly Circus proved somewhat disappointing. In fact, the outer façade of most of the buildings was of soot-coloured grey stone, which looked depressing. Uncertain about my next move, I started walking along the footpath of Regent Street—one of the roads radiating from Piccadilly Circus. I was surprised to see Pir Ibrahim coming slowly towards me on the same footpath. We gladly joined forces and he volunteered to take me for a ride on London's famed underground railway.

In the evening, we returned to the airport in a BOAC bus and boarded a Stratocruiser a plane bigger than the Argonaut with a cruising speed of 350 miles per hour. We were served dinner on the plane and in due course relaxed on our comfortable seats to rest. In those days, a direct hop from London to New York was considered risky; besides, the plane required refuelling. So, we had to cut our Trans-Atlantic flight in two, by a brief stopover at Reykjavik, capital of Iceland, at midnight. I shivered at the thought of having to disembark in freezing cold at the Reykjavik airport. However, the first thing we saw on the brilliantly lit tarmac was a young girl walking towards our plane in a flowered cotton frock! Mercifully, we were not required to get down at the airport and, after a brief stopover, left for America.

The customs officials at New York's Idlewild airport (now John F. Kennedy international airport) insisted on opening our suitcases and checking the contents. I carried with me a large sealed envelope given to me by Col. Hainsworth (chief engineer, NWFP), addressed to General Raymond A. Wheeler, engineering adviser of the World Bank (1949–56). The customs official demanded to know what was in the envelope. I told him it was an official document addressed to an official of the World Bank; its contents were unknown to me. He insisted on opening it. I said, 'By all means do so, but at your own risk.' He, thereupon, stayed his hand. Opening a little at a corner of the envelope, he espied the type written document, 'Oh well,' he said, 'it appears to be an official document,' and left it at that. Outside the airport, a secretary of the Pakistan embassy was waiting for us, with two hired taxis. He whisked us to LaGuardia airport (another of New York's airports) from where planes left for Washington DC, our final destination. Two hours later we arrived at the imposing airport of Washington DC. The secretary was waiting for us outside the airport with taxis. He proposed that we should first go to the Pakistan embassy to pay respects to the ambassador, and receive our weekly wages in dollars, before going to Fairfax Hotel where rooms were reserved for us. We met our ambassador, Mohammad Ali Bogra, from East Pakistan, who seemed to have nothing much to say. We next received our weekly allowance of $30 per day (the rate of exchange at that time was Rs 2.8 per dollar) and were driven to our hotel, escorted by our helpful guide, the secretary. Our rooms were on different floors of the twelve-storey Fairfax Hotel on Massachusetts Avenue. The daily rent per room was only $6, without meals. The secretary was apologetic about the 'poor' accommodation which had been hastily provided for us and promised to look for a better hotel, but we assured him that we were fully satisfied with it.

Negotiations with India

The conference room was on the second floor of a building in a busy quarter of Washington city. It was a large hall, with a long table in the middle. We met our Indian counterparts who were led by A.N. Khosla, as leader, and N.D. Gulhati, J.K. Malhotra, and others—all my acquaintances. Khosla had been junior to me by one year at Roorkee. Gulhati and Malhotra belonged to Lahore, and we had met several times during the Punjab Engineering Congress sessions, at Lahore, during pre-Partition days. We also met the three American observers, deputed by the World Bank. Their leader was General Wheeler, a remarkable man of rare wit and tact. We became very friendly later on. A large map of the Indus Basin was hung on one of the walls of the conference room. Being interested in geography by nature, I examined the map while my other colleagues were busy settling down. I was flabbergasted to see that the whole of the plateau of Rajputana was shown as commanded[1] by the Indus Basin rivers! I said in a loud voice: 'This map is wrong'. This pronouncement created a stir in the assembly. 'Mr Rahman says the map on the wall is wrong,' was the somewhat acid comment of the Americans. I pointed out to the members crowding round the map that only a small northern strip of the Bikaner state of Rajputana was commanded by Sutlej River; the rest of the vast territory of Rajputana was a high uncommanded sandy plateau.

The map was hastily removed from the wall. I wondered how a blatantly incorrect and pro-Indian map of the Indus Basin had come to be hung on the wall of the conference room, where the fate of the waters of the Indus Basin rivers was going to be debated on the negotiating table. I subsequently learnt, when it was too late, that the attitude of the USA government had been pro-Indian from the very beginning. The first day was spent in taking a combined group photograph of both the delegations, seated round the conference table. The actual business of the conference started the next day. According to an established convention, only the leaders of the two delegations spoke across the table. Other members passed on their written suggestions on slips of paper to their respective leaders, who included them, as they thought fit, in their talks. The three American delegates of the World Bank took no part in the discussions, as they were observers. At the very start of the meeting, Khosla, the leader of the Indian delegation, delivered a bombshell, by asserting that according to the agreement arrived at between

India and Pakistan the entire supplies of the Indus Basin rivers were to be treated as a common pool out of which both India and Pakistan would take their claims according to their needs. This meant that Pakistan's 'historic' rights in the supplies of the Indus Basin rivers, for its existing canals in the Punjab, Sindh, the NWFP, and Balochistan would have to be decided afresh. This very contingency was in my mind, when I had questioned the inclusion of the purely Pakistani rivers, in the chief engineers meeting at Karachi. Pakistan had consented to the 'common pool' proposal, on the tacit understanding that her 'historic' rights in the supplies in her existing canals were inviolable, while India had readily agreed to the proposal. The whole day was spent in Hamid, leader of the Pakistani delegation, trying to refute India's ridiculous claim. India's Khosla, however, remained adamant.

The next morning, we met early in the office of John Laylin, Pakistani's attorney in the canal water dispute with India. Mr Laylin asked each member to express, on a piece of paper, his written opinion about the Indian claim of the previous day. To his eternal credit, Dr S.M. Qureshi from Sindh was the only member who not only advocated total rejection of India's claim (which all the rest of us had also done), but also threatened to 'walk out', and return to Pakistan on the issue. Mr Laylin advised that, since it was imperative that the vote of the members be unanimous, he accepted Qureshi's view, while admiring his stand, and prophesying a great future for him. And so, it was decided that the delegation would threaten to break off negotiations, and return to Pakistan to consult their government, if the Indians persisted in their unreasonable stand.

When we met the Indians later in the day, Hamid announced that, in view of the fact that the Indian delegation refused to recognise Pakistan's 'historical' right on the supplies in Pakistan's existing canals in the Indus Basin, the Pakistan delegation had decided to suspend further negotiations in Washington and was returning to Pakistan for further deliberations with their government. Surprisingly, Khosla's attitude became suddenly very friendly. He gracefully acknowledged Pakistan's water rights in the existing canal systems. It was evident that the Indians were anxious that the negotiations should continue. The subject of annual withdrawals by Pakistani canals was acknowledged as a legitimate subject for 'study' by the two delegations. Surprisingly, the Indians possessed complete records of the withdrawals from Pakistani canals ever since the records had been maintained. The evaporation and absorption losses in the canals and the gains in the rivers through seepage were also parallel subjects of study. The same principles applied to the existing Indian canals in East Punjab and Rajasthan, which were fed from the rivers of the Indus Basin. The studies also included the extent of forest areas, in both the countries, in the Indus Basin and the calculation and tabulation of daily runoff in the Indus Basin rivers at Kalabagh, Mangla, Marala, Madhopur, Larji, and Bhakkar for the Indus, Jhelum, Chenab, Ravi, Beas, and Sutlej rivers, respectively. Both the delegations worked on the collection and analysis of data on specified subjects and then met on the negotiation table for discussions, which were usually characterised by arguments and wrangling. I was assigned the tedious but important job of recording the proceedings of the discussions, which I did in long hand. We worked on the data in our own office and were briefed by Mr Laylin, our attorney, before confronting the Indians on the conference table. Royce J. Tipton,[2] of Tipton & Kalmbach (consulting engineers from Denver, Colorado), and a vice president of the American Society of Civil Engineers, was engaged by the Pakistan government to provide expert advice on various technical issues. Mr Tipton and I worked on the same table in our office and consequently struck up a warm friendship.

Shortly after our arrival in Washington, Mr Laylin took us to call on the chairman of the Tennessee Valley Authority in Washington. There was a huge photograph of Grand Coulee Dam hanging on a wall of the chairman's office. The chairman was plainly unsympathetic to Pakistan's case, and did not conceal his critical views. We were stunned at his attitude; Mr Laylin's attitude, after the interview, was of smug satisfaction, which aroused a faint suspicion in my mind about his sincerity.

Hospitality and Social Entertainment

Mr Laylin took charge of entertaining our delegation. Mr and Mrs Laylin invited us to a comedy show at a theatre in Washington, where a clown kept hopping across the stage on one leg, with the other one in a horizontal position, crying, 'I like Ike'![3] The story concerned a lady ambassador of the US in a Latin American country; whose foreign minister soundly berated the US selfish foreign policy, with the result that the US government sanctioned aid worth $21 billion for the small Latin American country. The roars of irreverent laughter of the audience at the lampooning of President Dwight Eisenhower and the political leadership of the country were, to me, a refreshing example of the many freedoms enjoyed by ordinary Americans. On another occasion, Mr Laylin took us to lunch at the Metropolitan club in Washington with its imposing building that put our Government House, Peshawar, in the shade. We mingled with many American senators and members of the congress and were introduced to some who met us with courteous disinterest. Mr Laylin also took us for dinner to the posh Lafayette

Restaurant, opposite the White House. To our great relief, Mr Laylin always expected each one of us to pay for our own meals! He also invited us several times for lunch or dinner to his 350 acres ranch situated about 18 miles from Washington on the West Virginia road. At the ranch, he had a small but comfortable bungalow fitted with all sorts of modern conveniences. The large ranch was of rolling grassland, with clumps of tall pine trees and raised cattle. A tiny stream in a miniature valley within the ranch carried a small perennial flow. A small square swimming pool had been built across the tiny stream capturing part of the flow, with a small hut built nearby as a changing room. During our first visit to the ranch we saw a fashionably dressed man, cutting the pine trees with an axe. A large car was parked nearby. Mr Laylin informed us that it was his brother, to whom he had given the site of the clump of trees to build a house as a present. And he was cutting the trees himself and clearing the site, for building his own house unaided, in pioneer spirit.

Our colleague from Bahawalpur, Hasnain Ahmad, cut a sorry figure in our Washington circle with his lanky figure, ill-fitting suits, red fez, and awkward ways. He was generally avoided by the members of the delegation. However, sensing that I had forgiven and forgotten his offensive behaviour at Bahawalpur, he clung to me at every opportunity for moral support. Whenever I went out for a walk along the Massachusetts and Connecticut avenues or further on at the F Street on an afternoon after office, he would be waiting for me in the lobby, and would walk along with me without invitation. Even Mr Laylin noticed Hasnain's awkward behaviour, and proposed to invite him alone to tea, on a Sunday. He said he would also invite a few American couples to enable Mr Ahmad to shed his shyness. We all gladly fell in with the conspiracy. On the appointed day, we pretended to feel jealous about Hasnain being singled out for the invitation by Mr Laylin. When tea-time arrived Hasnain was reluctant to go, 'Why don't you come with me?' he asked me. 'But I am not invited,' said I. A taxi was called, and we bundled Hasnain into it. To our amazement, he returned to the hotel in a very short time. He explained that as the outer door of Mr Laylin's house was shut, he had knocked, but there was no reply and so he had returned. Simultaneously, there was a phone call from Mr Laylin saying that Mr Ahmad was awaited for tea, while the other guests were also waiting for the 'guest of honour'. We explained to Mr Laylin that Mr Ahmad had returned after unsuccessfully knocking at the closed front door of his house. Mr Laylin said, there was an electric call bell set in a conspicuous place at the outer door, which Mr Ahmad should have used. Anyway, he should come over soon because the tea was getting late, and the other guests were waiting.

But Hasnain refused to go back to Mr Laylin's house. So, we had to call and explain the situation and Mr Laylin's other guests had their tea without the pleasure of meeting Mr Ahmad.

At lunch break one day, we got down the elevator from Mr Laylin's office, situated on the eighth floor of the Union Trust Building in Washington DC, and were standing on the crowded footpath below waiting for a taxi. A bevy of young girls had emerged from the same building and were standing in a group a few feet away from us. An old, stocky-built American came slowly limping along the footpath, supported by a stout wooden staff. Seeing five elderly foreigners, he stood stock still in front of us. Looking at each of us in turn, and perhaps espying me to be the oldest member, he addressed me, pointing his staff at the group of girls standing close by, 'Sir, look!' The girl he was pointing at with his staff had golden blond hair, with the sun glistening through it and creating a golden aura around her hair. 'Sir,' I replied, 'she is indeed very pretty.' Satisfied, he remarked, 'That is the stuff we Americans are proud of,' and moved on.

At first, we only subsisted on 'ranch-style' eggs, bread, butter, and fish for our lunch in Washington's cafeterias; our dinners and afternoon teas were prepared by an African-American lady named Florence, whom we had engaged on a weekly wage of $28. In due course, we discovered some Jewish restaurants, where we had 'kosher' meat, and also a fake form of Pakistani *pulao*. An attractive waitress, who used to serve us at one of the Jewish cafes, was serving other customers on a particular evening. Pir Ibrahim, who had pretension about his looks, felt neglected and, to my embarrassment and chagrin, told her, 'Mr Rahman is upset because you are not serving us tonight.' She carelessly replied, 'The list of people dissatisfied with me would fill a volume,' and continued to serve other customers. On another occasion we ordered beef sausages at a Jewish restaurant. Finding the finely ground meat of the sausages insipid, all of us left the dish uneaten. The café happened to be empty of other customers except for an old lady sitting on a separate table. Noticing that all of us had left our food uneaten she rose and said, 'Oh you poor old gentlemen, you are going to go hungry; let me fix something for you.' She went into the kitchen. In a short while, a waitress brought us plates of 'ranch-style' eggs, bread, and butter. When we offered to pay extra for the eggs, bread, and butter, the old lady said that was 'on the house'. She was the owner of the café!

One morning, on entering the office, my secretary noticed my worried looks. 'Mr Rahman,' she said, 'you look glum today. Is anything the matter?' 'I have not heard from my wife for a month.' I replied. 'I left her in poor health, and I do not know where the Western Union is situated, so I could cable her.' She had her typewriter before her. 'If you just dictate your cable to your wife, I will phone Western Union to receive it. Provided your wife is prompt in reply, I shall bring the reply to you the day after tomorrow morning.' Promptly on the morning of the second day, she brought a cabled reply from my wife. 'What about the cable charges,' I asked. 'Oh, don't worry, the Western Union shall send you your bill at your hotel address.'

Mr and Mrs Tipton invited us for lunch at the Hotel Statler (now The Capital Hilton) in Washington. They also invited some of their friends. The hotel lobby was, as usual, crowded with other guests when our party arrived. An American casually asked me whether I knew Kashmir. On my reply in the affirmative, he asked me what sort of a place it was. 'Imagine an ellipse of snow-clad peaks, roughly a hundred miles in length and fifty miles in width, half of the oval turning rosy pink in the morning, the other half towards the evening. In spring in April, the snows travel down the slopes to within a thousand feet of the vale, revealing incredibly green lower slopes. The valley of Kashmir itself, viewed from the top of the 1000-foot high peak of the Takht-i-Sulaiman hill on the shore of the Dal Lake, presents a long vista of fruit trees smothered in light and deep pink and white blossoms, interspersed with a carpet of bright, yellow mustard, and green fields with rows of tender green wheat, with patches of bright grass, studded with red poppies and white lotus. And there is the sparkling Dal Lake, with its water depth transparent for a hundred feet, and its surface decked with lotus flowers yielding ambrosial lotus-honey!' I was suddenly aware of a hush in the lobby. The entire crowd in the lobby had surrounding us and was listening to my ecstatic description of Kashmir in pin-drop silence.

Both the Indian and Pakistani members of the canal water delegation were invited to lunch by Eugene Black, the third president of the World Bank (1949–1962). The house of the distinguished gentleman was an imposing two-storey building on Massachusetts Avenue, not far from our hotel. It was lavishly decorated with costly furniture, pictures, and rugs. It had a low ceiling, only about 10 feet high, both on the ground and the first floor. A low flight of stairs ascended from the ornate hall on the ground floor to the dining room on the first floor. A short electric elevator was also installed there along with the staircase. Mr Black insisted that his guests use the lift, instead of the stairs, and appeared upset when a few of us used the stairs instead of the lift.

There was neither any lady in the house nor any waiter. The food was laid ready to serve on the dining table, to which we were led directly. We were expected to help ourselves. Kosher meat had been provided, in consideration of the scruples of the Muslim members of the delegation. During the course of the lunch, Mr Black pushed a large oval dish, heaped with what looked like boiled rice, in our direction with the casual remark, 'wild rice'. We were expected to pass the dish around to the next guest, after helping ourselves. The dish went around the table without anyone taking a helping. I noticed that our host did not appear to be pleased that the *'piece de resistance'* on his menu was being rejected so casually. The boiled rice grains were large and round, and tasted faintly like sweet pudding and were, in fact, quite tasty. Noticing the expression on Mr Black's face, I said, addressing him, 'Did I understand you said, Mr Black, that this is really WILD RICE!' 'Yes! replied Mr Black, 'it is *real* wild rice; it was gathered in the lakes of Wisconsin. It is a rare treat.' 'Very delicious! May I have a second helping please,' I said to Mr Black, to whom the dish had reached back untouched. 'Oh, sure,' said Mr Black, pushing the dish in my direction. Taken aback, both Muslims and Hindus took large helpings and were effusive in their praise of the delicacy. After lunch, Mr Black made us use the lift down the 10-foot descent to the ground floor, escorting me along with him ahead of the others.

One afternoon, I managed to give the slip to Hasnain and went out for a walk alone down the Massachusetts and Connecticut avenues to the busy F Street. I was walking along the footpath of the F Street, ahead of a crowd of pedestrians when, on a street crossing, the traffic lights turned red, bringing the whole crowd to an abrupt halt. Being at the head of the crowd, I absent-mindedly stepped down from the footpath on to the road below. Suddenly, I was seized from the armpit and jerked back on to the footpath. A split-second later, a fast-driven car, coming from the back, swerved to the left, and thundered past; its left rear wheel passing over the very spot from which I had been pulled up. In utter bewilderment and relief at my miraculous escape, I looked back to find a couple of policemen, smartly clad in blue uniforms, looking down and smiling at me. By chance they were in the crowd and on seeing my imminent danger they had pushed forward and hauled me up. I was so shocked at the danger I had faced and subsequent relief that I forgot to thank my saviours!

We were invited to a number of receptions hosted by Pakistani diplomats during our two-month stay in Washington. We were invited by our ambassador, Mohammad Ali Bogra to a formal dinner at the Shoreham Hotel, in order to introduce our delegation to some state department officials. He had sent invitations to about a dozen state department officials, stating that the dress code was formal, about which we were again, discretely, reminded by his secretary. Some of us, who had not brought their formal suits with them, were advised to obtain them on rent from dealers of readymade suits. Hasnain accompanied me for renting a formal dress suit. The dealers had to cut the length of the trousers to fit my size. Our suits were lying on the counter in neatly packed bundles, marked 'Mr Rahman' and 'Mr Ahmad', when we went to receive them. We paid $7 each for rent for the suits for forty-eight hours and collected our bundles. I waited for the salesgirl to note down our address. She, however, waved us out of the shop, saying that it was not necessary; and that we were to return the suits after forty-eight hours.

About ten couples from the state department had been invited to the dinner. My seat on the longish dinner table was on the right of a young lady from the state department. Pir Ibrahim was seated on her left. Breaking the ice, she turned to Pir for conversation, but soon turned to me, and we were forthwith engaged in interesting conversation about social life in Washington. The young lady related how much she admired her many Pakistani friends in the diplomatic community and complemented Pakistani women on their good looks and elegant clothes. She particularly referred to a few ladies whom I had chanced to meet in recent diplomatic receptions. I related to her my great embarrassment in engaging in a long conversation with one of these ladies without recognising that she was the hostess. She admitted that the lady was very pretty. After a while she added, 'How would you feel, Mr Rahman, if Mrs Qureshi were a nurse and you were a convalescing patient in a hospital and she would bend low over you, and say, "how do you feel this morning, Mr Rahman?"' After dinner, she took a large cigar, and said, 'Mr Rahman you must be feeling outraged at my smoking a cigar!' 'I have ceased to wonder at anything,' I replied. 'After noting the perfect freedom enjoyed by women in the US!'

General Raymond A. Wheeler—engineering adviser, World Bank and leader of the party of three American 'observers', in the parleys—gave us (Muslims) a grand dinner in the luxurious Army and Navy Club in Washington. General Wheeler was another American, besides Mr Laylin, who had impressed me the most during my stay in the USA. He concealed his engineering talents under cover of his wit and cheerful disposition. He would readily tell a funny story that would be faintly off colour. Only on one occasion, I caught him off guard when under my persistent questioning, he described the sewerage system of Washington DC in somewhat deprecating terms, and his own part in its planning and execution. General Wheeler was particularly agreeable and friendly with me.

Our two-month long parleys with the Indians concluded without any tangible results except the boring task of conducting joint studies of withdrawal of Pakistani and Indian canals from the supplies of the Indus Basin rivers. It was further resolved that the combined delegations, comprising the American 'observers', the Indians, and Pakistanis would later undertake a joint physical inspection at the site of the Indian canals in East Punjab, and of the Pakistani canals in West Pakistan.

Return Journey Home

The return journey was uneventful. Mirza Abul Hassan Ispahani, Pakistan's high commissioner in London, had arranged accommodation in the luxurious St. James's Court. Khan Bahadur Sheikh Abdul Hamid had cabled Mr Blench, a retired pro-Pakistan chief engineer of the Punjab Irrigation Department, to meet our delegation in London, hoping to obtain from him an affidavit regarding the historic rights of Pakistan's canals in the waters of the Indus Basin rivers. On the evening of our arrival in London, Mr Blench met us at Trafalgar Square. Hoping to impress our delegation with the sights of London, Mr Blench pointed to the fountains of Trafalgar Square and said, in Urdu, *'Bahut Pani, Bahut Pani'* (much water, much water). He then took us to Piccadilly Circus, and pointed to the neon lights in various colours, with slogans advertising certain well-known products. Hamid, said, 'Rahman, tell Mr Blench about the lights of Broadway in New York!' Mr Blench looked visibly annoyed at the tactless remark. I said, 'Of course, the lights of Broadway are brighter but, after all, what is New York? It is but a precocious child of London, the mother of cities, and the birthplace of Western civilisation and democracy.' This saved the situation.

We left London on the evening of the second day and at midnight landed at Frankfurt airport in a heavy rainstorm. The huge hall of the airport was almost deserted, but the hospitable Germans had tables laid with cakes and coffee, and the famous German beer. The rainstorm, with a strong wind, was still raging outside. I felt apprehensive at the prospect of our having to cross the Alps in the storm and wished we would spend the night at Frankfurt airport. However, the storm subsided. A fellow passenger, obviously a clergyman, who was wandering disconsolately, stopped for a chat. He, too, was feeling apprehensive about our having to cross the Alps that night and said he wished we would spend the night at the airport. 'Exactly my sentiments,' I said. Thus, a kind of bond developed between us. In reply to his query, I told him that I was from Pakistan. 'Are you a Muslim?' he asked. And I said, 'Yes.' 'Oh, I want to know about Islam,' he said, 'will you kindly tell me about it.' I told him that both Islam and Christianity were revealed religions. According to our scriptures, we Muslims were commanded to treat Jesus Christ with great reverence, as a Messenger from God, but we did not believe in his divinity.

At that point, our leader Hamid burst in, and asked me in Urdu what we were talking about. I told him that we were discussing religion. He interrupted our conversation, and addressed the reverend gentleman, 'Sir, you know, the Indians have deprived us Pakistanis of our rightful share of canal water.' The reverend looked at Hamid with a confused expression and walked off. I told Hamid he should not have burst in while the reverend and I were discussing purely religious topics. His astonishing reply was that we should not miss any opportunity to carry on propaganda against the Indians! We reached Karachi the next morning. I experienced a strange feeling of extraordinary love for the soil of my homeland, on viewing the barren landscape around Karachi, as our plane slowly descended towards Karachi airport. At the airport, the customs officials refused to open and examine our suitcases out of respect for our mission and being senior officials. But I said, 'Please don't put my grey hair to shame as I have brought along some taxable articles such as costume jewellery.' My suitcase was sent to the customs warehouse, where I paid Rs 500 as customs duty, before recovering it.

A crowd of Hasnain's subordinates and friends were waiting to welcome him with garlands of flowers, when the Khyber Mail reached Bahawalpur railway station. I was looking at the scene of his triumphant return home from the adjoining compartment. Most of the crowd was known to me. Some had worked under me. One of the latter casually walked over in my direction and enquired what had brought me there. 'He also went with us to the USA,' said Hasnain who had been listening to our conversation; whereupon the person also put one string of faded flowers round my neck. The others looked on indifferently. Poor Hasnain! He did not live long to enjoy the glory of his visit to the USA and the fruits of his post as chief engineer of Bahawalpur. He died a few months later, of an undisclosed ailment. The next morning our branch train arrived at the Havelian railway terminus. My destination, Abbottabad, was 10 miles further on by road. My wife and children were waiting on the railway platform to welcome me back home with garlands in their hands.

25 Pakistan Delegation's Visit to India

As had been proposed at Washington DC, the combined canal water delegation, under the leadership of General Raymond Wheeler started the site inspection of Indian and Pakistani canals in East Punjab and West Pakistan, respectively. It was a sunny winter morning on 18 December 1952, when we left Lahore for Delhi on a Pakistani Dakota plane, and two hours later landed at New Delhi's Safdarjung airport, known officially as the Willingdon airport.[1] We were allotted spacious rooms in Old Delhi's famous Maidens Hotel, as guests of the Government of India. The food served was mainly Western, with the addition of typical Hindu dish of papadams. We were whisked away by train to Jalandhar—the then headquarters of the East Punjab government—and the Pakistanis and the Indians were put up in the Circuit House at Jalandhar. That evening, the East Punjab government arranged an official dinner for us. Noting that I was a Pathan from NWFP, the Hindu chief minister of East Punjab took me to a corner of the drawing room and said in confidence that the Government of India was prepared to extend all help to the people of NWFP and he personally would be pleased to be of any service to me! I later reported the conversation to Abdul Qayyum Khan, the chief minister of NWFP.

We made Jalandhar our temporary headquarters from where, in the company of the Indians and Americans, we made excursions by car to the small Shah Nahar Canal taking off from the Beas River; the Bhakra Dam and the connected Nangal Canal on the Sutlej River; the Sirhind Canal also on the Sutlej River and the Harike Barrage, on the confluence of Beas and Sutlej rivers; and the mammoth Rajasthan Canal (now Indira Gandhi Canal), taking off from the Harike Barrage. The trips to the small Shah Nahar Canal, dating from the Mughal times, and to the Sirhind Canal with its old-fashioned weir and canal headwork, built by the British during the early nineteenth century, were uneventful. The trip to the Bhakra Dam under construction, however, came as a painful revelation. The Rau Commission had allotted only 12 per cent of the annual runoff of the Sutlej River for the development of new areas in East Punjab, through storage of the summer

Sutlej floods in the proposed Bhakra Dam, whose height was to be restricted to 425 feet, and the dam was to start storage of the floods, annually, from 15 July onwards only. Moreover, the 'historic' rights of Pakistan's existing Sutlej valley canals had to be taken into consideration before deciding on the extent of development of India's new areas in East Punjab. In short, the question of allocation of Sutlej River supplies between India and Pakistan was understood to be still a matter of dispute.

India's Perfidy

I was expecting to see Bhakra Dam under construction, with the foregoing conditions in mind. However, I was shocked to see that the Indians were already halfway through with the construction of a concrete dam of stupendous dimensions. It was going to be the 'highest dam in the world', even higher than the famous Hoover Dam on Colorado River, in the USA. What was more, the Indians had secured the services of the designer and builder of the Hoover Dam itself at a colossal salary, exceeding that of even the president of India. It was evident that the Indians had confronted Pakistan with a *fait accompli* and had no intention of allowing even a single drop of perennial as well as flood water of the Sutlej River to go into Pakistan's Sutlej valley canal. General Wheeler, who was standing with me while we were looking at the under-construction Bhakra Dam, on seeing signs of confusion on my face, betrayed a smile, thus creating in my mind a suspicion that the Americans were well aware of the Indian plans and conspiracy against Pakistan.

Later, we inspected the new barrage, constructed on the Sutlej River, a few miles downstream of the Bhakra Dam, and its off-taking Nangal Canal with a capacity of 13,000 cusecs. The entire 40-mile length of the Nangal Canal was lined in concrete. The Nangal Canal was intended to utilise the storage of the Bhakra Dam, for its perennial discharge. Here was another accomplished fact staring at us. My suspicion about the Americans' collusion with the Indians in the canal water dispute deepened when, a few days later, we inspected the new Harike Barrage constructed by the Indians on the Sutlej River, a few miles downstream of the junction of the Beas River and the Sutlej at Harike, with its off-taking Rajasthan Canal, with a stupendous capacity of 30,000 cusecs. It meant that not a drop of water would go down the Sutlej River to feed Pakistan's Sutlej valley canals, except during floods exceeding 30,000 cusecs. At Harike, I was introduced to a Sikh minister of the East Punjab government. 'We know you already,' he said. 'Your former Sikh PA (personal assistant) at Bahawalpur has been sending long letters to the government, eulogising you for having saved his life in Bahawalpur during the disturbances following the Partition.' 'How is he getting on here?' I asked, politely. 'He is very corrupt,' was the Sikh minister's comment.

From Jalandhar, we moved to Amritsar where we stayed in the local Circuit House. Spencer & Co. had been given the catering contract for our tour of East Punjab. Their catering wagon followed us wherever we went. A number of taxis had also been requisitioned for us by the Indian government. Each taxi seated three people, an American, an Indian and a Pakistani. From Amritsar we drove most of the way along the Indian section of the Upper Bari Doab Canal main line, to its old-fashioned headworks at Madhopur, and back. The canal was closed due to the annual winter closure and was observed to have been heavily silted. Addressing my Indian counterpart in the car, I remarked that Kennedy had formulated his famous silt theory on that very canal which, in Kennedy's time, neither silted, nor scoured. He, however, expressed his ignorance about this and did not seem to know the cause of silting or its remedy. I did not enlighten him that the deposit of heavy bed silt was the natural consequence of the channel attempting to conform to Lacey's dimensions, by reducing its depth, enlarging its bed-width, and modifying its slope.

We next inspected the proposed Larji Dam site on the Beas River, situated a few miles short of the famous Indian hill resort of Kullu. The long route from Amritsar lay via Gurdaspur and Pathankot, through the Kangra valley, along and parallel to the foothills of the Dhauladhar range of the inner Himalayas, past Palampur, Jogindernagar, Mandi to Larji. Along the way to Palampur, I saw the bushes of the much-advertised tea plantation experiment in the Kangra valley. At Palampur we stopped our cars to allow a batch of beautiful Gaddi[2] highlander girls to pass. They are a Mongolian race, with an admixture of Caucasian blood, which reminded me of the Hazaras of Quetta valley. There was a good deal of rice cultivation in the terraced fields at Palampur, which were watered from small perennial watercourses fed from the Dhauladhar snows higher up. N.D. Gulhati, a senior member of the Indian delegation, wistfully asked me to suggest a remedy for the dwindling discharges of the 'cools' (the snow-fed water courses in question). I did not have the heart to tell him that the diminution in the 'cool' discharges was the natural consequence of the ruthless cutting down of the luxuriant forests which used to cover the middle and lower slopes of the Dhauladhar range at the turn of the century. I have a vivid memory of those forests from the times when I was living in Dharamshala, as a teenager, in 1906.

We had picnic lunch at the Larji Dam site, thanks to the ever-present Spencer's catering wagon. From Amritsar, we drove to Simla, via Chandigarh, the new headquarters of East Punjab, then under construction. The Muslim members were lodged in a hotel at the far end of the Simla Mall, which was formerly owned by a Muslim who had migrated to Pakistan. The objective of the visit to Simla was not clear. Perhaps the American 'observers' wanted to visit the former summer capital of the British government in India. At Simla, the East Punjab government invited the combined delegations to a formal lunch at the Grand Hotel, situated on top of a bluff at the near end of the Mall. A number of Indian ministers and secretaries of the East Punjab government and their wives attended the lunch. It was an imposing show, some forty people sat on a long table laid in the middle of the wide dining hall of the hotel which had high glazed windows all around. It was snowing outside, but the hall was comfortably warm with the log fires burning brightly in two huge fireplaces situated at the opposite ends of the hall.

As few days later, the Indian and American members of the combined canal water delegation arrived in Pakistan for inspection of West Pakistan's irrigation works. We started our inspection with Sindh's Ghulam Mohammad Barrage, under construction on the Indus River, near the town of Hyderabad. Chief Engineer T.A.W. Foy[3] was present at the site to provide us with information. The members evinced keen interest in the pre-stressed concrete T beams, being fabricated for the road bridge connected with the barrage. We also saw the concrete vibrators at work on the lining of the Fuldi Canal. I ruefully pondered the fact that while we were building the Ghulam Mohammad Barrage on the purely Pakistani Indus River, the Indians were building their Bhakra Dam on the disputed Sutlej River, disregarding the effect it would have on Pakistan's eight Sutlej valley canals that were already drawing 8 MAF of water annually from the Sutlej River.

I did not accompany the combined delegation during their inspection of the Sukkur Barrage, and the irrigation works of West Punjab, most of which I had already visited. I returned to Peshawar, and reported to the chief minister of NWFP, Khan Abdul Qayyum Khan, about the Government of India's lavish hospitality extended to us, during our tour of East Punjab, and suggested that the NWFP with its proverbial hospitality should not lag behind. Khan Abdul Qayyum Khan took the hint, and when the combined delegation reached Peshawar, he arranged a formal dinner at the Government House, Peshawar, hosted by the governor of NWFP, and a public dinner himself, attended by hundreds of prominent citizens of NWFP and the tribal territory. In due course, the combined delegation also inspected the Kurram Garhi project, under construction on the Kurram River in Bannu district. I had to admit that I was the 'father' of the Kurram Garhi canal scheme. General Wheeler, in particular, was amazed to learn that I was the sole planner and designer of the scheme and had personally prepared the working drawings of the various works, assisted only by a single draftsman. In the USA, a whole team of engineers and draftsmen would be needed to prepare a scheme of comparable size. In the visitors' book, kept at the site of the work, General Wheeler wrote some very complimentary remarks about me.

A few days later, during the second fortnight of January 1953, we again went to Delhi by air, to continue our dialogue with the Indians, which was interrupted by our return from Washington. We arrived at New Delhi on a Saturday. As the next day was Sunday, the banks would only open on Monday. The problem was that Hamid, our leader, was short of Indian money. Luckily, during our circuit around the shops of Connaught Circus, we came upon Kirpa Ram & Brothers, formerly of Peshawar. The manager met me cordially, and gladly cashed the Rs 500 cheque for Hamid. Our parleys were held in a spacious room of the new Secretariat Building of New Delhi. This time the Muslims were lodged in the Swiss Hotel, while the Americans were put up in Hotel Cecil. Soon after our arrival, the manager of the Swiss Hotel informed me that my pre-Partition friend, R.B. Nanak Chand, advocate of Mardan, had invited me to dinner on the following day. But on the morning of the appointed day, the manager again informed me that Nanak Chand had cancelled my dinner invitation.

The highlight of our stay in Delhi was the Independence Day celebrations by the Government of India for which we also received invitations. There was a grand display of India's military might on land, sea, and air, watched by large Indian crowds. There was a long march past of units from the land and naval forces, followed by an impressive procession of weaponry, in the shape of tanks, guns and armoured vehicles, with frequent flypasts by the air force. We watched moving pageants depicting the typical occupations of the people of the numerous Indian provinces in Delhi's national stadium in the afternoon. It was depressing to note that the ten centuries of Muslim rule in India were represented in this pageant by a solitary female singer of the Muslim era.

We also attended the grand garden party held in the beautiful grounds of the palatial residence of the president of India (the pre-Partition viceroy's palace of New Delhi). There I met Brigadier Sarup Narain, formerly a colonel and officer commanding at Peshawar, and his wife. Mrs Sarup Narain remarked that they wished to invite me to dinner but were unable to do so for obvious reasons. We also saw Maulana Abul Kalam Azad and Sheikh Abdullah, seated in the shade of an ornamental bush, engrossed in deep conversation. At my request, an Indian executive engineer of the buildings and roads branch undertook to show me the twenty-five new townships which had sprung up between New Delhi and the Qutub Minar. I was distressed to see that the spacious areas of the royal cemetery of Nizamuddin Auliya and the mausoleum of Emperor Humayun had been walled in to make room for the new townships. The executive engineer's own residence was in one of the new townships. I was surprised to see that it consisted of only two small living rooms, with a bathroom and a kitchen in the back veranda; a small walled back courtyard, and a narrow front veranda with a narrow space separating it from the roadway. What a contrast to the imposing official residences of executive engineers in Pakistan.

My colleagues got heavy woollen overcoats tailored in India. I purchased four woollen blankets from the famous Lal Imli Woollen Mills (Kanpur). There was a dearth of black peppers and white cardamom in Pakistan. So, we persuaded our friend, the Sikh deputy superintendent of police, who had been deputed by the Government of India for our protection, to procure a pound of each item for all five of us, and he willingly obliged. So, when we moved into the lounge of the Safdarjung airport, our suitcases filled it with the aroma of cardamoms. Our 'protector', the Sikh deputy superintendent of police, followed us right into the departure lounge to see that we safely carried away the contraband stuff. At the Lahore airport, our own customs officials pounced upon our odorous suitcases. I told the custom officials that they should feel ashamed of themselves for demanding duty on small quantities of pepper and cardamom, which the Indians knowingly allowed us to carry away. On hearing this, they did not question us further about the pepper and cardamom.

In 1954, Royce J. Tipton came to Pakistan as a guest of the Government of Pakistan. He contacted me by phone as soon as he came to Peshawar. Since he was the guest of the NWFP government, I invited him to tea at my house. He said that he would be pleased to come, provided Mrs Rahman personally received him. He complained that, in America, they introduced their wives to us, but when they came to Pakistan, we said our wives were in purdah. I assured him that Mrs Rahman would be pleased to receive him. In any case, Mr Tipton was a few years older than me, and was a very dignified person. I invited Mr Tipton to my son's posh official residence, where he also met my daughter-in-law. My son was an executive engineer at Peshawar PWD. With characteristic American candour, Mr Tipton declared that of all the women he met in Pakistan, Mrs Qureshi and Mrs Rahman were the most impressive.

During the same year (1954), the World Bank announced its plan,[4] allotting the three eastern rivers of the Indus Basin, namely, the Sutlej, the Beas, and the Ravi to India, and the three western rivers, the Indus, the Jhelum, and the Chenab to Pakistan. From the pre-emptive activities of the

Indians at the Bhakra Dam and at Harike, below the junction of the Sutlej and the Beas, one may well have expected the impending award. Pakistan, finding itself in a weak position, had to accept the award perforce, through which India secured about 25 per cent of the resources of the Indus Basin, against only 12 per cent allotted by the Rau Commission. Since the sources of the Jhelum River lay wholly in the valley of Kashmir, and those of the Chenab and the Indus lay in the eastern parts of Kashmir, the Kashmir dispute could automatically be resolved in favour of Pakistan, if Pakistan could stake a successful claim, through the World Bank award, of the three western rivers of the Indus Basin right up to their sources. Following this 'brain-wave', I wrote to Sheikh Abdul Hamid at Lahore and to Chaudhry Mohammad Ali, finance minister (1951–5), at Karachi, to urgently develop a case based upon these new realities which logically linked the rights to the waters of the three rivers allocated to Pakistan to their basin which included the disputed territory of Kashmir. I further suggested that the Pakistan government should, in pursuance of the World Bank award, and the various UN resolutions on Kashmir, make out a case with the UN and the World Bank and also pursue it politically with the US government for their support for a comprehensive resolution of the linked issues of Kashmir and Indus Basin waters. As expected, Hamid ignored my letter, but Chaudhry Mohammad Ali promptly replied, to say that it was a brilliant idea worth pursuing and advised me to get ready to leave for Karachi for further discussions, while he would discuss the matter with the central minister of industries. However, later in another letter, he expressed his helplessness to proceed further because of internal differences of opinion and 'because he was not on speaking terms with Fazlur Rahman, the minister of industries.'

At this time, Khan Abdul Qayyum Khan was apparently becoming too powerful in NWFP, so Governor General Ghulam Mohammad made him central minister of industries, food, and agriculture and transferred him to Karachi, with the option to choose his own successor, as chief minister, from amongst the heads of departments of the NWFP. Khan Qayyum selected Sardar Abdur Rashid Khan, inspector general of police, NWFP, to succeed him; however, Sardar Rashid soon developed a strong antipathy for his former mentor. So, later when Khan Qayyum came on tour to Peshawar, as minister of industries, Sardar Rashid forbade government officials of the NWFP to meet him, and had the Peshawar's Circuit House, where Khan Qayyum was staying, watched by plainclothes policemen to see which of the local officials visited him. Both Khan Qayyum and Sardar Rashid were my friends and I met Khan Qayyum at Peshawar and told him about my correspondence with Chaudhry Mohammad Ali. Khan Abdul Qayyum did not have positive things to say about Chaudhry Mohammad Ali, accusing him of being self-serving and political. He told me that there was considerable merit in my proposal and that he would personally follow it up on his return to Karachi. But fate had decided otherwise. When the Khyber Mail bearing Khan Qayyum reached Lahore railway station, dismissal orders by Governor General Ghulam Mohammad were handed over to him, and he was told to step down from the train, as he was under arrest and was permitted to live only within the precincts of Lahore, until further orders.

One evening, in the course of my usual walk along the Saddar road at Peshawar, I stopped at a small marble dealer's shop. In a corner of the shop, there was a large stack of marble tablets, each bearing the following identical legend:

The Foundation Stone of......was laid by Khan Abdul Qayyum Khan, Chief Minister of NWFP, on......

In reply to my enquiry, the dealer explained, 'You know, since our popular Chief Minister Khan Qayyum used to place orders for immediate supply of marble foundations stones for opening ceremonies for the numerous public works constructed under his orders, I had these tablets made for meeting the government's urgent demand. But now, alas, he has been transferred to Karachi and so all these tablets are a total waste.'

26 Warsak and Mangla Dams

In the ten years or so since the Partition, momentous changes had taken place in the administrative structure and social fabric of the new state. Pakistan was conceived and born with a spirit of great idealism which, however, did not last long. The moral fabric of the country fell into a sharp decline. Col. Muirhead had lamented, albeit hypocritically, that 'The moral fibre had snapped after independence.' However, some Britishers, too, were making hay during the sunset days of the Raj. A new breed of national politicians and officials who had assumed control of the country after independence set new standards for loot and pillage of our national resources.

One factor was the sudden and unexpected elevation of a junior cadre of relatively inexperienced government officials to senior positions of responsibility and authority, which they could not have dreamt about, even months before Independence. Faced with a shortage of qualified professionals to fill positions of leadership in the country, a free-for-all took place and important positions were filled without regard to either competence or integrity. The British had run their administration by maintaining their preponderance through hard work, discipline, and integrity, irrespective of the official ranks or positions they held. On retirement from service in India, they returned to England to live on their modest pensions in relatively straitened circumstances. Retiring Indian officers followed the same pattern. In those times, one's personal prestige and position carried more weight than one's net worth which, in the case of landlords and *banias* for instance, was far more than that of government officials. Landed property was considered to have lesser social value than official status in the social hierarchy; even large landowners preferred to have their sons appointed to junior 'officer' positions in the government.

This sudden elevation to positions of importance introduced an interesting change in the mindset of the engineers of the PWD. They were now the 'officer' class and the new 'bosses' and, therefore, their role was to manage and oversee, rather than work on the drawing boards and construction sites. The tight norms and procedures that were routinely followed in the day-to-day work under the British, now became lax and ineffective. This happened at the tremendous cost of high level of professionalism that existed among engineering professionals working in government departments during the

British period. Alongside this new mindset was the gradual but pervasive growth of corruption, reaching the highest echelons. All of this resulted in changing the culture of official dealings from one based on formal decorum and propriety to one based on interpersonal relationships and informality. This negatively affected the discipline of the department. Another setback was the departure of many qualified Hindu engineers who had established a name for themselves due to their hard work and professionalism.

Based on my personal experience, I believe that Pakistan missed a splendid opportunity to build a superior cadre of highly skilled professional engineers in all technical specialisations, when our engineers took a back seat and preferred a passive managerial role instead of taking active part in gaining expertise in design and construction work from the foreign experts who were involved in gigantic Indus Basin Replacement works, which included some of the world's largest dams such as Mangla and Tarbela and link canals, and irrigation networks, that provided one of the world's largest field laboratories for honing their skills. Of course, this was not across the board and there were notable exceptions among Pakistani engineers like S.S. Kirmani,[1] who was quickly recruited by the World Bank in a senior position. The greatest beneficiary of these projects was our skilled labour, who became experts in operating heavy construction machinery and earned a well-deserved place in the future building and construction booms in the Middle East and elsewhere.

Furthermore, with the departure of hundreds of thousands of rich and moneyed Hindu and Sikh property owners, large urban and rural properties became available, ostensibly for distribution to refugees who had left their properties in India, but in many instances were occupied by local elites. I recall that, in 1946, shortly after my retirement, there was a massive exodus of Hindu and Sikh gentry from Peshawar. They belonged to the rich class and lived in posh residential houses in Peshawar cantonment and city. Many of them were my friends and came to me beseeching me to buy their bungalows at throwaway prices. I refused all these offers for two reasons. First, I didn't feel that it was right to take advantage of a friend in distress; and second, I had a conviction that the Pakistan experiment was not going to last long, and I may land in a situation where I may have to return the property I bought to its claimant in the future.

Some years later, in 1952, I was allotted a very large and spacious house on the Mall, Peshawar, situated on many acres; it belonged to a Hindu who had been assassinated by his Pathan servant. The family had fled, leaving the premises replete with carpets, furnishing and even expensive crockery and cutlery. I requested the tehsildar of the evacuee department to carefully make an inventory of all the items in the house, to which he had wryly replied, 'That isn't necessary as you are not going to be asked to account for the contents.' He meant that since I had taken possession of the premises, all of it was mine for good. My wife and I consulted on the matter and decided to surrender the house with all its contents to the government after a stay of three months and made our own living arrangements. And so, I was back in Peshawar without a house of my own.

Finally, in 1956, I decided to build my own house in the University Town, Peshawar, on an acre-and-a-half of land which I had purchased earlier. This was my second venture at housebuilding, having completed our spacious house in Darul Khair, Abbottabad, in 1928. It was an exciting moment when my wife laid the foundation stone. It, however, turned out to be a great misadventure because, having become used to living in spacious houses and being somewhat out of tune with the times, I started construction on a large old-style house, situated in the middle of extensive lawns with an impressive driveway. The house depleted our savings and placed us in considerable financial stringency, since I was then without a job.

Warsak Dam

In September 1956, I received a telegram from Mohammad Azam Khan, chief engineer, Warsak Dam Project Organisation (WDPO) of the Government of Pakistan, and a former colleague, whether I would like to work as senior consulting engineer with the Canadian contractors of the Warsak Dam Project. I wired back my acceptance and took up the job. There had been a long controversy about the alternative hydel power schemes, namely, the Mianwali hydel scheme from the Indus River, advocated by the West Punjab, and the Warsak hydel scheme on the Kabul River, backed by the NWFP government. And there was also the problem of finding a foreign donor to sponsor and finance these projects. Finally, the Government of Canada undertook to finance the Warsak project, as a gift to Pakistan. H.G. Acres and Co., Niagara Falls, Ontario, were appointed as the consulting firm for the design of the project.[2] They selected the Warsak site on the Kabul River, for a concrete gravity dam.

The site selected for the dam was a narrow section of the gorge of the Kabul River situated about 5 miles upstream of the weir-less head of the existing Kabul River canal. The rock at the dam site consisted of hard Gneiss.[3] The width of the Kabul River gorge at the proposed roadway level on top of the proposed central spillway was 300 feet. The designed height of the dam, from the foundation level to the base of the spillway, was only 100 feet because, with that height, the level of the headed-up back-water upstream reached the gorge just short of the boundary of Afghanistan. It was, in fact, primarily a hydroelectric project, designed in addition to command an area of about 100,000 acres in Peshawar district under a high-level canal, running on high ground parallel to the existing Kabul River canal along its bank. The head of the high-level right bank canal was made through a high inlet built in the body of the dam, followed by a three-and-a-half-mile long concrete-lined tunnel through the hill side. The small gorge storage behind the dam was proposed to be utilised for a low-level left bank canal.

Mr Blakeman was the chief engineer of the consulting engineers and was my direct boss. My job, as senior design engineer, was to design various structures, as aids to the construction work of the main dam. Fair drawings, from my calculations and rough drawings after discussion with Mr Blakeman, were sent to the Canadian and Pakistani draftsmen in the drawing office and were, thereafter, passed on to the Canadian engineering construction crews for implementation on the ground. The Warsak Dam project turned out to be a bonanza for the Canadian consultants and contractors. Their original estimate of a cost of Rs 70 million, swelled to Rs 140 million in the second revised estimates. The work was completed at a total cost of Rs 400 million under the third revised estimate. Even for that colossal cost, the work proved immediately productive, owing to the quick sale of hydel power, and early development of the new areas commanded by the right bank high-level canal, as well as the left bank canal.

My design work comprised all sorts of structures in *reinforced concrete* steel and timber, such as *reinforced concrete* bridges on the Joe Sheikh Canal,[4] falling within the project area; elevated drinking water tower and water supply for the Warsak colony; elevated *reinforced concrete* towers housing giant concrete mixers; a steel gantry for handling steel penstock tubes each weighing 15 tons; steel dowels for connecting concrete lining of power tunnels to virgin rock; form work of the powerhouse, and a roadway over the central spillway; cableway foundations along the hill sides; temporary gates for draft tubes, etc. In fact, I had to design all sorts of odd structures, with all kinds of materials, for the construction equipment. I worked on the Warsak Dam construction for a period of five years, from 1956 to 1961, the last year as employee of the WDPO of the Government of Pakistan, for writing the history of the Warsak Dam Project.

The construction work on the Warsak Dam was carried out on stereotyped lines, starting with the construction of a diversion tunnel along the gorge, through a rocky flank, for passing the winter discharge of the river, followed by a pair of cofferdams, enclosing and dewatering the area of the proposed dam for excavating the dam foundations. The Canadians, apparently, did not know the art of constructing diversion weirs, submersible during floods in a river. They, accordingly, started building the cofferdams with huge dry boulders, each weighing up to five tons, blasted from the rock of the surrounding slopes. Some of the Punjab irrigation chief engineers, who happened to visit Warsak, predicted that cofferdams built in dry boulders would not stand the floods in the river. Mr Blakeman, called me and said, 'Mr Rahman, you are our irrigation expert; we look to you to comment on the objection of the irrigation chief engineers.' I unhesitatingly replied, 'Mr Blakeman, your cofferdams in dry boulders will be washed away like butter, under the floods.' Later, I reinforced my remarks, by calculations of the velocities which would be generated along the downstream slopes of the cofferdams under floods. My calculations were tested by the consulting engineers on models, at Niagara Falls, Canada, and were verified as correct. Consequently, heavy reinforcements in stone masonry in cement mortar, topped by a substantial casing of reinforced concrete, were specified on top of the boulder cofferdams by the now thoroughly impressed consultants.

The armoured cofferdams withstood the summer floods in the river. By the next year's flood season, the concreting of a substantial portion of the dam had been completed, leaving a 50 feet wide gap in the middle to allow the summer floods to pass. Concreting had also been completed to a height of 50 feet on both sides of the gap. I was told to design a 65 feet span suspension bridge across the gap, to enable the labourers to work on both sides during the flood season. Mr Blakeman warned me, 'Mr Rahman, a faulty design of a single bolt and nut in the suspension bridge may cause it to collapse, resulting in fatal casualties of Canadian and Pakistani labour.' The beauty of the situation was that the designs were forthwith turned into structures, so that one could easily see the merits and demerits of a design. Sometimes the Canadian labour would push me, saying, 'Hey, when will you finish with the design?'

Early in 1957, Prime Minister Huseyn Shaheed Suhrawardy,[5] a somewhat elderly widower, with a weakness for the fair sex, made several visits to Warsak, for dancing the whole night with the Canadian ladies at Warsak, after sneaking out of the Government House, Peshawar, where he was staying. Later in the same year, President Iskander Mirza (23 March 1956–27 October 1958) paid an official visit to Warsak, accompanied by Pakistan's First Lady, Begum Naheed Iskander Mirza. As usual, the senior engineers stood in a row, for formal introductions, and a handshake with the president. I, however, being the senior-most engineer stood at the head of the row. The president had only a suspicion of the smile for me, while shaking my hand.

Of course, Iskander Mirza knew me very well. As assistant commissioner, Tank, he had introduced his first wife to me, while I was touring his area, as executive engineer, D.I. Khan. I was then his senior by one step. And later, we were on friendly terms, when he was deputy commissioner, Abbottabad and Peshawar. Early in 1959, President Ayub paid an official visit to Warsak. Seeing me standing at the head of the row of the consulting engineers, he shouted, 'Khan Bahadur Sahib, *Assalam-o-Alaikum*!' And he came swiftly towards me and enquired about my welfare. The next day in office, McAlister, head of the drawing branch asked me, 'Is President Ayub your friend?' 'Used to be,' I briefly replied.

In my work in the design office, I was assisted by Amin Ismail, a bright young Pakistani engineer with foreign qualifications in structural design. He was from Peshawar and showed great promise. Many years earlier, his father had also served with me as an SDO in D.I. Khan. One day Ismail informed me, to my great disappointment, that he was planning to leave. My dissuasion and predictions of a bright future for him in engineering did not seem to change his mind. He subsequently resigned and, I was informed, abandoned his promising engineering career to become an artist. Some years later, I learnt with great pleasure, that he had achieved fame and fortune as 'Gulgee' (1926–2007), one of Pakistan's foremost painters.

One day, when Mr Park, assistant to Mr Blakeman, and I happened to be sitting in the drawing office, a Canadian welder, working at the Warsak Dam, came to see Mr Park, and familiarly called him a 'son-of-a-bitch', to Mr Park's visible embarrassment. The welder asked Mr Park for the services of an engineer to interpret the drawings of a suction pipeline for welding the pipes for him. The line ran around a building at the Jamrud railway station, for automatic unloading of cement from the railway wagons to the waiting trucks on a newly built highway between the Jamrud railway station and the Warsak Dam. Mr Park said he was sorry, but no engineer could be spared at the moment for the job. The welder looked disappointed. Then he looked at me, and said, 'Who is that guy sitting over there?' Mr Park replied, 'No, no, no, he is our senior adviser and a retired chief engineer and cannot be spared for your job.' The welder retorted, 'He might as well be the president of Pakistan himself. I must have that guy for my job, if I have to finish it on time.' So, Mr Park was compelled to nominate me for the job. And for the next month, I had to go on a truck full of red-necked construction workers from my house at Peshawar, directly to Jamrud, to guide them in the pipe welding job!

It has been a great irony that while some junior Pakistani engineers were working with the contractors and the consultants employed on specified construction jobs on the dam, and had acquired useful skills and experience, the Pakistan government engineers, working under the WDPO at Warsak, were not directly employed on any construction job on the dam. The Canadian consultants and contractors agreed that Pakistani engineers of the WDPO should do the job of rock cutting of the three-and-a-half-mile long tunnel at the head of the high-level canal, under the guidance of the Canadian experts. Thus, the Pakistani engineers did the blasting work, while the Canadians gave them the hole-boring points on the rock face, for each blast. The whole tunnel was bored by the Pakistani engineers. To celebrate the completion of the tunnel, a tea party was arranged on the hillslope on a level spot downstream and of the tunnel, in honour of Lt.-Gen. Mohammad Azam Khan,[6] martial law administrator, who rode through the completed tunnel on a trolley pulled by a small diesel engine. On emerging out of the tunnel, the General thundered, 'Tremendous! Most creditable!' The concreting of the tunnel was carried out by the Canadians themselves. When Mohammad Azam Khan (chief engineer, WDPO) was transferred, he appeared in a fix finding a suitable successor to hand over the charge to as he wanted someone who would remain loyal to him. He found a suitable successor in Mr Quraishi, a chief engineer of the B&R department, who knew very little about irrigation or dam construction, but boasted about his public relations capabilities in dealing with the Pakistan government and the Canadian consultants and contractors.

In February 1961, Queen Elizabeth II of England and Prince Philip, paid an official visit to the Warsak Dam. Invitations were issued to a select few Canadians and Pakistanis to meet the distinguished guests. It being a rainy day, seating arrangements were made in the powerhouse of the dam. The guests sat in the form of a horseshoe, while an ornate chair was provided for the Queen. We sat in hushed silence waiting for the Queen's arrival, who was inspecting the dam, high up. Presently she looked down at the interior of the powerhouse, from a loft opening near the ceiling. She was dressed in white, and looked enchantingly beautiful, framed like a picture in the loft's rectangular opening. Finally, she came down. We all silently stood up, as a mark of respect. Chief Engineer Quraishi, as the host, respectfully offered her a cup of tea which she gravely accepted but without paying much attention to him. After tea, the Queen and Prince Philip left for Peshawar, while we remained seated. It was reported that Prince Philip visited the University of Peshawar the next day and was warmly welcomed by the students of the Islamia College.

Small Dams Organisation

Before joining the Mangla Dam contractors, I served in the Small Dams Organisation. How I landed in that position? It so happened that while en route to Peshawar from Abbottabad I was visiting my old friend Ambassador Abdur Rahman at his roadside village in Dervish when, to my surprise, President Ayub Khan, a close relation of my host, unexpectedly dropped in. He was on his way to his village Rehana, which was close by. President Ayub met me with his usual cordiality and during the course of our conversation enquired as to what I was doing. I told him about the recent completion of my assignment at Warsak, where we had briefly met when he visited the dam for its opening ceremony and distribute medals to the Pakistani engineers. He mentioned how impressed he was with the construction of Warsak Dam, which to him was a great achievement. He said that Pakistan should build many more dams, particularly small dams to harness the water resources of streams in mountainous regions, such as those in Hazara division. He said he had already discussed his ideas with the governor of West Pakistan (1960–6), Malik Amir Mohammad Khan, Nawab of Kalabagh for setting up an organisation for the construction of small dams and suggested that with my extensive experience I could make an important contribution to its success. I was pleasantly surprised by President Ayub's off-hand offer of an interesting assignment and, on return to Peshawar, wrote to him, in March 1961, expressing my interest in joining the newly set up organisation called the Small Dams Organisation.

I received a prompt reply from him suggesting that I should see him at Rawalpindi. During the interview at Rawalpindi, he told me to meet him at Lahore where he would personally introduce me to the Nawab of Kalabagh, the governor of West Pakistan, at the Government House. When I reached there, the president and the governor had gone to have tea with my former friend, Justice M.R. Kayani, the outspoken chief justice of West Pakistan (1958–62). On their return, President Ayub introduced me to Nawab of Kalabagh. The nawab said he would appoint me as director, Small Dams Organisation under the charge of Pir Mohammad Ibrahim (part of the canal water delegation) whom, he presumed, I knew quite well. However, it was not until November 1961 that Pir Ibrahim personally came to Peshawar to ask me to take up the job as director. A house was requisitioned for me by the government in Satellite Town, Rawalpindi. I was pleased to learn that my neighbour was the additional district magistrate of Rawalpindi, who had formerly served under me at Mianwali as land acquisition officer. My other neighbour was a captain in the army, who called on me, and introduced himself with the surprising information that he had nine daughters, most of them were adults, but still unmarried.

The Small Dams Organisation was a newly constituted department, with Pir Ibrahim, who had retired from the irrigation department, appointed as its first chief engineer. He had an influential General as his near relative. With all his self-proclaimed accomplishments, Pir Ibrahim knew little about dam design and construction. So, the burden of the survey, design, and construction of the first small dam, on a small perennial stream at Misriot, a few miles from Rawalpindi, fell on me. The rest of the year was spent in surveys of small dam sites in the Jhelum, Campbellpur (now Attock), Peshawar, Mardan, and Hazara districts.

Pir Ibrahim wanted to construct a masonry dam on the Mung Nullah in the Haripur tehsil of Hazara district, situated a few miles from President Ayub's native village. His boss, General Haq Nawaz Khan, also approved of the proposal. I, however, had strong objections to it because the Shale[7] rock at the dam site had wide fissures, which would speedily deplete the reservoir storage. The geologist employed by the Small Dams Organisations also agreed with me. I, too, stood my ground with the result that General Haq Nawaz, during his inspection of the dam site, said that, if we two old men persisted in the quarrel, one of us would have to quit. He further declared that the leakage in the reservoir could be prevented by spreading a waterproof plastic sheet on the reservoir bed.

Since I refused to acquiesce to the above impracticable suggestion, I was the one who had to quit. A masonry dam was subsequently constructed at the Mung site after I left. Its leaky reservoir remains empty to this day! Before the one-month notice period of quitting my job was over, I received an interesting offer from the Associated Consulting Engineers (ACE) Limited in Karachi, offering me an attractive assignment of chief consultant and engineering adviser of a hydroelectric project in Karnaphuli in the Chittagong Hill Tracts of East Pakistan. My wife, however, vetoed the offer.

Mangla Dam

By now, although fairly advanced in years, my professional reputation as well as the demand for my services had spread, especially among foreign engineering consultant and construction firms, involved in many important projects in the Indus Basin. I accepted the position of a senior design engineer with the Mangla Dam contractors. John Stevenson Eilers, the design office manager, was a youngish American, about forty years of age. The engineering design branch at Mangla, where I worked, was on a grander scale than at Warsak. It was equipped with many advanced drawing gadgets and appliances. The majority of the thirty or forty people working in the design office were either Americans or other foreign nationals, such as Germans, Italians, etc. There were about half a dozen Pakistanis as well. Another important difference was that, unlike at Warsak, we had to prepare our own fair drawings, based on our calculations. Amendments to the drawing sometimes necessitated a discussion with the boss, and changes had to be expeditiously incorporated in the fair drawings, after duplication in a machine, within a few minutes.

The summer and winter office hours were from 8 a.m. to 5 p.m., with one hour's break, between 12 noon and 1 p.m. for lunch. We took lunch in a cheap Pakistani restaurant run at the dam site. At lunch time, it was crowded with Pakistanis of all categories working on the dam. There were two coffee breaks at 10 a.m. and 3 p.m., when large cups of coffee were served at our worktables. I was offered accommodation in the posh American colony, built by the American Mangla Dam contractors themselves, complete with poultry and vegetable farms, and equipped with a fine hospital, a bowling alley, and several large department stores, where custom-free American goods (for Americans only) were available. But, in order to protect my teenage daughter from the somewhat permissive atmosphere prevailing in the American colony, I declined the offer, and decided to live in a small house in the Jhelum cantonment, 16 miles away from the dam site. I left my house daily at 7 a.m. in the morning and walked the one-mile distance from my house to the bus stand in the Jhelum city, where I boarded a bus, hired by the Mangla Dam contractors, and returned in the afternoon on the same bus arriving at my house at 6 p.m., making it an eleven-hour working day.

The Jhelum River at Mangla formed an S-meander curve through the last range of low-lying hills, before emerging into the wide Jhelum valley. The 380-feet high earth dam was constructed on dry ground, in the upper limb of the S-curve, just avoiding the winter channel of the river which hugged the concave rim of the upper limb. A diversion tunnel which also ultimately served as the power tunnel was constructed through the hillside flanking the concave rim of the upper portion of the S-curve, with the mouth of the tunnel in line with the dam leaving the river channel clear. The spillway was located through the convex portion of the lower limb of the S-meander curve. During my one-year service with the Mangla Dam contractors, I carried out four main tasks, namely, preparing working drawings of the foundation drains under the sloping double spillways and the powerhouse; the survey, calculations, and preparation of longitudinal sections of the Jhelum River at Mangla, showing flood levels of discharges varying by stages from 5000 to 500,000 cusecs for the design of the 'closure dam' to be subsequently constructed for diverting the river into the power tunnel, and completing the portion of the dam across the dry river channel. My last chore was the design of a log boom in the river, upstream of the dam, to intercept floating debris during the flood season.

Apart from my transport problem from Jhelum to Mangla, life at Mangla was not unpleasant. John Eilers was a very competent engineer and had engaging manners. I invited him to dinner at the Jhelum club. He met my wife and two teenage children. In return, he invited me, my wife, and children to afternoon tea, and dinner on a Sunday. He came to Jhelum cantonment driving his station wagon and took us to Mangla. We met his wife and four daughters, ranging in ages between two and eleven. Both husband and wife expressed very positive feelings about Pakistan and how much they were enjoying their stay here. My wife and I were greatly taken by their warmth and hospitality. In summer, some teenage boys came from the US to join their parents at Mangla. They were sent daily to the design branch to sit there on the vacant tables until lunchtime, to be kept busy at some useful work. One day, one of the boys came and sat by me and looking wistfully at me said that if I were to visit the US, I would be welcomed by his people. I was amazed at the boy's fond, impulsive tone. He was an utter stranger to me, and I was at a loss as to what he had seen in me which had so impressed him.

After my first year of service with the Mangla Dam contractors was over, my second eye was ripe for a cataract operation. My first eye had been successfully operated upon, three years ago, at Lahore, by a Pakistani doctor. Consequently, I had difficulty in turning out fair drawings of the standard of neatness demanded by the Mangla Dam contractors; I, therefore, politely quit the job and left for Lahore to have my second eye operation.

27 The Last Lap

During the 1960s, interesting new developments were taking place in the overall paradigm of planning and execution of engineering projects in Pakistan. The role, responsibilities, and functions of traditional engineering departments, such as the public works, irrigation, public health engineering, etc., underwent a sea change. The execution of gigantic dams, such as Tarbela, Mangla, and Warsak as well as the Indus Basin replacement works, occupied the Water and Power Development Authority (WAPDA)—a large autonomous organisation supported by major internationally renowned firms of engineering consultants and contractors as principle agents for undertaking these projects. This was a logical development because traditional government departments like the PWD and Irrigation lacked both the know-how and the capacity to design and construct them. This diminished their role and importance and also, inevitably, the professional skill and expertise that was traditionally associated with their engineering staff. Furthermore, their subsequent role generally got confined to the execution of smaller projects and maintenance works.

However, the completion of the large Indus Basin projects did not herald a reversal of fortune for these departments. The large foreign consulting and contracting firms were replaced by a number of Pakistani consulting firms and contractors. They were now in competition with the traditional departments and, in due course, monopolised the engineering design and construction supervision market in the country. Many of these firms were set up by former chief engineers, who exercised considerable influence over their former departments, and the government, as well as with international donors. Sheikh Abdul Hamid, my former colleague and delegate in the canal water dispute negotiations, had set up a firm of Associated Consulting Engineers and was doing a thriving business. He had secured the contract for preparing a feasibility study and preliminary designs of a surface drainage sewage scheme for Islamabad and requested me to supervise the study. I hesitated but decided to accept it, thinking that it would be a useful contribution to Pakistan's new capital city.

The developed and underdeveloped sectors of Islamabad city are a compact block of land about 100 square miles situated on a sloping well-drained ground, along the foothills of the Margalla spur of the Murree

Range. The area is interspersed with seven natural ravines, having their sources in the slopes of the Margalla spur facing Islamabad. All seven nullahs have a small perennial discharge running through their deep and narrow beds during the dry season. The soil of Islamabad consists of fine, loess-alluvium interspersed with stray lenses of gravel within the subsoil which are water-bearing aquifers in the area. The natural nullah draining the Islamabad area upstream coalesces in the Satellite Town area at various points lower down into a single channel of the large Leh Nullah.

Hamid had arranged a retired officer of the Punjab irrigation department to undertake the survey work, who did a rather mediocre work with the assistance of some surveyor and overseers. He also personally lacked the experience to design urban drainage works, so I was given the responsibility to correct the survey loopholes, carrying out the preliminary design and writing the feasibility report. The work was finally completed to Hamid's satisfaction and presented to the chairman of the Capital Development Authority (CDA) by him, with due ceremony.

Whilst working on the surface drainage project, I had the opportunity to study the drinking water supply situation in Islamabad and arrived at the conclusion that drinking water for the city from the existing and potential sources was highly inadequate and that this showpiece capital city was likely to face serious drinking water shortages. The three drinking water sources at that time were: the Saidpur spring at Saidpur village situated at the foot of the Margalla spur; the Rawal Dam reservoir, constructed within the Islamabad city area with its source in the Murree hills; and the tube wells. The Saidpur spring had a perennial but inadequate water supply; the Rawal Dam reservoir was technically a defective structure, having a wide surface, subject to excessive evaporation losses, and a shallow depth. Besides, the source of the Rawal stream in the low Murree hills received comparatively little snow. The potential tube well supply in the Islamabad–Rawalpindi area was also uncertain, being dependent on stray pockets of gravel aquifer of limited extent in the loess-alluvium subsoil. I pondered over the question whether the planners of Pakistan's new capital failed to plan for providing an adequate water availability.

The official proposal in the field was to supplement the existing inadequate water supply in Islamabad–Rawalpindi by the construction of a Rs 60 million Simly Dam on the Soan Nullah in the Murree foothills. But, in my view, the proposed Simly Dam suffered from the same disadvantages as the Rawal Dam. So, I started investigations for a dependable, preferably perennial, source of water supply for the nation's capital. The only feasible alternative appeared to be at the Haro River at Khanpur village which was, however, situated on the opposite side of the Margalla spur. The record of water discharge of the Haro River showed that at Khanpur it had a minimum discharge of 32 cusecs, equivalent to 17 million gallons of water per day, and enough to cater to the needs of over 400,000 inhabitants of Islamabad at the rate of 40 gallons per person per day.

My scheme consisted of a diversion weir on the Haro River upstream of Khanpur village and a reinforced concrete pipeline along the top of the winding ridge and a one-mile long concrete-lined tunnel emerging in Islamabad at a height of 2000 feet, the highest point in the inhabited area. The scheme also envisaged that the proposed dam at Khanpur, which was located downstream of the Khanpur village, would take care of the traditional irrigation water needs and rights of cultivators though the five water courses taking off from the river below the village. I believed that the scheme was practicable, cheaper, and sounder than the Simly Dam scheme. A few years later, on learning that the government was planning to construct a direct storage dam on the Haro River at Khanpur, I succeeded in locating an indirect storage site for the dam which would have served the purpose far better. This scheme was presented to and was received by the chairman CDA with due ceremony but was, apparently, shelved.

In 1968, my interest and expertise in water supply and drainage projects came to the attention of S.S. Moreno, chief engineer and manager of Parsons Corporation, a leading American Engineering firm, who had been hired by the Government of West Pakistan for the planning and implementation of water supply and sewerage projects of all major cities of West Pakistan. He requested me to visit him to discuss what he termed as the rather vexing problem of the Greater Quetta Water Supply Scheme. The availability of drinking water in the arid areas of Balochistan, including the Greater Quetta Basin, was a growing problem. The drying up of karez and sinking aquifers in tube wells was depleting Quetta's meagre water resources. Mr Moreno enquired whether I would undertake a six-month investigative study to evaluate the water resource potential of the Greater Quetta Basin for planning a water supply project for the city. I accepted his challenging offer and was given *carte blanche* to work on the study, using all available government historical data and the extensive library of the Parsons Corporation, that contained the entire proceedings of the American Society of Civil Engineers and many other works on water supply, sewerage, hydrology, and climatology.

I submitted my report to Mr Moreno, after the prescribed six months, which astonished him considerably and which he shared with the firm's headquarters for double-checking. My analysis and conclusions were corroborated through computer modelling at their US headquarters. My conclusions, supported by voluminous data, tables, charts, and calculations, were as follows: after allowing for various annual withdrawals of water by various means—the Quetta Water Supply, by karez, tube wells, shallow wells, annual flood, and dry weather runoffs of the Sar-i-Ab (Sariab) and Bareli nullahs, by springs and evaporation losses from land and water surfaces, and transpiration from the cropped area and sublimation from snow surfaces in winter—about 20 per cent of the amount of average rain and snowfall in the Quetta Basin (with an area of 320 square miles) was still unaccounted for and was probably represented in deep percolated pools, residing in strata that were below the maximum depth of 600 feet currently reached by tube wells. Tapping this water source, described by a senior geologist as 'fossil water', required new and imaginative engineering and financial resources. This report added greatly to my personal prestige as an engineer, but did not provide for a comprehensive solution to the problems of adequate drinking water for Quetta, as there was no follow up.

In 1972, when I had crossed my eightieth birthday, I was requested by senior WAPDA officials to review the design parameters of the Gomal Zam Dam prepared by the Yugoslav firm Messrs Energoprojekt, which had prepared a feasibility study for the dam in 1963 for WAPDA. The Government of Pakistan had approved the construction of a concrete dam in 1963 and subsequently, in 1964, preliminary work had started on the construction of a colony and access infrastructure such as roads and services. Further work had been stopped in 1965 due to the India–Pakistan War and the project had languished since then with conflicting views on the suitability of the dam site, the height of the dam, the size of the reservoir, and the comparative cost and benefits of irrigation as against power generation.

After carefully reviewing the reports and the various options under consideration, I drove to D.I. Khan to visit the dam site and see for myself the progress that had been made. Staying overnight at the D.I. Khan Circuit House, I left the next morning in the company of WAPDA engineers for the project site. We left the main D.I. Khan road and headed west towards Waziristan. I was overwhelmed with a wave of nostalgia, remembering my earlier visits under the protection of armed convoys of Scouts, in the strict discipline of young British officers, plainly clad in native clothing and barking short orders in Pashto. We were now traversing the same track, on a motorable road, with only two armed scouts for protection. The spectacular site of the Gomal Zam gorge and the sharp rocky cleavage of the dam site ascending upwards into the blue sky roused the same emotions as when I had seen it first. We arrived at the modest rest house in the colony which was showing signs of dilapidation because it had been left abandoned for long.

I felt further depressed with the thought that there had been no progress in over ninety years since this scheme was first conceived. Much time and effort had been invested by numerous engineers for its implementation. The brave Wazir and Mahsud tribes and the people of D.I. Khan had been denied the benefit of developing their parched lands and the electrification of their villages and cottage industries. I could barely understand the rationale of the British and the colonial policy for not being able to see any direct benefit from investing in this wild and deficient borderland. The failure of the Pakistan government to do so, however, defied my comprehension. A tall Mahsud who was assigned the task to guard me and was following me all around, looked at me quizzically and enquired in Pashto, 'Sir, when will they complete this project?' I looked at him and found myself completely at a loss for an answer. I finally mumbled, 'Soon.' He looked hard at me before responding, 'In Sha Allah!'

Notes

Chapter 2 — Boyhood in Abbottabad

1. Chapter 68: 'Fall in the East,' *The Decline and Fall of the Roman Empire* by Edward Gibbon.

Chapter 3 — Sialkot - Dharamshala - Abbottabad

1 Dharamshala is a popular hill town located in the state of Himachal Pradesh. It is situated in the upper reaches of the Kangra Valley at an elevation of 1,457 meters above sea level.

Chapter 4 — College Days

1. Chaudhry Zafarullah Khan later became foreign minister of Pakistan, a one-time president of the UN General Assembly, and a judge of the International Court of Justice at The Hague.

Chapter 7 — Homeward Bound

1. *Note for the reader*: Our father's first marriage was to Khanum Begum, who was his first cousin, in 1909. He was eighteen years old and she was sixteen. She passed away in 1930. He then married our mother Iqbal Begum (Bibi Gul), who was 24, in 1935. She passed away in 1976.

Chapter 8 — Mardan, 1917

1. Khan Abdul Ghaffar Khan was born on 6 February 1890 in Utmanzai, Hashtnagar, Frontier Tribal Areas of Punjab Province, of British India, nicknamed as Bacha Khan (King of Chiefs). A close friend of Mahatma Gandhi, Bacha Khan was nicknamed the 'Frontier Gandhi'. He founded the Khudai Khidmatgar (Servants of God) movement in 1929. Members of the movement were known as 'Red Shirts' because of the red uniforms they wore.
2. Following the outbreak of rioting and violence in Amritsar in April 1919—which included the killing of four Europeans and the beating of Miss Marcella Sherwood, supervisor of the Mission Day School for Girls in Amritsar—General Reginald Dyer who had arrived in Amritsar on 11 April to take command issued a proclamation on 12 April restricting movement, imposing a curfew and outlawing all assemblies. The

13 April 1919 was Baisakhi, the Sikh New Year's Day. A large number of people who were probably unaware of the proclamation congregated at Jallianwala Bagh.
3. On 9 April 1919, Miss Marcella Sherwood was assaulted and beaten up by elements of a mob in a narrow Amritsar street called Kucha Kurrichhan. General Dyer designated the spot where she was assaulted as sacred. Under his orders, all Indians who crossed that Amritsar street had to crawl a distance of some 200 yards on all fours, their bellies to the ground. The order took effect from 19 April 1919 and stayed until 25 April 1919. Source: Sanjeev Aga, 'Crime and Punishment: Indian state faces assault, though from corruption, not from mobs,' *The Economic Times*, 25 December 2012.
4. The Secretary of State for India, Edwin Montagu, established an inquiry committee to look into the Jallianwala Bagh incident. So, on 14 October 1919, the Government of India announced the formation of the Disorders Inquiry Committee, later dubbed the Hunter Committee/Commission after its chairman, Lord William Hunter. The Hunter Commission members were as follows: Chairman Lord William Hunter, former Solicitor-General; W.F. Rice, additional secretary to the Government of India (Home Department); Justice G.C. Rankin, judge of the High Court, Calcutta; Maj. Gen. Sir George Barrow, commandant of the Peshawar Division; Sir Chimanlal Setalvad, vice-chancellor of Bombay University and advocate of the Bombay High Court; Pandit Jagat Narayan, lawyer and member of the Legislative Council of the United Provinces; and Sardar Sahibzada Sultan Ahmad Khan, lawyer from the Gwalior state.
5. Later Sir Olaf Caroe, governor of NWFP and author of the well-known book *The Pathans*.

Chapter 10 — Dera Ismail Khan (D.I. Khan)

1. From 1913 to 1945 James Glasgow Acheson served in the Indian Civil Service (Political Department from 1920). His service was in the United Provinces; Delhi/Simla (twice officiating in the Government of India as Foreign Secretary); then in Balochistan and the North-West Frontier Province; his final post was as Political Resident in Kashmir 1943–5. He was appointed CIE in 1929 and Knight Bachelor in 1945.
2. Sir Ambrose Dundas (1899–1973), KCIE, CSI, joined the Indian Civil Service in 1922, and remained in the ICS until 1947. He served as the last British governor of North-West Frontier Province from 1948 to 1949.
3. Embankments made in steps of stone slabs along the river bank.

Chapter 12 — Across the Frontier into Balochistan

1. A system of irrigation. An underground irrigation tunnel bored horizontally into rock slopes.

Chapter 13 — Government of India's Inspection

1. Jamsetji Nusserwanji Tata (1839–1904) was an Indian pioneer industrialist who founded the Tata Group, India's biggest conglomerate company. Named the greatest philanthropist of the last century by several polls and ranking lists, he also established the city of Jamshedpur.

2. Tata Iron and Steel Company (TISCO) was founded by Jamsetji Nusserwanji Tata and established by Sir Dorabji Tata on 26 August 1907. It was the first steel plant in India and was set up at Jamshedpur.

Chapter 14 — Players of the Great Game

1. Mullah Powindah born Mohiuddin Mahsud (d. 1913) was a religious leader in the Pashtun tribe of the Mahsuds.
2. https://en.wikipedia.org/wiki/Edward_Noel_(Indian_Army_officer).

Chapter 15 — Political Turbulence

1. https://en.wikipedia.org/wiki/Satyendra_Prasanna_Sinha,_1st_Baron_Sinha.

Chapter 16 — Farewell to D.I. Khan

1. An ill-defined medical condition characterized by lassitude, fatigue, headache, and irritability, associated chiefly with emotional disturbance.
2. Zilladar means an officer exercising control over a canal in respect of proper distribution and regulation of water and is responsible for the assessment of canal revenue.

Chapter 17 — Bannu

1. Sir Mian Fazl-i-Husain (1877–1936), KCSI, was an influential politician during the British Raj and a founding member of the Unionist Party of the Punjab.

Chapter 18 — The Second World War, 1939–1945

1. On 30 September 1938, Neville Chamberlain announced 'peace for our time' on the steps of 10 Downing Street, straight after returning from Munich, where he and government leaders from France, Italy and Germany had signed an agreement over the division of Czechoslovakia in the hope of averting war.
2. Designed by the famous Sir Pierre Louis Napoleon Cavagnari, the British representative to Kabul, who was killed in 1879. In 1861 he was appointed an assistant commissioner in the Punjab region of British India, and in 1877 became deputy commissioner of Peshawar (now in Pakistan) and took part in several expeditions against the Pashtun tribes.
3. From 22 March to 11 April 1942, Sir Stafford Cripps, a member of the War Cabinet, was dispatched to India to discuss the British government's draft declaration on the Constitution of India with representative Indian leaders from all parties. The Cripps Mission failed, and the issue of India's constitution was postponed until the end of the war.

Chapter 19 — Peshawar

1. Mir Jafar and his fellow conspirators took no active role in the Battle of Plassey (1757). Nawab Siraj-ud-Daula was defeated and killed in battle, and the whole of Bengal had thereby come under the sway of the British. Mir Jafar was installed afterward as the nawab of Bengal.
2. The Spin Ghar or Safed Koh meaning both White Mountains, or sometimes Selseleh-ye Safid Kuh meaning White Mountain range, is a mountain range in South Asia to the south of the Hindu Kush. It ranges from eastern Afghanistan into Khyber Pakhtunkhwa, Pakistan.
3. R.G. Kennedy, executive engineer, Punjab PWD.
4. Gerald Lacey was professor of civil engineering (1915–17, 1928–32, 1945) and the last British principal (1945–6) of the Thomason College of Civil Engineering, Roorkee.

Chapter 20 — Retirement and the Partition of India

1. Apollo is the Greek god of the sun, light, music, truth, healing, poetry, and prophesy, and one of the most well-known gods in Greek mythology.
2. He was the father of Lt. Gen. Habibullah Khan Khattak, Aslam Khattak, Yusuf Khattak and Begum Kulsum Saifullah Khan.

Chapter 21 — Bahawalpur

1. Sir Richard Marsh Crofton (1891–1955), CIE, was an English officer in the British Indian Army who served as the first Prime Minister of Bahawalpur between 1942 and 1947.
2. Master Tara Singh (1885–1967) was the foremost Sikh leader of pre-Partition Punjab. Born into a Hindu family in Rawalpindi district, he was influenced and impressed by the teachings of Sikhism and converted to a Sikh at the age of twelve under Sant Attar Singh.
3. Gurdwara Panja Sahib is situated at Hasan Abdal, 48 km from Rawalpindi in Pakistan. This is one of the most holy places of Sikhism because it marks the spot where the founder of the faith, Guru Nanak Dev visited and instilled an important lesson for his adherents. Still visible is the sacred rock with the hand print of Guru Nanak.
4. Sir Yadavindra Singh (1913–1974) was Maharaja of Patiala from 1938 to 1974.
5. Madrasi also spelled as Madrassi, is a term used as a demonym and a regional slur for people from southern India. In earlier usage it was a demonym to refer to the people of Madras Presidency; however, this use of the term is now outdated.
6. There are two powers in the world; one is the sword and the other is the pen. There is a great competition and rivalry between the two. There is a third power stronger than both, that of the women. (Jinnah)
7. Born in 1899 and trained at the Royal Military Academy at Woolwich, Jefferis was commissioned into the Royal Engineers in the summer of 1918. In 1920, Jefferis was sent to India, and it was here that his career took shape. During the Waziristan campaign of 1922, he built roads and bridges through supposedly impassable mountains, connecting the strategically important settlements of Isha and Razmak. This work was carried out under regular fire from enemy snipers, demonstrating Jefferis's courage

in the face of adversity, for which he was awarded the Military Cross. Though he continued to demonstrate a genius for bridge building, Waziristan brought out a new side of Jefferis. He became interested in guerrilla warfare and the use of explosives to destroy the very feats of engineering he was used to building. By the start of the Second World War, Jefferis was a major. He took part in the Norwegian campaign of 1940, using his specialist knowledge to destroy bridges ahead of the Germans as the Allies retreated north. Jefferis's weapons, which ranged from mines to mortars to specialist ammunition, were soon in use all over the world, from resistance operations in Europe to anti-submarine warfare in the Pacific. Military and political figures increasingly recognized his special genius and came to consult with him on armaments. In recognition of his work, he was made a Commander of the British Empire in the 1942 New Year's Honours list, a prestigious reward for a man whose work was largely secret. In late 1945, with the Second World War finally over, Jefferis was made Chief Engineer to the Indian Army. He was also promoted to acting major-general and made a Knight Commander of the British Empire. He spent five years in India and Pakistan, returned to England as a brigadier, and served as an aide to King George VI from 1951. Retiring from service in 1953, he was given the honorary rank of major-general. Millis Jefferis died in 1963.
8. Lt. Col. Sir Arthur John Dring (1902–91), KBE, CIE, JP, was Bahawalpur's second prime minister, and last prime minister of British origin, and served from 1948 to 1952.

Chapter 22 — Balochistan

1. Hamun/Hamoun are marshlands with lagoons that vary strongly in water level over the seasons and years.

Chapter 23 — Irrigation Advisor, NWFP

1. David E. Lilienthal was an American businessman and government official, who was codirector (1933) and first chairman (1941) of the Tennessee Valley Authority (TVA) and first chairman of the Atomic Energy Commission (AEC).
2. Sir B.N. Rau was an Indian civil servant, jurist, diplomat and statesman known for his key role in drafting the Constitution of India.
3. The Government of India Act of 1935 gave greater autonomy to provincial governments in a number of areas, including irrigation. The Indus Commission, commonly called the Rau Commission, was convened in 1941 to consider the province of Sindh's complaints against Punjabi river control structures, existing and planned.

Chapter 24 — Canal Water Dispute Negotiations

1. Command area means an area irrigated or capable of being irrigated.
2. During his fifty-year career, Colorado civil engineer Royce J. Tipton (1893–1967) designed numerous large water infrastructure projects in the US West and around the world. As an accomplished and respected consultant, he helped negotiate several interstate water compacts and international treaties.

3. 'I Like Ike' was the campaign slogan of Dwight David Eisenhower during his 1952 election campaign for the presidency of the United States.

Chapter 25 — Pakistan Delegation's Visit to India

1. Used extensively in the Second World War, what used to be known as Willingdon Airfield started operations in 1929 during the British Raj. Once serving as Delhi's main airport until 1962, it is now called Safdarjung airport, and has been officially closed since 2002.
2. The Gaddi is a semi-pastoral Indo-Aryan ethno-linguistic tribe living mainly in the Indian states of Himachal Pradesh and Jammu and Kashmir.
3. Thomas Arthur Wyness Foy (1895–1971) was Educated at Truro College, Cornwall, he studied engineering at Birmingham University before and after the First World War, taking his BSc (Eng.) degree in 1920. From 1914–18 he was on active service with the Royal Fusiliers, the Royal Engineers, and the Royal Flying Corps (later the RAF). In 1920 he joined the Indian Service of Engineers and served in the Punjab PWD (Irrigation Branch) until 1947, when he became chief engineer to the Government of Sindh on the Lower Sindh Barrage, a position he held until his retirement eight years later. During his first five years in the Punjab he was partly responsible for the construction and design of the Islam Barrage over the Sutlej River under E.R. Foy and F. Burkitt; on this river he was also engaged on the Ferozepur Barrage. Appointed executive engineer in 1925, he embarked on the construction of the Bikaner Canal, under T.B. Tate. Foy was next engaged on maintenance and administration of the Sirhind Canal and reconstruction of Rasul Weir on the Helum River, which had suffered serious damage in the record flood of 1929. For a time, he worked as under-secretary to the Punjab government before becoming chief engineer to Bikaner state on the state-controlled Bikaner Canal. From 1940–45 he was superintending engineer under A. St. C. Lyster on the bitter dispute which developed between the Punjab and Sindh over distribution of the waters of the Indus and its tributaries between the two provinces. This resulted in the setting up of the Rau Commission, on which Foy was primarily responsible for the preparation of the Punjab case in its technical aspects. He represented the state of Bikaner before the committee presided over by Sir Frederick Anderson, and was also in charge of the Sirhind Canal. He carried out substantial repairs to Sulemanki headworks on the Sutlej River. After further legal work on the water dispute he was included in the deputation to the Government of India on the subject; later he visited Afghanistan to advise on certain dams and irrigation works. Under A.N. Khosla he was responsible for the preparation of projects for dams in the Himalayas, and under F. Haigh was in charge of the Thal Project (completion of Kalabagh headworks, diversion of river and construction of a concrete-lined canal of considerable proportions). In 1945, Foy was appointed chief engineer and secretary to the Punjab Public Works Department (Irrigation Branch) and for the next two years was in charge of the completion of the Thal Project on the Indus River, began construction of Bhakra Dam and of the Nangal Project (his own conception). He was also responsible for the design and construction of earth dams for storage near Delhi. In 1947, he became chief engineer and secretary to the Government of Sindh, in charge of the design and construction of the Lower Sindh Project (comprising the Kotri Barrage, 3000 feet long, and three earthen channels totalling 37,000 cusecs, designed to divert the waters of the

Indus). During the difficult changeover period he was in the Punjab as chief engineer on contract in West Pakistan. His career in India ended with his retirement in 1955, but for a further fourteen years he continued to work as a consultant. Foy was made a Companion of the Order of the Star of India in 1947. He received a knighthood in 1956. He was a member of the Indian Government (Retired) Association, and was elected direct to the senior grade of membership of the Institution in 1955. Source: Obituary, Thomas Arthur Wyness Foy (1895–1971), *Proceedings of the Institution of Civil Engineers*, vol. 51, no. 2, 1972, pp. 429–31, https://doi.org/10.1680/iicep.1972.5979.

4. India and Pakistan submitted their plans in October 1953. The Indian Plan allotted to India all the three eastern rivers (Ravi, Beas, Sutlej) plus 7 per cent of the three western rivers (Indus, Jhelum, Chenab). The Pakistani Plan allotted to Pakistan all the three western rivers plus 70 per cent of the eastern rivers. World Bank realised, 'the problem could not be solved solely by technicians...[its] representative [felt] the responsibility to put forward a proposal...to serve as the basis of the comprehensive plan.' (World Bank's Press Release No. 380, 10 December 1954) World Bank presented its own plan on 5 February 1954 based on the general principle that with the exception of local uses in Kashmir, the three western rivers would be reserved exclusively for the use and benefit of Pakistan and the three eastern rivers would be reserved entirely for India. Source: Muhammad Nasrullah Mirza, 'Indus Water Disputes and India–Pakistan Relations,' Doctoral Dissertation, M.Phil., Department of Political Science South Asia Institute, University of Heidelberg, Federal Republic of Germany, 2016, p. 114.

Chapter 26 — Warsak and Mangla Dams

1. Syed S. Kirmani was born on 1 July 1921 in Guntur district of Madras and was educated at Andhra Christian College. He graduated in civil engineering with honours from the College of Engineering, Guindy, Madras and joined the irrigation branch, Punjab Public Works Department in 1944 as an assistant engineer. Luckily, he was assigned to the Central Design Office where his special talent and intellect began to take shape. His first important assignment was the design of Rasul Hydroelectric Project and the Ravi Syphon for the design of which he worked for about a year in 1949. At the young age of 29 he was promoted as Director, Design, for the planning of Irrigation and Hydroelectric Power in the former Punjab. In 1954, he represented Pakistan as an official delegate to the Third International Irrigation and Drainage Conference and the same year he was promoted as superintending engineer. This was a chance for him to show his intellect and the deep understanding of the problems of irrigation, so that in December 1954, he was sent to America as a member of the Pakistan Water Delegation in connection with the Indus Water Dispute, being resolved under the good offices of the International Bank of Reconstruction and Development. On return to Pakistan after a brief stay in the Dam Investigation Circle, he was sent by the irrigation department to West Pakistan, WAPDA as Director, Planning and Investigation. He was again called upon for a brief visit to USA as a member of Pakistan Water Delegation for Indus Water Dispute, and on return in 1960, was appointed Director Indus Basin Projects. In March 1961, he was awarded the title of Sitara-e-Quaid-i-Azam and the next year in May 1962, he was put in as Director General Indus Project and later on in August of the same year promoted as chief engineer. In November 1967, he was again awarded

the Sitara-e-Imtiaz. Source: 'Engineering News,' Vol. 12, no. 4, Dec. 1967, *Quarterly Journal of the West Pakistan Engineering Congress.*
2. The contract for the construction of the actual dam was awarded in August 1955, to a well-known Canadian construction firm. Angus Robertson (Overseas) Limited did the work on behalf of Angus Robertson Limited of Toronto, Ontario.
3. Gneiss is a high-grade metamorphic rock, meaning that it has been subjected to higher temperatures and pressures than schist. It is formed by the metamorphosis of granite, or sedimentary rock. Gneiss displays distinct foliation, representing alternating layers composed of different minerals.
4. Construction of Joe Sheikh inundation canal was carried out from the right bank of Kabul River during the reign of Mughal Emperor Aurangzeb and to irrigate areas around Peshawar.
5. H.S. Suhrawardy was the fifth prime minister of Pakistan, 12 September 1956–17 October 1957.
6. Lt. Gen. M. Azam Khan (1908–1994) was the general officer commanding of Lahore Garrison and was appointed martial law administrator when Governor General Ghulam Mohammad called upon the army to bring the anti-Ahmadiyya movement under control in 1953. He successfully restored law and order situation in Lahore. Azam Khan gave laudable leadership during the flood in the Punjab in 1953. He was one of the three senior generals of the Pakistan Army who helped Ayub Khan in staging his coup d'état and declaring martial law in 1958. Ayub Khan took him into his cabinet as senior minister on 28 October 1958 and placed him in charge of the department of refugee rehabilitation.
7. Soft finely stratified sedimentary rock that formed from consolidated mud or clay and can be split easily into fragile plates.

Index

A

Abdul Wadud, Miangul, 94, 202
Acheson, J.G., 98, 100, 102, 103
Afghan War, 77, 78, 90, 119, 131, 136, 138, 144, 145
Afridi, 73, 76, 94, 128, 129, 130, 145, 204
Afridi, Sardar Mughal Baz Khan, 171
Agra, 19, 45, 66
Ahmad, Mirza Bashiruddin Mahmud, 24, 236
Ahmad, Mirza Ghulam, 15, 16, 23, 24, 236
Ahmadiyya movement, 15, 23, 24, 236
Aibak, Qutubuddin, 38
Akali Sikhs, 211
Akbar Allahabadi, 12
Akbar, Emperor, 18, 19
Alexandra Bridge, 3
Ali, Chaudhry Mohammad, 243, 265, 266
Ali, Choudhry Rahmat, 157
Ali, Maulana Mohammad, 31, 160
Ali, Maulana Shaukat, 31
Aligarh College, 27, 28, 29, 30, 31, 33, 146, 157, 162, 193, 234, 235
All India Competitive Examination, 27, 31, 32
Aligarh Muslim University, 27
Aligarh, 26, 27, 28, 29, 30, 31, 33, 40, 146, 157, 162, 193, 234, 235
All-India Muslim League, 192
Amir of Bahawalpur, 3, 4
Amritsar, 75, 214, 261
Arabia, 86, 150, 155, 246
Ashoka, 83
Assam, 44, 47, 48, 50, 51, 57, 66, 72, 192, 195, 218
Ataturk, Mustafa Kemal, 150, 155
Atkinson, Lt. Col. E.H., 34, 67
Aurangzeb, Emperor, 66, 179
Azad, Maulana Abul Kalam, 157, 263

B

Babar Ghundai, 121, 122, 133, 134, 138
Babar, Emperor, 121
Badragas, 117, 118, 131
Badshahi Masjid, 37, 39
Bahawalpur, 3, 37, 154, 210–12, 215–19, 220, 221, 223–9, 242, 243, 247, 252, 258, 260, 261
Baildars, 73, 74
Bajaur, 2
Balochistan, 15, 105, 113, 114, 123, 128, 129, 131, 133, 137, 138, 141, 142, 150, 195, 219, 227, 228, 231–8, 250, 280
Bannu, 11, 98, 100, 116, 120, 140, 148, 152, 159, 175–9, 180, 184, 188, 189, 208, 238, 240, 262
Barra Bazaar, 118
Beas River, 18, 19, 243, 251, 259, 260, 261, 264, 265
Bhakra Dam, 259, 260, 262, 265
Bhatia, 117, 118
Bhittani, 115, 116, 118, 119
Bihar, 58, 65, 72, 84, 160
Bilot, 97, 100, 101, 106–8, 152, 164, 165
Black, Eugene, 254, 255
Bogra, Mohammad Ali, 248, 255
Bolan River, 232
Bombay Mail, 72
Bombay Technical Institute, 18
Bombay, 18, 65, 66, 72, 139, 158, 170, 181, 211, 216
Brahmaputra River, 47
Brave Company, 5
British Empire, 77, 242
British India, 2, 66
Bruce, Charles Edward, 103–5, 166
Buddhist, 76, 83–5, 175, 201
Buner, 72, 83, 89, 90–4
Burkitt, F.A., 69, 156, 161, 162, 164–6, 171, 179, 184, 194

291

INDEX

C

Calcutta, 36, 46, 48, 55, 57, 58, 61, 117, 118, 215
Cantonment Military Hospital, 103
Cantonment, 10, 14, 16, 17, 30, 32, 36, 41, 102, 116, 117, 120, 148, 149, 188, 190, 268, 276, 277
Capital Development Authority (CDA), 279, 280
Carne, F.W., 72
Caroe, Olaf, 82, 84, 204
Cavagnari House, 188, 191
Chamberlain, Neville, 187
Charsadda, 167, 169, 172, 185, 191, 223, 241
Churchill, Winston, 88, 202, 228
Civil Disobedience Movement, 157–9, 160, 193
Companiganj rest house, 52, 55
Company Garden, 9, 10, 186
Comrade, 31
Constantinople, 13
Cripps, Sir Stafford, 193
Crofton, Sir Richard, 154, 210, 218
Cunningham, Sir George, 176, 189, 195, 208
Curzon, Lord, 12, 13, 36

D

Darya Khan, 161
Delhi Durbar, 12, 36, 39
Delhi, 5, 12, 13, 19, 36, 38, 39, 45, 66, 142, 181, 215, 216, 218, 259, 263
Deoband, 43, 44
Dera Ghazi Khan, 108, 176, 177
Dera Ismail Khan (D.I. Khan), 95–9, 100–3, 105, 106, 108, 109, 110, 114, 115, 117, 118, 120, 128, 131, 132, 134, 135, 138, 139, 140, 142, 144, 146–8, 150–4, 156, 161–9, 182, 216, 227, 231, 272, 281, 282
Dewadand, 59, 60, 62
Dharamshala, 16, 17, 261
Dhauladhar range, 17, 19, 261
Din, Khan Bahadur Maulvi Ahmad, 148, 149
Dundas, Ambrose, 105, 106, 167–9, 194, 227, 228
Durand Line, 69, 126, 148, 186, 197

Durand, Sir Mortimer, 148, 237
Durrani, Ahmad Shah, 2, 170
Dyer, General Reginald, 75, 76

E

East Pakistan, 248, 275
East Punjab, 203, 207, 214, 215, 217, 218, 220, 221, 223, 226, 250, 257, 259, 260, 261, 262
Eden Gardens, 57
Edward VII, King, 3, 11, 12
Edward VIII, King, 86, 87
Empress Bridge, 3
Empress Road, 211, 212

F

Farrukh Siyar, Emperor, 59
Fazal-e-Rahman, Sahibzada, 88
First World War, 31, 34, 67, 69, 72, 73, 76, 86, 94, 150, 155, 227
Fort Sandeman, 114, 128, 129, 130, 135–7, 141, 237
Frontier Crimes Regulations, 157

G

Gandapur, Sardar Aurangzeb, 192, 207, 208
Gandhara, 83, 91
Gandhi, M.K., 74–6, 86, 95, 157–9, 160, 176, 186, 192, 195, 208, 214
Ganges River, 35, 43, 46
Gauhati, 47, 48
George V, King, 13, 17, 36, 37
George VI, King Emperor, 187
Ghilzai, 118, 223
Gibbon, Edward, 13
Gomal, 113–15, 117–19, 123–6, 128, 131–3, 135, 136, 138, 139, 140–2, 281
– Gomal Tangi dam, 139, 141
– Gomal Tangi gorge, 140
– Gorge, 115, 135, 136, 140
– Pass, 118, 123
– River, 115, 117, 119, 123–6, 129, 133, 135, 138, 139, 140, 141
– Torrent, 113, 114, 115
– Valley, 117, 135, 139, 140

INDEX 293

Gomti River, 45
Government College, 2, 20, 21, 24–8, 224
Gracey, General Sir Douglas, 236
Grand Trunk Road, 58
Great Depression, 163
Gul Kachh, 113–115, 119, 120, 123–126, 128, 133, 138, 140–142, 169, 194, 195
Gurkha, 10, 17, 19
Gurmani, Nawab Mushtaq Ahmad, 218, 219, 224, 225, 226

H

Hamzakot, 83, 85, 91–4
Harvey, W.B., 70, 71, 72
Hazara, 5, 59, 70, 77, 86, 160, 199, 202, 274, 275
Howrah railway station, 57
Humayun, Emperor, 38, 39, 58, 264
Husain, Khan Bahadur Nabi Bakhsh Mohammad, 226, 227
Hyderabad, 65, 66, 69, 70, 180

I

India, 2, 5, 11, 12, 14, 17, 25, 28, 31, 32
Indian Imperial Service, 35
Indian National Congress, 74, 86, 95, 143, 147, 159, 160, 176, 193, 195, 196, 204, 214
Indian Service of Engineers (ISE), 11, 35, 67, 113, 161, 179, 209
Indraprastha, 38
Indus Basin River system, 242, 249, 250, 256, 257
Indus River, 1, 2, 4, 80, 95–7, 99, 100, 106–9, 110, 111, 114, 132, 148, 152, 161, 164, 165, 175, 189, 219, 227, 231, 232, 262, 269
Iqbal, Dr Mohammad Allama, 24, 25, 30, 65, 157, 159, 160, 180, 185
Iraq, 86
Irwin, Lord, 142, 159
Islamgarh, 2
Islamia College Peshawar, 88, 162, 164, 273

J

Jagannath, 71–4, 79, 184
Jahanara Begum, 39

Jahangir, Emperor, 123
Jahanzeb, Miangul, 202
Jalalabad, 86
Jalbai, 78, 79
Jallianwala Bagh massacre, 75, 76
Jammu and Kashmir state, 3
Jammu, 3, 17
Jandola, 114, 115, 116, 118, 119, 128, 131, 194
Jefferis, General Millis, 227, 228, 232, 233, 237
Jinnah, Quaid-i-Azam Mohammad Ali, 157, 163, 192, 193, 207–9, 214, 218, 221, 223, 229, 235
Jumna River, 35, 36, 38

K

Kabul, 76, 77, 145–7, 198, 269, 270; River, 269, 270
Kalabagh, 100, 108, 109, 110–14, 124, 251, 274
Kangra, 17–19, 25, 261
Karakul, 117, 129
Karamar hill, 83, 84, 86, 93
Karimganj, 48, 49
Kashmir, 17, 37, 44, 85, 178, 179, 208, 211, 217, 218, 222, 223, 254, 265
– Disputed territory, 265
– Maharaja of, 17, 222
– Resident of, 17
– Smast cave, 85
Khairpur railway station, 3
Khajuri Kachh, 114, 128, 131, 135, 136, 138, 139, 140, 141
Khalassis, 58, 60, 61, 63, 109, 116, 120, 121
Khan, Abdul Ghaffar, 70, 95, 147, 159, 172, 185, 186, 196, 223
Khan, Abdul Ghani, 172
Khan, Abdul Latif, 144
Khan, Abdur Rahman, 80
Khan, Amir Ahmad, 137
Khan Sahib, Dr (Abdul Jabbar Khan), 95, 147, 175, 176, 180, 181, 185, 186, 187, 192, 195, 196, 203, 204, 208, 209, 214, 241
Khan, Ayub, 235, 236, 272, 274, 275
Khan, Chaudhry Nasrullah, 2
Khan, Chaudhry Zafarullah, 2, 23, 224, 236

INDEX

Khan, Faqir Mohammad, 95–8, 169, 170, 181, 182, 183
Khan, Fazal Din, 2
Khan, Ikramullah, 164, 165, 166
Khan, Jaffar, 136
Khan, Khair-ud-Din, 8
Khan, Khan Bahadur Saadullah, 168, 169
Khan, Khan Bahadur Sharbat, 130, 136
Khan, King Amanullah, 127, 144, 145, 146
Khan, Lt. Gen. Mohammad Azam, 273
Khan, Malik Amir Mohammad (Nawab of Kalabagh), 274
Khan, Maulana Zafar Ali, 30, 31
Khan, Mir Alam, 168, 169
Khan, Mohammad Aslam, 80, 206
Khan, Mohammad Din, 26
Khan, Mohammad Said, 172, 189, 240
Khan, Mohammad Umar, 84
Khan, Nadir, 77, 145, 146
Khan, Nawab Mohammad Akbar, 87
Khan, Nawabzada Liaquat Ali Khan, 235
Khan, Nisar Mohammad, 182
Khan, Sahibzada Sir Abdul Qayyum, 157, 160–2, 173, 175, 176, 192, 223, 229, 241, 242, 259, 262, 265, 266
Khan, Sardar Abdur Rashid, 265
Khan, Sheikh Mahbub Ali, 144–7, 169, 188, 189, 190, 196, 204, 207
Khan, Sir Syed Ahmad, 28
Khan, Sir Zafarullah, 2, 23, 236
Khan, Sultan Mohammad, 2
Khanum Begum, 65, 156
Khasadars, 74, 123, 138, 148, 149
Khattak, Khan Bahadur Quli Khan, 208
Khilafat Movement (1919–22), 86, 90, 95
Khilji, Sultan Alauddin, 66
Khudai Khidmatgar, 147, 158
Khushalgarh, 1, 189
Khyber Mail, 244, 245, 258, 266
Khyber Pass, 76, 118, 199, 204,
Khyber, 76, 118, 176, 199, 204, 244, 245, 258, 266
Kirmani, S.S., 268
Kohat, 1, 5, 77, 130, 145, 187, 188, 189, 190, 191, 192
Kurram, 110, 112, 150, 175, 180, 189, 190, 197, 198, 199, 207, 238, 240, 241, 262
– Agency, 150, 197, 207

– Garhi Scheme, 238, 240, 241, 262
– River, 110, 112, 175, 180, 189, 190, 197, 198, 240, 241, 262
– Valley, 198, 199

L

Lahore Airport, 264
Lahore High Court, 30, 172
Lahore Mall Road, 74
Lahore Railway station, 24, 212, 266
Lansdowne Bridge, 2, 4
Larji Dam, 261
Lawrence, Col. T.E., 150, 155
Laylin, John, 250–3, 256,
Leeper, 116, 120, 121, 126, 169, 194
Lilienthal, David, 242, 243
Linlithgow, Lord, 181
Lucknow, 45, 153, 157; High Court, 157

M

Mafrurs, 90
Mahmud of Ghazni, Sultan, 19, 25
Mahsud, 115–19, 120, 128, 129, 131, 132, 136, 138, 139, 140, 141, 145, 148, 149, 159, 182, 183, 222, 282
Maira, 70–3, 78, 79, 82, 83, 85, 90, 184
Malakand, 88, 181, 182, 194, 199, 200, 201, 202, 204, 206, 207; Fort, 88, 200, 206,
Malik(s), 120, 121, 124–7, 132, 133, 134, 149, 177, 182, 204
Mandi, 15, 16, 70, 203, 261
Mangla Dam, 191, 251, 268, 274, 276–8
Mardan, 68, 69, 71, 72, 73, 77, 81, 82, 84, 85, 87, 88, 90–4, 108, 166, 167, 169, 170–4, 180, 181, 183, 187, 194, 196, 199, 263, 275
Mary, Empress, 17, 36
Maynard, Sir Herbert John, 75
Mehmed VI, 86
Miami Conservancy Project, 114
Mianwali, 109, 110, 269, 274
Minhajuddin, 95, 162
Minto Circle hostel, 27
Mir Jafar (Sayyid Mir Mohammad Jafar Ali Khan Bahadur) 196
Mirza, Iskander, 196, 271, 272

Mohmand: agency, 171; tribe, 2; tribesman, 69, 70, 72, 77
Montagu, Edwin Samuel, 87, 147, 160
Montagu-Chelmsford Reforms (1918), 87, 160,
Moon, Penderel, 218, 220, 221
Mountbatten, Lord Louis, 208, 214, 221, 224
Muhammadan Anglo-Oriental (MAO) College, 26, 27, 28
Muirhead, 227, 229, 232, 237, 238, 239, 267
Mullick A.P., 49, 51, 57
Murree: hills, 278, 279; road, 7
Murtaza Post, 117, 118, 135, 140
Muslim League, 157, 159, 192, 193, 204, 207, 214
Mutiny (1857), 5, 28

N

Nazimuddin, Khawaja, 29, 230, 234
Nehru, Pandit Jawaharlal, 180, 203, 204, 208, 215, 218, 219, 222
New Delhi, 146, 154, 198, 215, 216, 218, 259, 263, 264
Niamatullah Committee, 157
Nizam of Hyderabad, 37
Nizamiyya University, 44
Noel, Lt. Col. Edward, 150–5, 162, 199
Non-Cooperation Movement, 76, 86
North Western Railways (NWR), 2, 43, 45, 212
Nur-ud-Din, Maulana, 23, 24
Nushki, 15, 230, 231
NWR Bridge, 3, 15

O

Oram, Arthur, 107, 131–6, 139, 140–2, 154, 156, 157, 161, 162, 173, 188, 192, 208
Osmania University, 65

P

Paharpur Canal, 100, 107, 108, 115, 150–2, 161, 164, 165
Pakistan Army, 227, 228, 232, 236
Palestine, 86

Panipat, 2, 170
Parachinar, 150, 190, 197, 198, 199
Pawindah, 117, 118, 125, 126, 128, 139, 149
Peshawar Fort, 209
Peshawar Mall, 238, 268
Punjab Engineering Congress, 74, 184, 249
Punjab Irrigation Department, 75, 98, 109, 111, 114, 162, 223, 225, 243, 257, 279

Q

Qadian, 24, 236
Qissa Chahar Dervish (*Tales of the Four Dervishes*), 4
Qissa Khwani Bazaar, 87, 121, 158
Quit India, 192, 193, 195, 196

R

Radcliffe, Cyril, 224
RAF (Royal Air Force), 76, 190, 191
Ram, Kirpa, 95, 96, 98, 100, 156, 157, 175, 176
Rau Commission, 259, 265
Rawal Dam, 279
Razmak, 116, 117, 120, 148, 149, 204
Reading, Lord, 128, 130
Red Fort, 12, 37, 38, 39
Red Shirts, 70, 95, 147, 186, 196, 223, 241
Risalpur, 76
Robertson, A.N.M., 98, 102–5
Robertson, T.E., 3, 4
Rohri, 2
Rohtas, 58, 59, 63
Roorkee, 21, 26, 27, 32–6, 39, 40–3, 48, 65, 67, 95, 157, 162, 249

S

Safed Koh, 197, 199
Saidpur, 279
Sandes, Captain, E.W.C., 21, 39, 42
Sapru, Sir Tej Bahadur, 43
Sarwakai, 119, 120, 121, 131, 132, 133, 138, 139, 142, 182
Satwari Gardens, 17
Second World War, 188, 195, 199, 208, 210, 226, 242

Shah Jahan, Emperor, 38, 39
Shah, Aitibar, 73, 74, 90
Shah, Makhdoom Miran, 223, 225
Shah, Pir Jamaat Ali, 23, 24
Shahbaz Garhi, 83, 93
Shahin-Badshah graveyard, 6
Shewa, 84, 92, 93, 170
Sialkot, 2, 3, 6, 7, 15, 16, 18, 19, 23, 44
Sibi, 136, 228, 232, 235, 237, 238
Simly Dam, 279
Simon Commission, 147
Simon, Sir John, 147
Singh, Tara, 212, 214, 215, 218
Sir Syed Court, 27, 234
Siraj-ud-Daula, Nawab, 196
Smith, Mr, 68, 76–79, 80–82, 85
Sone, 58, 59, 60, 65
Stein, Sir Aurel, 14
Stoddard-Harvey module outlet, 71
Sukkur, 2, 4, 100, 108, 128, 136; Barrage, 208
Sulaimankhel, 117, 119, 120, 123, 124, 125, 237, 238
Suri, Sher Shah, 58, 64
Surma River, 48, 49, 50, 54
Swabi, 68, 78, 80, 90, 169, 180, 183
Syed Mahmud Court, 27, 28
Sylhet, 44–6, 48, 49, 50, 51, 52, 54–8
Syria, 86

T

Taj Mahal, 66; Palace, 65
Tank, 114, 115, 116, 118, 131, 139, 140, 159, 272
Tarbela Dam, 191, 268, 278
Thal Canal, 110, 112
Thomason Civil Engineering College, 21, 26, 32, 33, 157
Treaty of Sevres, 86, 87
Tughlaq, Sultan Mohammad, 66

U

University of Peshawar, 130, 273
University of the Punjab, 26, 30

Upper Swat Canal, 70, 71, 73, 81–5, 89, 90, 91, 94, 169, 180, 181, 184, 200
Utmanzai, 70, 185

V

Victoria, Queen, 5, 11, 178
Viqar-ul-Mulk, Nawab, 28, 33

W

Wadia, D.N., 139, 140,
Walker, Samuel, 105, 106, 108, 110–14, 123, 140, 142
Wana, 116, 120, 122, 123, 125, 131, 133, 134, 204
Ward, Sir Thomas, 68, 69
Warsak Dam, 269, 270, 271–4, 276, 278
Water and Power Development Authority (WAPDA), 278, 281
Waziristan, 77, 78, 113–19, 120, 123, 125, 126, 128, 129, 131, 132, 135, 136, 138, 140–3, 147–9, 159, 175, 182, 190, 194, 195, 204, 241, 281
– North Waziristan, 77, 116, 148, 159, 175, 190
– South Waziristan Scouts, 115, 116, 120, 128, 135, 136, 169, 194, 195
– South Waziristan, 77, 113, 114, 116, 119, 120, 122, 123, 131, 132, 135, 138, 141, 148, 149, 182
– Waziristan Survey Division, 113, 129, 131, 142
West Pakistan, 95, 257, 259, 262, 274, 280
West Punjab, 214, 215, 218, 262, 269
Wheeler, General Raymond A., 248, 249, 256, 259, 260, 262, 263
Willingdon, Lord, 181
World Bank, 242, 248, 249, 254, 256, 264, 265, 268

Y

Younghusband, Sir Francis, 17, 178
Yousufzai, Najib Khan (Najib-ud-daula), 170
Yusufzai, 83, 88

Z

Zhob, 114, 128, 129, 130, 131, 135, 136, 138, 139, 140, 141, 228
- Gorge, 129
- Levies, 128, 129, 131, 136
- River, 114, 128, 129, 130, 131, 136, 138, 139, 140, 141
- Torrent, 114
- Valley, 114, 128, 129, 131, 135, 136

Ziarat, 233–7
Ziauddin, Dr, 28, 29, 33, 157